Society of Earth Scientists Series

Series editor

Satish C. Tripathi, Lucknow, India

The Society of Earth Scientists Series aims to publish selected conference proceedings, monographs, edited topical books/text books by leading scientists and experts in the field of geophysics, geology, atmospheric and environmental science, meteorology and oceanography as Special Publications of The Society of Earth Scientists. The objective is to highlight recent multidisciplinary scientific research and to strengthen the scientific literature related to Earth Sciences. Quality scientific contributions from all across the Globe are invited for publication under this series.

More information about this series at http://www.springer.com/series/8785

Sunil Bajpai · Satish C. Tripathi
Vandana Prasad
Editors

The Indian Paleogene

The Society of Earth Scientists
2008

Springer

Editors
Sunil Bajpai
Birbal Sahni Institute of Palaeosciences
Lucknow, Uttar Pradesh
India

and

Indian Institute of Technology
Roorkee, Uttarakhand
India

Satish C. Tripathi
Geological Survey of India
Hyderabad, Andhra Pradesh
India

Vandana Prasad
Birbal Sahni Institute of Palaeosciences
Lucknow, Uttar Pradesh
India

ISSN 2194-9204 ISSN 2194-9212 (electronic)
Society of Earth Scientists Series
ISBN 978-3-030-08456-1 ISBN 978-3-319-77443-5 (eBook)
https://doi.org/10.1007/978-3-319-77443-5

Printed on acid-free paper

This Springer imprint is published by the registered company Springer International Publishing AG part of Springer Nature
The registered company address is: Gewerbestrasse 11, 6330 Cham, Switzerland

From Series Editor

The Paleogene, comprising the early part of Cenozoic Era, was the most dynamic period in the Earth's history with profound changes in the biosphere and geosphere. The period began with post-K/T mass extinction event at ~65.5 Ma and ended at ~23 Ma, when the first Antarctic ice sheet appeared in the southern hemisphere. The early Paleogene (Paleocene–Eocene) has been considered a globally warm period, superimposed on which were several transient hyperthermal events of extreme warmth. Among them, the Paleocene–Eocene Thermal Maxima (PETM) boundary interval is the most prominent extreme warming episode of 200 Ka duration. PETM is characterized by 2–6‰ negative carbon isotope excursion in both terrestrial and marine sediment records, worldwide. The event coincides with the benthic extinction event (BEE) in deep sea and larger foraminifera turnover (LFT) in shallow seas. Rapid ~6°–8° warming of high latitudinal regions led to major faunal and floral turnovers in continental, shallow-marine, and deep-marine realms. Emergence, dispersal of mammals of modern aspect including artiodactyls, perissodactyls, and primates (APP), and evolution and expansion of tropical vegetation are some of the significant features of the Paleogene warm world.

In the Indian subcontinent, the beginning and end of the Paleogene are marked by varied events, which resulted in shaping various physiographic features of the subcontinent. Indian subcontinent lays within the equatorial zone during the earliest part of Paleogene. Carbonaceous shale, coal, and lignite deposits of Paleogene age (~55.5–52 Ma) of the western and northeastern margins of the Indian subcontinent are rich in their fossil content and provide information on climate as well as evolution and paleobiogeography of tropical biota. Indian Paleogene deposits of India-Asia collision zone also provide information pertaining to the paleogeography and timing of collision. Indian Paleogene rocks are exposed in the Himalayan and Arakan mountains; Assam and the shelf basins of Kutch-Saurashtra, western

Rajasthan; Tiruchirapalli-Pondicherry and Andaman, and though aerially limited, these rocks preserve very important geological records. I hope this concise account of Indian Paleogene will provide valuable insight.

Satish C. Tripathi
Series Editor

Contents

About the Editors

Prof. Sunil Bajpai is a geologist with main research interests in vertebrate paleontology, biostratigraphy, paleobiogeography, paleoecology, and related fields. He studied at the Centre of Advanced Study in Geology, Panjab University, Chandigarh, and subsequently joined the Indian Institute of Technology Roorkee in 1996 as a faculty in the Department of Earth Sciences. In 2013, he moved to Lucknow as Director of prestigious Birbal Sahni Institute of Palaeosciences (BSIP), an autonomous research institute under the Department of Science and Technology, Government of India.

He has authored over 115 scientific publications, and in recognition of his contributions, he was elected Fellow of the Indian Academy of Sciences and the National Academy of Sciences. He has also been the recipient of the National Geosciences Award of the Government of India and the Ramanna Fellowship of the Department of Science and Technology. Among important international fellowships/assignments held by him are the Smithsonian Institution Fellowship at Washington DC, USA; Commonwealth Fellowship at the University of Wales and the University College London; CNRS Professor at the Natural History Museum, Paris; JSPS Fellowship in Kyushu, Japan. Currently, he is the Chief Editor of the journal *Palaeobotanist* and is also in the editorial board of *Current Science, Indian Journal of Geosciences*. As Director of BSIP, he holds the position of the President of the Palaeobotanical Society of India.

Dr. Satish C. Tripathi is Director, Geological Survey of India. He has contributed extensively on the sedimentary environment, facies analyses, sequence stratigraphy and paleontology of Cretaceous basin of Son-Narmada rift valley, central India and Cretaceous–Paleogene basin of Lesser Himalaya. He is presently carrying out detailed geological investigation in the Tethyan sequence of Higher Himalaya and flysch–molasse sedimentation along ITSZ. He has contributed 36 research papers in international and national journals. He is Founder Editor of the e-Journal Earth Science India-www.earthscienceindia.info and General Secretary, The Society of Earth Scientists. He has organized several international and national conferences and field workshops and is Series Editor of this Society of Earth Scientists Series with Springer.

Dr. Vandana Prasad is Scientist at Birbal Sahni Institute of Paleosciences, Lucknow, India, and is internationally recognized for her outstanding contributions in understanding the vegetational history and climate change during late Cretaceous–Paleogene and Quaternary period of Indian subcontinent. She established the oldest (65 Ma) record of grasses in India. Extending this work, she showed that the rice tribe evolved in India. Her studies on early Paleogene extreme global warm climate have helped in understanding the vegetation turnover due to long-term effect of elevated CO_2 in tropical ecosystem. A very significant finding of her provides evidence of rain forest vegetation in the southern Western Ghats as Gondwana relics of 55 Ma. She used palynofacies analysis to understand the process-driven changes for the establishment of sequence stratigraphy in Vastan lignite mine succession. Utilization of proxies such as grass phytoliths and marine dinoflagellate cysts in solving climate issues are her very significant contributions in Quaternary studies. She has also evaluated the role of mid-Holocene climate change on the rise and fall of Harappan civilization in Gujarat. Her researches in solving various regional and global issues will have long-term relevance in the area of earth science. She has been a member of various scientific advisory bodies including those of DST, MoES, etc.

Paleogene Stratigraphy of India: An Overview

D. S. N. Raju

Abstract The Paleogene stratigraphic record in India left imprint with many clues towards understanding the global and Indian plate events. The KTB and the beginning of the Paleogene is heralded by the Deccan volcanism, which is considered to be the main killer responsible for mass extension. Towards the end of the Paleogene there was a global drop in sea-level and major hiatus which left marks in the Indian basins. Within the limits of the Paleogene a number of major geological events are recognizable. They are: (1) Maximum flooding during zones P4/P5 (Late Paleocene), (2) Paleocene-Eocene Thermal Maxima (PETM), the impacts of which are not fully explored in Indian strata, (3) Late Paleocene-Early Eocene sedimentary succession of India occupies an important position as a major source rocks and to a lesser extent reservoirs of hydrocarbons. Case examples are: (a) Bombay Offshore Panna Formation deposited in several cycles in a marginal marine setup, (b) Facies comparable to the Panna Formation extended up to Kerala-Konkan basin where these are referred to Kasargod Formation. (c) Cambay Basin: Olpad/Cambay Shale deposited in lacustrine to marginal marine environment is the source for the sandstone reservoir in several fields, (d) Tura Formation in NE Region, (4) Zone P13 (Middle Eocene) maximum flooding surface recognizable and valuable in correlation, (5) Himalayan orogenic movement-I (41.3–42 Ma), (6) Late Eocene drop in sea-level left record in several basins except in Cambay Basin and NE Indian basin where there was sea-level rise due to local tectonic subsidence greater than sea-level drop, (7) drop in paleotemperatures during the Early Oligocene documented globally and in the Kutch basin. (8) Extinctions close to the Paleocene-Eocene boundary and top of the Eocene. Further studies by Research scholars require access to deep well samples of the ONGC and other oil companies. Lithostratigraphy of Paleogene strata of India from exposures and subsurfaces, paleoshelfs and bathyal settings is well recognized/established. All the formations are briefly described and further references are being provided for more detailed account. Sedimentary basins covered in this paper include: Himalayan foreland, NE Region, Andaman, West Bengal, Mumbai offshore, Cambay, Kutch and

D. S. N. Raju (✉)
Sri Sannidhi Grand, Flat no. B1, Prabhu Gardens, JN Road, 2nd Floor, Rajahmundry 533103, Andhra Pradesh, India
e-mail: rajudsn1@gmail.com

S. Bajpai et al. (eds.), *The Indian Paleogene*, Society of Earth Scientists Series,
https://doi.org/10.1007/978-3-319-77443-5_1

Jaisalmer basins. Preserved record in exposed sections is incomplete-punctuated by hiatuses. Diachronos nature of the limits of the formations on basin scale is not fully appreciated by workers in India.

Keywords Paleogene stratigraphy · PETM · Foraminiferal markers · Indian basins

Introduction

The Cenozoic (originally Cainozoic) Era (Phillips 1841) derives its name from the relatively new (Kainos) biota, compared to the Mesozoic Era. The Cenozoic Era is subdivided into the Paleogene (Palaios = old, genes = birth), the Neogene and the Quatenary periods. The term Tertiary (Ardunio 1759) has been traditionally used to indicate the lives interval between the Mesozoic and the Quaternary and therefore comprise the Paleogene and the Neogene. The use of the term Tertiary is discouraged. However, the term Tertiary still widely used, in particularly since the ratification of the term Quaternary as a period (Head et al. 2008).

Naumann (1866) combined his "Paleogene Stufe," the Eocene and the Oligocene, as opposed to the "Neogene Stufe of Hornes" (1853) which included not only the Miocene and the Pliocene, but also fauna of the Pleistocene. The term "Nummulitique," which have been employed as an equivalent of Paleogene mainly by French speaking stratigraphers such as Renevier (1873), Haug (1908–1911) GTS (2004).

The Paleogene system is subdivided into three series: The Paleocene, Eocene and Oligocene, referring again to the evolution of the biote (Eos = down and oligos = little). The three Paleogene series have been formally subdivided into nine stages as decided by the International Subcommission on Paleogene Stratigraphy at the 1989 International Geological Congress in Washington (Jenkins and Luterbacher 1992). The Paleocene Series is further subdivided into the Danian, Selandian and Thanetian stages, the Eocene series into the Ypresian, Lutetian, Bartonian and Priabonian stages and the Oligocene Series into the Rupelian and Chattian stages, all stages being listed successively from old to young (GTS 2012).

- All Paleocene stages (i.e., Danian, Selandian and Thanetian) have formally ratified definitions, and so have the Ypresian and Lutetian Stages in the Eocene, and the Rupelian Stage in the Oligocene.
- The Bartonian, Priabonian and Chattian Stages are not yet formally defined.
- After the global catastrophe and biotic crisis at the Cretaceous- Paleogene boundary, stratigraphically important marine microfossils underwent new evolutionary trends, and on land the now flourishing mammals offer a potential for stratigraphic zonation.
- During the Paleogene the global climate, being warm until the late Eocene, shows a significant cooling trend culminating in a major cooling event in the beginning of the Oligocene, preparing the conditions for modern life and climate. Orbitally

tuned cyclic sedimentation series, calibrated to the geomagnetic polarity and biostratigraphic scales, have considerably improved the resolution of the Paleogene time scale.

Indian Stages Biochrons

Paleocene

GTS 2012 recognized 9 stages in the Paleogene. For Indian basins, 7 stages and 19 biochrons are recognizable. Comparison/correlation of Indian biochrons and stages with standard stages and zones is shown in Figs. 1 and 2.

Meghalayian Stage (2.72 My)

Interval from the LAD of *Racemiguembelina fructicosa*, *Globotruncana*, *Globigernelloides*, *Rugoglobigerina* (66.04 Ma) to FAD of *Morozovella angulata/Igorina pusilla* (62.29 Ma) (Early Paleocene).

1. **MEG-I**: Definition: defined as the interval from the LAD of *Racemiguembelina fructicosa*, *Globotruncana*, *Globigernelloides*, *Rugoglobigerina* (66.04 Ma) to FAD of *Globanomalina compressa* (63.94 Ma). Duration: 2.1 My (Early Danian). Datums within this biochron MEG-I are Planktic foraminifera 4 FADs.
2. **MEG-II**: Defined as the interval from the FAD of *Globanomalina compressa* (63.90 Ma) to FAD of *Morozovella angulata/Igorina pusilla* (62.29 Ma). Duration: 1.61 My (Early to Late Danian). Datums within this biochron MEG-II are planktic foraminifera: 3 FADs.

Mawsmaian Stage (6.19 My)

Interval from FAD of *Morozovella angulata* (62.29 Ma) to LAD of *Globoanomalina pseudomenardii* (60.73 Ma). Duration: 1.56 My (Middle Selandian).

MAW-I: Defined as interval from FAD of *Morozovella angulata* (62.29 Ma) to LAD of *Globoanomalina pseudomenardii* (60.73 Ma). Duration: 1.56 My (Middle Selandian). Foraminiferal datums within this biochron MAW-I are Planktic foraminifera 5 FADs and 2 LADs.
MAW-II: Defined as an interval from FAD of *Morozovella aequa* (57.79 Ma) to FAD of *Globoanomalina pseudomenardii* (60.73 Ma). Duration: 2.94 My (Late Selandian). Datums within this biochron MAW-II are planktic foraminifera: 1 FAD and 1 LAD.

Khasian Stage (4.53 My)

KHA-I: Defined as an interval from FAD of *Globanomalina pseudomenardii* (60.73 Ma) to LAD of *G.pseudomenardii* (57.10 Ma). Duration: 3.63 My (Middle Thanetian). Datums within this biochron KHA-I are planktic foraminifera: 2 FADs and 3 LADs.

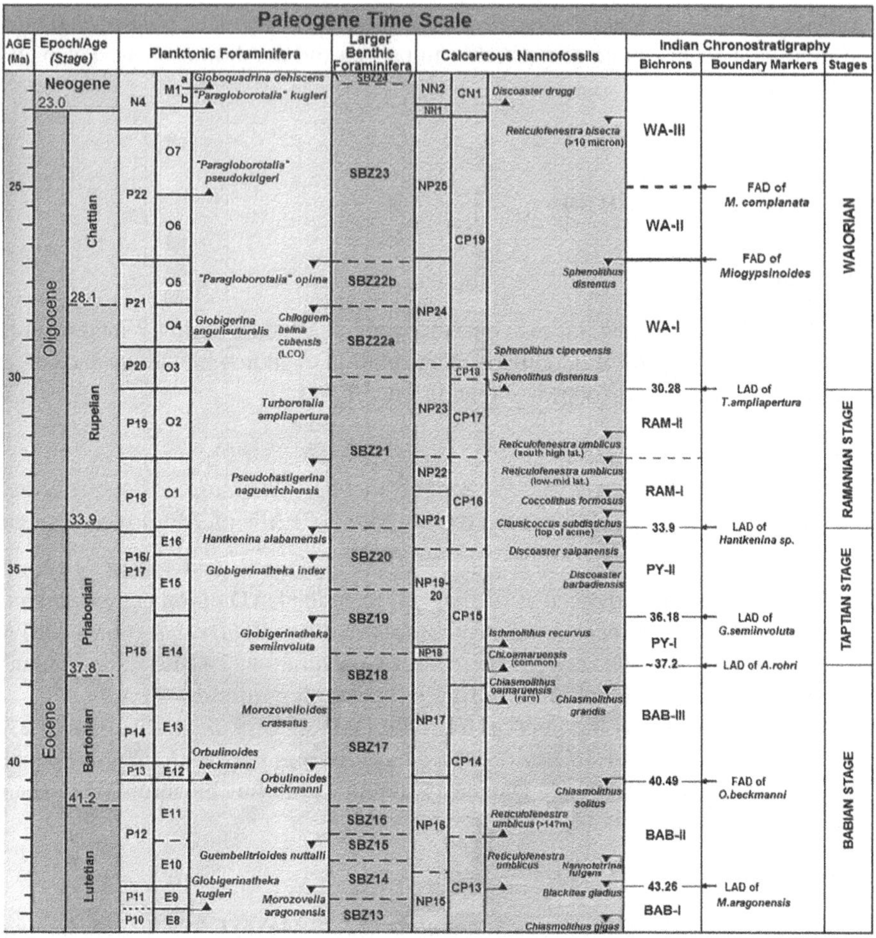

Fig. 1 Comparison/correlation of Eocene-Oligocene global standard stages, zones and Indian biochrons, stages and boundary marker

KHA-II: Defined as interval from LAD of *Globanomalina pseudomenardii* (57.10 Ma) to LAD of *Morozovella velascoensis* (55.20 Ma). Duration: 1.9 My (Late Thanetian). Datums within this biochron KHA-II are planktic foraminifera: 1 LAD and 4 FADs.

Kakdian Stage (6.28 My)

1. **KAK-I**: Defined as an interval from LAD of *Morozovella velascoensis*-(55.20 Ma) to FAD of *Morozovella aragonensis* (52.54 Ma). It occupies the lower part of the Ypresian. Duration: 2.34 My. Datums within this biochron KAK-I are Planktic foraminifera: 2 FADs and 4 LADs, Nannoplanktons: 2 FADs and 1 LAD.

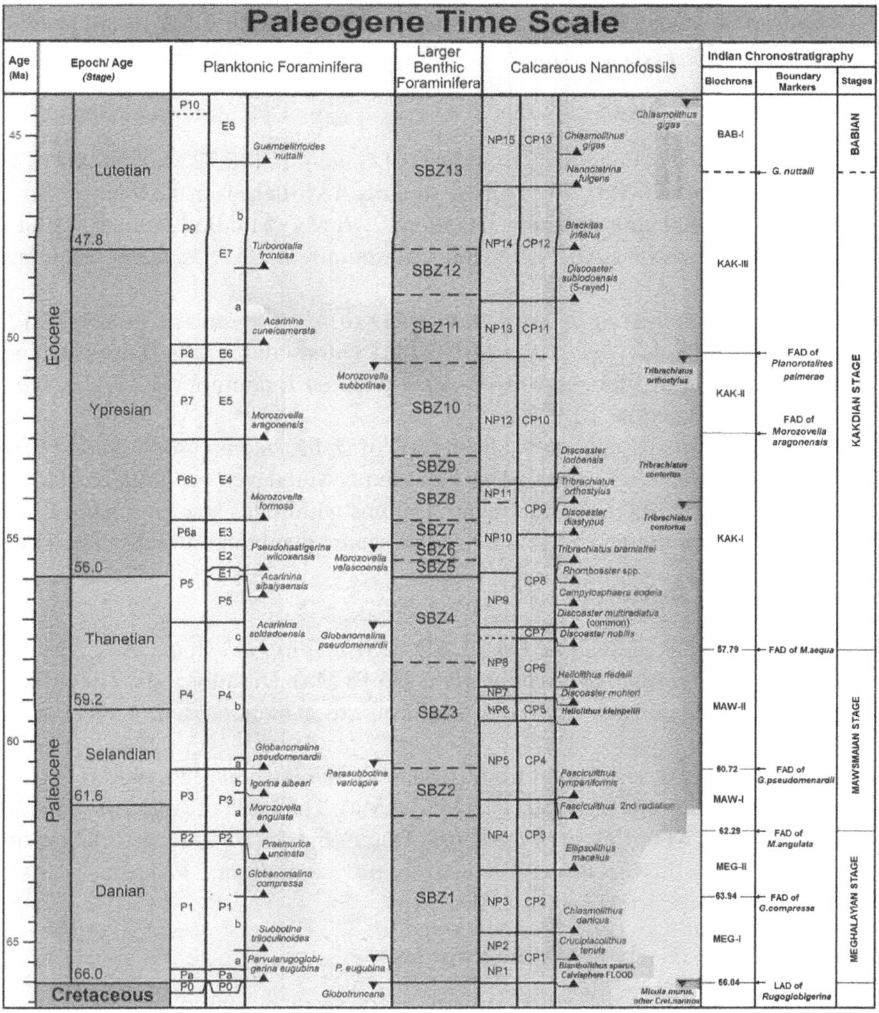

Fig. 2 Comparison/correlation of Paleocene-Eocene global standard stages, zones and Indian biochrons, stages and boundary markers

2. **KAK-II**: Defined as an interval from FAD of *M. aragonensis* (52.54 Ma) to FAD of *Planorotalites palmerae* (~50.20 Ma). This biochron occupies the middle part of Ypresian. Duration: 2.34 My. Datums within this biochron KAK-II are Planktic foraminifera: 1 LAD and 1 FADs, Nannoplankton *Tribrachiatus orthostylus* LAD 50.47 Ma
3. **KAK-III**: Defined as an interval from the FAD of *P. palmerae* (50.20 Ma) to the FAD of *Guembelitrioides nuttalli* (45.72 Ma). This biochron represents the upper Ypresian. The upper limit is close to the LAD of *Acarinina soldadoensis*

angulosa. Duration: 1.6 My. Datums within this biochron KAK-III are planktic foraminifera: 1 FAD, Nannoplankton *Discoster sublodoensis* FAD at 49.43 Ma.

Babian Stage (8.52 My)

1. **BAB-I**: Defined as an interval from FAD of *G. nuttalli* (45.72 Ma) to LAD of *M. aragonensis* (43.26 Ma). The interval is highly fossiliferous in the Cauvery Basin in wells Adiyakkamangalam-1. Duration: 2.46 My (Middle Lutetian). Datums within this biochron BAB-I are planktic foraminifera: 2 FADs, nannoplanktons: 2 FADs and 1 LAD.
2. **BAB-II**: Defined as an interval from LAD of *M. aragonensis* (43.26 Ma) to FAD of *Orbulinoides beckmanni* (40.49 Ma). Duration: 2.77 My (Late Lutetian). Datums within this biochron BAB-II are planktic foraminifera: 2 FADs and 1 LAD, nannoplankton: 1 LAD.
3. **BAB-III**: Defined as an interval from FAD of *O. beckmanni* (40.49 Ma) to LAD of *Acarininarohri* (~37.2 Ma). This biochron is well developed in Indian basins. Duration: ~3.29 Ma (Late Bartonian). Datums within this biochron BAB-III are planktic foraminifera: 1 FAD and 6 LADs, nannoplanktons: 1 FAD and 1 LAD.

Taptian Stage (3.30 My)

1. **PY-I**: Defined as an interval from LAD of *Truncorotloidesrohri* (~37.2 Ma) to LAD of *Globigerinatheka semiinvoluta* (36.18 Ma). Duration: ~1.02 My (Early Priabonian). Datums within PY-1 biochron are nannoplankton: 1 FAD and 1 LAD.
2. **PY-II**: Defined as an interval from LAD of G. *semiinvoluta* (36.18 Ma) to LAD of *Globorotalia cerroazulensis* (*s.l*) (~33.9 Ma). The LAD of *Pellatispira* is very close to the LAD of *G. cerroazulensis*. Duration: ~2.28 My (Late Priabonian). Datums within this biochron PY-II are planktic foraminifera: 3 FADs and 5 LADs, nannoplanktons: 1 FAD and 1 LAD.

Ramanian Stage (Early to Middle Rupelian)

RAM-I: (Early Ramanina, Early Rupelian): Defined as an interval from the FAD of *N. fichteli* to FAD of *Lepidocyclina isolepidinoides*. Nannoplankton assemblage referable to zone NN23 was recorded by Dr. Jafar and others from the uppermost part of this zone (personal communication).
RAM-II: (Upper Ramanian, Middle part of Rupelian): Defined as an interval from FAD of *L. isolepidinoides* to LAD of *N. fichteli. Nummulities clypeus* is common in this zone. The definition and subdivision of this Ramanian stage is based on larger

foraminifera. However, based on subsurface data in the Cauvery Basin, a linkup with planktic foraminifera is attempted.

(a) The LAD of *N. fichetli* is close to the LAD of *Globigerina ampliapertura* (30.28 Ma).
Datums within this biochron RAM-I are Planktic foraminifera: 1 LAD and 1 FAD, Nannoplanktons: 2 FADs and 3 LADs.

Waiorian Stage

WA-I (Early Waiorian, latest Rupelian to Early Chattian): Defined as an interval from LAD of *N. fichteli* to FAD of Miogypsinidae (*M. bermudezi* in Kutch Basin). In Kutch the FAD of Miogypsinidae coincides with FAD of *Globigerinoides primordius*.

WA-II: (Middle part of Chattian): Defined as an interval from FAD of *M. bermudezi* to FAD of *M. (Miogypsinoides) complanata*. It is characterized by *M. bermudezi* and transitional forms from *M. bermudezi* to *M. gunteri* like forms with primitive lateral chambers. It is differentiated from younger intervals with *M. gunteri* in the absence of *M. (Miogypsinoides) bantamensis* which has a distinctly larger protoconch.

WA-III (Upper part of Chattian): It is characterized by an interval from FAD of *M.(M.) complanata* to FAD of *Spiroclypeus ranjanae. Lepidocyclina praemarginata-morgani* occurs in this interval.

WA-III b: Defined as an interval from FAD of *S. ranjanae* to a level of *M. gunteri* with mean value of X = 11. Datums within this biochron WA-III are planktic foraminifera: 5 LADs and 7 FADs, Nannoplanktons: 2 LADs.

Limitations of Bio- and Chronostratigraphy

Biochronology rests on the principle of irreversible evolution in biota and establishment of ranges of species particularly their FADs (First Appearance Datums) and LADs (Last Appearance Datums). A case example is given below. *Fasciolites elliptica (=Alveolina elliptica)* was described by Sowerby (1840) from Kutch. The type level is from planktic foraminiferal zone P13. (Bartonian, Middle Eocene) Its range in Kutch is from P13 to P14. Govindan (2013) has shown the range of *A. elliptica* from P10 to P11 (Lutetian); such interpretations will lead to wrong correlations.

Resolution Achieved in Biochronostratigraphy Through Various Tools/Methods

Biostratigraphy based on planktonic foraminifera, nannoplankton, spore-pollen, acritarchs and dinoflagellate cysts are most useful and actually applied constantly in hydrocarbon exploration in India.

Some Inquiries and Requisites of Basic Research in India

What is the use of bio- and chrono-stratigraphic classifications if they are not useful in deciphering the geological evolution of a basin or the Earth as a whole, or evolutionary trends in biota or if not applicable towards economic benefits of mankind and other life?

In GTS 2012, standard Cretaceous-Cenozoic chronostratigraphic classification, several of the standard stage boundaries are located within foraminiferal or nannoplankton zone.

Resolution Achieved so Far

Late Cretaceous (Duration 34.5 My)

The divisions are: 5¼ polarity chrons according to Miller et al. (2005) and 3½ chrons by Gradstein et al. (2012); 15 planktic foram zones; 20 nanno zones; 9 cycles of global sea-level fluctuations. The resolution through planktic foram zones is about 2 My while by nanno plankton is 1.5 My.

Paleocene (Duration: 10 My)

The divisions are: 5 polarity chrons according to Miller et al. (2005) and 4½ chrons by Gradstein et al. (2012); 10 planktic foram zones (including subzones); 9 nannoplankton zones; about 5 cycles of sea-level (Miller et al. 2005) and 2 ½ major cycles according to Haq et al. (1987). The resolution achieved through plankticsis about 1 My. 22 dinoflagellate cysts datums are recognized by GTS 2012. Resolution can be 0.5 My.

Early Eocene (Duration: 8.2 My)

The divisions are 3¾ polarity chrons, 5 1/3 planktic foram zones and 4½ nanno zones. Resolution is about 1.4 My. 20 dinoflagellates cysts datums are recognized by GTS 2012. Resolution can be 0.3 My. Haq et al. (1987) has shown 2 major cycles and a relatively high sea-level?

Middle Eocene (Duration: 10 My)

The divisions are: 5 polarity chrons, 5½ planktic foram zones; 6½ nanno zones (including subzones); 6½ pulses in sea-level according to Miller et al. (2005) and 6 pulses by Haq et al. (1987).The resolution achieved is about 1.8 My. 20 dinoflagellates cysts are recognized by GTS 2004. Resolution can be 0.56 My.

Late Eocene (Duration: 3.9 My)

The divisions are: 3 polarity chorns, 2½ planktic foram zones; and 3½ nanno zones. The resolution achieved is about 1.6 My. 7 dinoflagellates cysts datums are recognized by GTS 2012. Resolution can be 0.47 My.

Oliogocene (Duration: 10.9 My)

The divisions are: 7½ polarity chrons, 5½ planktic foram zones; 4½ nanno zones; 6 sequence chornostratigraphic units in Europe; 7 larger foram zones (5 zones plus 2 subzones) in India. Seven parasequences are recognizable in Kutch. Resolution is about 1.5 My. 16 dinoflagellates cysts datums are recognized by GTS 2012. Resolution can be 0.68 My.

Global Sea-level Changes and Indian Record

Figures 3, 4 and 5 show standard chronostratigraphy, Indian stages and biochrons, relative sea- level curves for Indian basins and comparison with global sea- level curves of Haq et al. (1987) and Miller et al. (2005).

Paleocene (duration: 10 My)

A major rise of sea-level during P4/P5 with a peak in P5 is well recognized in Rajasthan, Bombay offshore, Kutch offshore, Cauvery, KG and Mahanadi. At some localities in Bengal Basin, Late Paleocene is represented by non-marine while a regressive phase recognized in KG basin during latest Paleocene–earliest Early Eocene. A minor hiatus was recognized by Narayanan and Raju (1996) at the top of Paleocene in SM-2-1, Bombay offshore.

Early Eocene (Duration: 8.2 My)

In the Cauvery Basin two distinctly major cycles are recorded as in PTK-1 and MNRG-1 where relative fluctuations of an order of 100 m are known. Miller also showed 4–6 pulses within 2 major cycles. Interestingly about 6 minor cycles are recognizable in Kutch, Cambay and Rajasthan basins. Sea regressed from several areas of Kutch, Cauvery and KG basins during latest Early Eocene. In Bombay offshore, Narayanan and Raju (1996) mentioned about probable hiatus during P9-lower P10.

Case example: Late Paleocene-Early Eocene

Late Paleocene-Early Eocene stratigraphic succession of India occupy an envious position as a major source rocks and to a lesser extent reservoir of Hydrocarbons. They are the source rocks for the major oil and gas accumulation in Bombay offshore giant and major oil fields. Cambay Basin, Barmer Basin, Assam and Krishna-Godavari Basin.

Globally significant event of the Paleocene-Early Eocene Thermal Maxima (PETM) and higher temperature are well recognized. Global environmental events such as PETM have had profound effects on evolution in geological past and must be considered when modeling of the history of life.

Bombay Offshore: The Panna formation (also referred as Panna basal clasts) deposited in several cycles in marginal marine setup are the source rocks for several reservoirs: (1) Late Paleocene- Early Eocene sands in Panna Formation; (2) Middle Eocene to Early Oligocene Paleocene-Early Eocene sands Oligocene carbonates of Mukta Formation and Early Miocene carbonates of Bomay formation

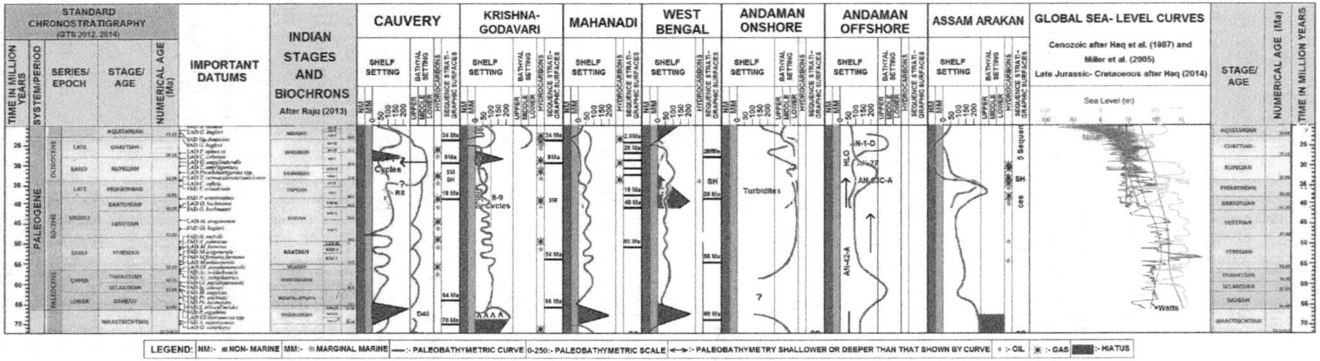

NOTE: 1. PALEOBATHYMETRIC CURVES AS SHOWN IN THIS TABLE ARE GENERALISATIONS VALID FOR MAJOR PART OF THE BASIN CONCERNED. 2. HYDROCARBONS OCCURENCES ARE TO BE READ AGAINST TIME SCALE.

SH= SUPPORTS HAQ et al. (1987); SM= SUPPORTS MILLER et al. (2005); SS= SUPPORTS SAHAGIAN & JONES, (1993); PALEOBATHYMETRY DEEPER THAN THAT SHOWN IN CURVE; DS= DROP IN SEA- LEVEL; AH& AM= DATA AGAINST HAQ AND MILLER; D60= DROP IN SEA- LEVEL BY 60m; RS= RAPID SUBSIDENCE; MFS= MAXIMUM FLOODING SURFACE

NOTES: CURVES ARE SHOWN HERE ARE GENERALIZATIONS FOR EACH BASIN TAKING ACCOUNT OF IMPORTANT UNITS OF STRATIGRAPHY PARTICURLARLY SOURCE OR RESERVOIRS OF HYDROCARBON.

Fig. 3 Record of relative sea level change, comparison with global sea level curves, sequence stratigraphic surfaces (ONGC) hiatuses and hydrocarbon occurrences of East Coast basins in India. Compiled by D. S. N. Raju

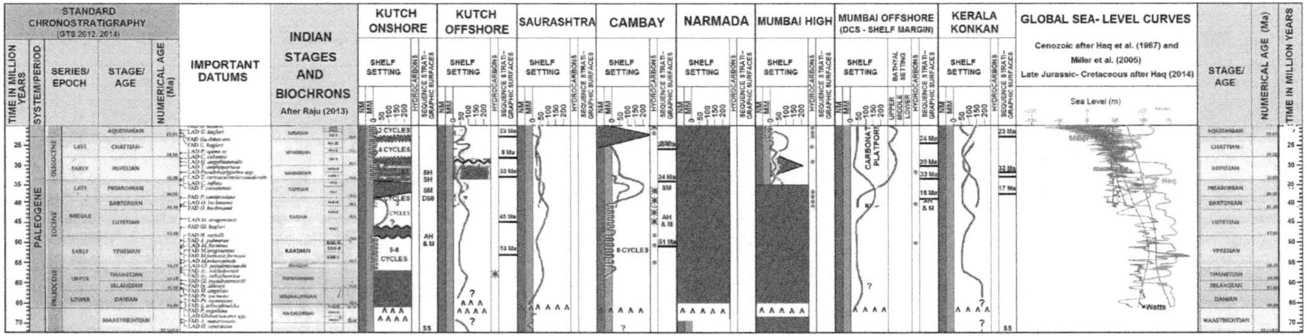

LEGEND: NM:- NON- MARINE MM:- MARGINAL MARINE —:- PALEOBATHYMETRIC CURVE 0-250:- PALEOBATHYMETRIC SCALE ⟷:- PALEOBATHYMETRY SHALLOWER OR DEEPER THAN THAT SHOWN BY CURVE •:- OIL ✳:- GAS ▬:- HIATUS

NOTE: 1. PALEOBATHYMETRIC CURVES AS SHOWN IN THIS TABLE ARE GENERALISATIONS VALID FOR MAJOR PART OF THE BASIN CONCERNED. 2. HYDROCARBONS OCCURENCES ARE TO BE READ AGAINST TIME SCALE.

SH= SUPPORTS HAQ et al. (1987); SM= SUPPORTS MILLER et al. (2005); SS= SUPPORTS SAHAGIAN & JONES, (1993); PALEOBATHYMETRY DEEPER THAN THAT SHOWN IN CURVE; DS= DROP IN SEA- LEVEL;
AH& AM= DATA AGAINST HAQ AND MILLER; D60= DROP IN SEA- LEVEL BY 60m; RS= RAPID SUBSIDENCE; MFS= MAXIMUM FLOODING SURFACE
NOTES: CURVES ARE SHOWN HERE ARE GENERALIZATIONS FOR EACH BASIN TAKING ACCOUNT OF IMPORTANT UNITS OF STRATIGRAPHY PARTICURLARLY SOURCE OR RESERVOIRS OF HYDROCARBON.

Fig. 4 Record of relative sea level change, comparison with global sea level curves, sequence stratigraphic surfaces (ONGC) hiatuses and hydrocarbon occurrences of West Coast basins in India. Compiled by D. S. N. Raju

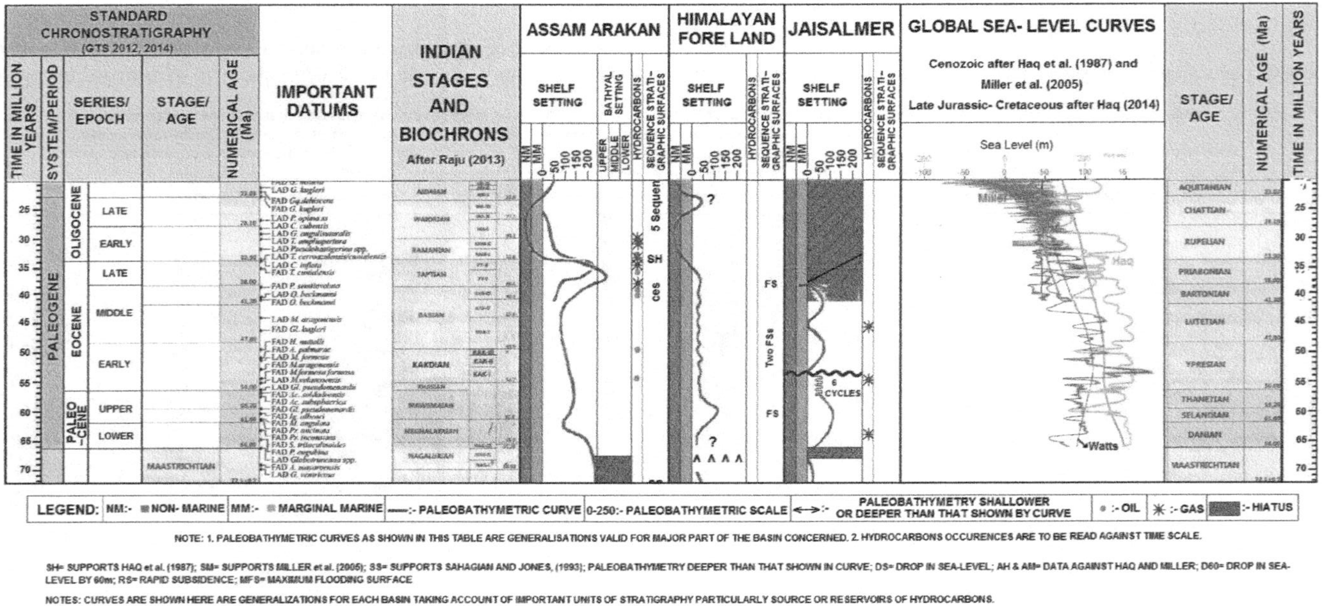

Fig. 5 Record of relative sea level change, comparison with global sea level curves, sequence stratigraphic surfaces (ONGC) hiatuses and hydrocarbon occurrences of North Eastern basins in India. Compiled by D.S. N. Raju

(Llll). Chakravarthy et al. (Unpublished Report) have identified altogether 12 prominent lithological episodes in the area. Out of them 8 are fining up cycles, 2 major coarsening cycles and 2 moderate cycles. There are 5 layers of coal and thin beds of open marine limestones in the Panna formation.

Cambay Basin: Olpad/Cambay Shale deposited in lacustrine to marginal marine rock environment are the source rock for the sandstone reservoir in of kadi and Kalol formations and Cambay Shale is the source for Middle Eocene to Early Oligocene Ankleswar Formation and Dadhar Formation.

The Strata of Vastan open cast Lignite Mine in the south-eastern part of Cambay Basin were studied in great details because of discovery of many vertebrate fossils and microfossils in an exposed column of on 32 m thick.

Sahni et al. (2006, Fig. 2) recorded six lignite seams, seven marine fish layers and inferred five cyclothems. There are at least seven marine incursions in a marginal marine to shallow marine set-up.

Raju et al. (2005) recognized five to six third order cycles in the Cambay Shale encountered in deep wells of Southern part of the Cambay Basin.

Grover et al. (2004) recognized seven sequences in deeper part of Mehsana block of Northern Cambay Basin. As a whole six to seven third order cycles/sequences were recognized in the Early Eocene of Cambay basin.

Barmer Basin: The Akli Formation of the Late Paleocene-Earliest Eocene, exposed in Lignite mines of Barmer district, was studied in detail due to the occurrence of vertebrate fossils. In Giral Lignite Mine, 10–11 lignite seams were recorded. Rana et al. (2005) have dealt with the summary of the Akil Formation.

Tura Formation: In the Tura Formation hydrocarbon were discovered recently by the ONGC. Near the top of the formation there are claystones and shales and occasional coal seams of Paleocene-Early Eocene age. Foraminifera in this formation include *Miscellania miscella*. The inferred environments of deposits vary from fluvio-deltaic to shallow marine (Pandey and Dave 1998).

Case example: Late Eocene-Earliest Oligocene

Miller et al. (2005) recognized two major fluctuations in Sea-level with amplitude of about 50 m but with an overall rise during Late Eocene when compared to Middle Eocene. Haq et al. (1987) predicted a drop in global sea-level by amplitude of 50/60 during the Late Eocene. Interestingly two cycles of Sea-level are recognized in PY structures, offshore of Cauvery basin.

In the exposed section at the location of Ramania dam, Kutch a drop in Sea-level by on an amplitude of 40/50 m is recognizable band on foraminifera. At this location carbonates of equivalent to zone P14 (may be a part) of Middle Eocene are overlain by strata bearing *Nummulites fitcheli* of Early Oligocene with a disconformity (duration of hiatus 3.5 ma). In the Berwali *nadi* section which is located some 20 km. west of the Ramania dam represents slightly basin ward facies. Shukla (1996) found about 1.5 thick strata of Late Eocene.

In major part of Mumbai offshore, the Bassein Limestone of Middle Eocene is overlain by the Early Oligocene strata with *N. fitcheli*. However, at a few areas along depressions, Late Eocene strata are present. However, a drop in sea-level along western onshore and offshore is well established.

On the other hand, stratigraphic record suggests a relative rise in Sea-level almost through out the basin. In the Cambay basin in the subsurface successions there was a rise of more than 200 m from the late Middle Eocene to the Late Eocene.

A similar deepening (rise of Sea level) took place during the deposition of the Disang Formation (group of some authors) across a long belt of Assam-Arakan, NE India. Examples of the Cambay Basin along the Western part of India and Assam-Arakan across NE India demonstrate that the Sea level rise in these basins was due to higher/rapid rate of tectonic subsidence compared to drop in eustatic seal level.

Early Oligocene

Haq et al. (1987) estimated a major rise of Sea-level in this interval. Indian data from Cauvery Basin, Bengal Basin and some areas of Bombay offshore support Haq et al. (1987). But the rapid rise due to tectonic subsidence appears more in these three Indian basins. Miller estimated a relative drop. The present author visualizes another possibility that estimate of major/abrupt rise in some areas of India could be due to migration of bathyal Uvigerinids (on whose basis the paleodepths were estimated) to shallower waters during cold (low) temperatues of Early Oligocene or paleowater depth. Estimates based on modern Uvigerinids may not be fully valid for Oligocene. Narayanan and Raju (1996) estimated hiatus of a duration of 2.5–5 My covering the interval of zones P_{20}/P_{21}. This event matches with Sea-level drop estimated by both Haq et al. (1987) and Miller et al. (2005).

Late Oligocene

Haq et al. (1987) estimated a major relative drop in sea-level with three pulses. There is relatively a major drop in India basins of KG, Mahanadi, Cambay and Bombay offshore. In Kutch 3–4 pulses are recognized but the drop when compared to the Early Oligocene is very minor. Hiatus immediately after LAD of *Nummulites fitcheli* (approximately close of zone P_{21}, *G. opima* zone) may match with the estimate of Miller et al. (2005). The Late Oligocene transgression (Waiorian) after P_{20}/P_{21} hiatus is prominent in Kutch and Bombay offshore.

Case example: Oligocene

Pekar et al. (2004) recognized eight cycles of Sea-level change within Oligocene. Interesting observations/results from India are: (i) seven to eight para sequences/cycles are recognizable in the Oligocene exposures in Kutch. (ii) Bhandari (personal communication) recognized seven zones/biochorns (including two sub zones) within the exposed succession of Kutch. He also recognizes eight ostracod zones in the exposed Oligocene strata of Kutch as well as in the subsurface successions of Mumbai-4 and AKM-1, 6–7 cycles are recognized.

Exposures in Kutch and continuously cored Oligocene strata of Mumbai offshore should be play grounds for future research.

Middle Eocene (Duration: 10 My)

Both Haq and Miller estimated a relatively high sea-level during zones P10–P11-lower part of P12. This interval is represented by a major hiatus in Kutch outcrop area while there are evidences of a lower sea-level in Bombay offshore and Cambay Basin. Miller et al. (2005) estimated a relative fall of sea-level during zones upper P12, P13 and P14. There is a maximum flooding during P13 in Kutch and Rajasthan. Haq showed a relatively high sea-level during the entire Middle Eocene. Zone P13 represents a regional MFS in Rajasthan, and Kutch.

Late Eocene (Duration: 3.9 My)

Miller et al. (2005) recognized two cycles within this interval with an overall rise in sea-level compared to zones P13 and P14 of Middle Eocene. Our data of Cambay Basin, Assam and Nagaland support Miller. On the other hand, Haq et al. (1987) showed a relative drop in sea-level of an order of 70–80 m. Indian records from Kutch, Bombay offshore, KG Basin support (Haq et al. 1987). These are clear evidences that tectonic subsidence versus rate of sediment accumulation played a major role. An interesting observation by Narayanan is: Eocene-Oligocene marine succession is continuous in the Murud depression while a hiatus of ~3 My across Eocene-Oligocene boundary in BHE-1 within the Bombay Offshore.

Early Oligocene (Duration: 5.8 My)

Haq et al. (1987) estimated a major rise of sea-level in this interval. Indian data from the Cauvery Basin, Bengal Basin and some areas of Bombay offshore support Haq et al. (1987). But the rapid rise due to tectonic subsidence appear more in these three Indian basins. Miller et al. (2005) estimated a relative drop. Uvigerinids (on whose basis the paleodepths were estimated) to shallower waters during cold (low) temperatures of Early Oligocene or paleo water depth estimates based on modern Uvigerinids may not be fully valid for Oligoene. Narayanan and Raju (1996) estimated hiatus of a duration of 2.5–5 My covering the interval of zones P20/P21. This event matches with sea-level drop estimated by both Haq et al. (1987) and Miller et al. (2005).

Late Oligocene (Duration: 5.1 My)

Haq et al. (1987) estimated a major relative drop in sea-level with 3 pulses. There is relatively a major drop in Indian basins of KG, Mahanadi, Cambay and Bombay offshore. In Kutch 3–4 pulses are recognized but the drop when compared to Early Oligocene is very minor. Hiatus immediately after LAD of Nummulites fichteli (approximately close of zone P21, *G. opima* Zone) may match with the estimate of Miller. Late Oligocene transgression (Waiorian) after P20/P21 hiatus is prominent in Kutch and Bombay offshore.

Systematic Stratigraphy

(Abbrevations given in the paper are: AU: Author; HOLO: Holo stratotype/type section; LB: Lower boundary, UP: Upper boundary; LITH: Lithology, THICK: Thickness; ENVIRON: Environment of deposition)

Zanskar-Spiti Basin

Zanskar marine sediments of Latest Precambrian to Early Cambrian and Early Paleoecene to Middle Eocene are known. There is a hiatus from Middle Eocene to Pleistocene. Three formations are briefly described below.

Stumpata Formation (*Au*: Gaetani and Garzanti 1991)

LB: Normal contact. **UB**: Normal contact with overlying Dibbling Formation (Gaetani and Garzanti 1991). **LITH**: White, brown-weathered quartz arenite, increasing in thick towards the East **THICK**: 10–12 m. **BIOTA**: Devoid of fauna **AGE**: Early Paleocene. **ENVIRON**: Transition zone, mouth bar deposits of a fluvial dominated delta.

Dibling Formation (*Au*: Gaetani and Garzanti 1991)

HOLO: Not defined, *Hypo*: Dibling village section (Spanboth Valley, Zanskar); **LB**: Normal contact with underlying Stumpata Formation. **UB**: Normal contact with overlying Chulungla Formation. **LITH**: Silty marls, dark marly packstone to wackstone, calcite balls with wackstone **THICK**: 100 m thick **BIOTA**: Rotallids; *Lockhartia* sp., miliolids, *Chrysalidinae*; *Daviesina danieli*; *Operculania* sp., *Ranikothalia* sp., *Furcoporella diplopora*; *Daviesina langhami*; *D. khatiyahi*, *Ovulites margaritula* etc. **AGE**: Paleocene **ENVIRON**: Open shelf carbonate sequence.

Chulung La Formation (*Au*: Gaetani et al. 1986)

LB: Conformable contact with Dibling Formation. **UB**: Its upper contact is unconformable **LITH**: Nummulitic limestone, red shale/siltstone/slate and impure fine grained sandstone, volcanic arenite. **THICK**: 160 m thick (>160–200 m) **BIOTA**: *Morozovella velascoensis* and *M. aragonesis* **AGE**: Early Eocene. **ENVIRON**: The sequence testifies to the progradation of a fluvial-dominated deltaic system in a shallow lagoon. Sedimentary features indicate tropical climate and high sedimentation rates.

Lesser Himalaya

Proterozoic sediments are established. There is a major hiatus from Cambrian to Maastrichtian, Maastrichtian to Pleistocene events are well known. Three formations are briefly described below.

Dagshai Formation (*Au*: Medlicott and Blanford 1879)

HOLO: Dagshai hill in H.P. **LB**: Subathu Formation. **UB**: Kasauli Formation/Dharmsala, Up. Murree. **LITH**: Red shale and sandstone intercalations. **BIOTA**: Ill preserved plants, Frag. Vertebrates, Bones. **AGE**: ?Late Eocene-Oligocene, Lr. Miocene. **ENVIRON**: Brackish continental water interval.

Subathu Formation (*Au*: Medlicott 1864)

HOLO: Subathoo H.P. *Hypo*: Various parts of Lesser Himalayan region; **LB**: Kakara Formation. **UB**: Dagshai/Dharmsala/Murree. **LITH**: Calcareous shale splintery firzile grey green khaki colure with minor limestone and sandstone. **THICK**: Maximum thick 410 m. at Subathu type area minimum thick 92 m at Kalakot (Jammu area). *LV*: Outcrops of Subathu Formation are exposed at various localities from Garhwal to Jammu area. **BIOTA**: 8 foraminiferal zones are recognized. **AGE**: Ypresian—early Lutetian. **ENVIRON**: Shallow marine.

Kakara Formation (*Au*: Juyal and Mathur 1990; Kakara original Au: Srikantia and Bhargava 1967)

HOLO: (Kakara Village) Tal Valley. **LB**: Unconformable contact with the underlying Cambrian/Pre Cambrian rocks of the Lesser Himalayan formations, i.e. Simla Group in type section, Arki, Blaini in Halog, Krol in Mangarh-Bagar, Dharamkot Limestone in Bilaspur (H.P.) (H.P.), and Sriban Limestone in Jammu area. **UB**: Subathu Formation. **LITH**: Grey limestone in Garhwal area **Age:** Maastrichtian to Paleocene. Carbonaceous Shale local coal (Kalakot) with lenses of limestone. **THICK**: 60 m in Garhwal area. Thickness varies in other area. **BIOTA**: Biozones and ostracods in Garhwal area and age diagnosis foraminifers and ostracodes in other area. **AGE**: Palaeocene–Maastrichtian. **ENVIRON**: Marine.

Ganga, Punjab and Purnea Basin

Proterozoic to Early Paleozoic sediments are known in the subsurface. Late Paleozoic to Eocene. One formation is described. Late Eocene to Oligocene are recognized in subsurface in well Matera-1, U.P.

Matera Formation (*Au*: Shukla et al. 1993). The name is proposed after Matera, a village NW of Bahraich U.P. These sediments have been encountered in Matera-1 whereas in all the other wells drilled in the Ganga basin, these Oligo-Miocene (?) sediments have not been encountered so far.

HOLO: Matera-1, Latitude; 270° 43′ 22″N. Longitude: 810° 36′ 51″E. Drilled depth: 3927 m. The top of the formation is at 3400 m (Fig. 24). **THICK**: 400 m **LB**: It is underlain unconformably by Bahraich Group. The contact being a pronounced unconformity can be easily identified in the type section. **UB**: The formation is overlain conformably by Lower Siwalik. **LITH**: Three lithountis are identified in the type section. Litho unit I (3280–3400): It consists of dark grey shale which is moderately hard and silty in nature and shows splintery characteristics. Occasionally siliceous matter has also been reported. It is noncalcareous and pyretic in nature. Litho unit II (3190–3280 m): It is a thick sandstone unit with thin intercalations of siltstone and claystone at places. Litho unit III (3000–3190): It is composed of intercalation of sandstone, claystone and siltstone and is considered to be a transition between Lower Siwalik and Matera Formation (core analysis report C C # 4). **BIOTA**: The palynomorph assemblage recorded in core no 5 and 6 is characterized by *Polypodisporites* sp., *Trialata* sp., *Inverturopollenites* sp., *and Caryapollentes* sp., **AGE**: Oligo-Miocene (?) age is assigned to the formation on the basis of palynofossils and heavy mineral analysis. **ENVIRON**: The sediments were deposited in a mixed braided/anastamosing and possibly meandering fluvial system.

Himalayan Foreland Basin

Kakara, Subathu and Dagshai are well known and details on biostratigraphy and sequence stratigraphy are published by Bhatia et al. (2013). Litho-bio- and chronostratigraphy is shown in Fig. 6.

Subathu Formation (*Au*: The term Subathu was first used by Medlicott (1864) to a group of pre-Siwalik Tertiary rocks occurring in Simla Hill area, after the small town Subathu, southwest of Simla. Medlicott's Subathu stage that formed the lowest unit of his "Sub- Himalayan Series" Sirmur Series included Subathu, Dagshai and Kasauli stages and the name "Subathu" was restricted to the lower most sequence. Since then, the term Subathu has been in wide usage for designating the Paleocene/Eocene sediments of the Himalayan foothills). **THICK**: Repetition of green and red coloured rocks in Lower Tertiary sequence is mainly due to thrusting and faulting due to which calculation exacts formation thick is difficult. The thickness of the Subathu Formation however, varies from 180 to 840 m in Himanchal Pradesh while in Jammu and Kashmir its thick ranges from 35 to 225 m. **LB**: It is noted that Subathu rests unconformably on the pre-Tertiary rocks. **UB**: There is however, a great deal of controversy on the upper contact of Subathu with the overlying Dharmasala. According to Bhatia and Bhargava (2006) the concept of a greater than 10 My hiatus in the foreland basin based on 40 Ar/39 Ar dares of a single detrital muscovite grains from the supposed Dagshai arenite is no longer tenables. **LITH**: The Subathu is characterized by olive green-dark grey, splintery shale, dark grey, fossilferous, carbonaceous shale and calcareous shale, red shale in basal and upper parts, with bluish grey to grayish, very hard and compact fossiliferous limestone. Limestones are best developed in the northwestern areas and gradually diminish into the southeast. The characteristic heavy minerals include rutile, zircon and tourmaline. **BIOTA**: The faunal content of Subathu outcrop has revealed the dominance of genera *Nummulites*, *Assilina*, *Rotalia*, *Lockhartia*, *Dictyconoides*.

Paleogene sediments are widely distributed from Pakisthan to Nepal in the Himalayan Foreland Basin (Fig. 7). The development of Paleogene succession on the Indian shelf is related to gradually accelerating to transgression of sea which seems to have reached its acme during Middle Eocene when extensive carbonate facies developed all around the Indian shield are shown in Fig. 8, Pandey and Dave (1996).

Shimla Hills: Subathu Formation

Faunal succession: A detailed account on larger foraminieral succession, paleoenvironmental setting and sequence stratigraphic frame work of the Paleogene of Himalayan Foreland Basin- Shimla Hills was presented by Bhatia et al. (2013). The data are shown in chart.

AGE: Considerable paleontological and palynologic work has been done in recent years to give an idea about the age of Subathu. Bhatia et al. (2013) assigned Late Thanetian to Early Priabonian age for the Subathu Formation. **ENVIRON**: The

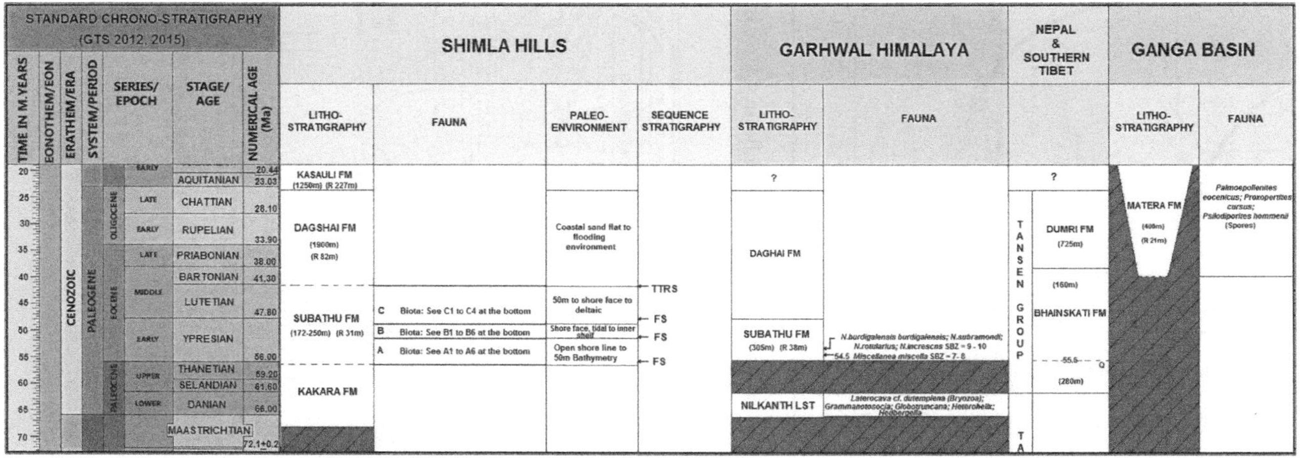

A BRIEF ON EACH FORMATION :

SHIMLA HILLS: Subathu Formation: Succession A: A2 Biota: Daviesina garumnensis (Tambareau), D. ruida (Schwager), D. khatyahi Smout, D. langhami Smout, D. danieli Smout, Lockhartia hunti Davies, Rotalia trochidiformis Lamarck, Pseudohastigerina micra(cole),Alveolina minervensis Hottinger and echinoid spines SHIMLA HILLS: Subathu Formation: Succession ***A: A3 Biota: N. burdigalensis burdigalensis de la Harpe, N. partschi de le Harpe, N. rotularius Deschayes, N. subramondi, Assi lina spinosa Davies, A. plana Schaub, Dasycl adacean algae, Cyanopolia sp, Distichoplax sp, Echinoid, Cyamidia numulitica and several species of ostracoda.*** **SHIMLA HILLS: Subathu Formation: Succession** ***A: A4 Biota: Cordiopsis subathooensis and Turritella subathooensis, Nimmulites rotularius Deschayes, Assilina plana Ishatherium subathuensis, Himalayacetus subathuensis.*** **SHIMLA HILLS: Subathu Formation: Succession** ***A: A5 Biota: Selechi Galeocrdo sp, Galeorhinus sp, Reptilia Pristichampsus sp; Batoidea (Rays and skates)- including Subathurura casieri (n.gen., and n. sp.), Dasyatis rafinesquei (n.sp.), Holostei- Pycnodus bicolestata; Telostei- Kankotodus capettai***

SHIMLA HILLS: Subathu Formation: Succession ***B: B2 Biota: Unidentifiable oysters, Cordiopsis subathuensis d' Archaic and Haime and bine fragments of selachians.***

SHIMLA HILLS: Subathu Formation: Succession ***B:B3 Biota: Nummulites rotularius and Assilina sp.***

SHIMLA HILLS: Subathu Formation: Succession ***B:B4 Biota: Nummulites planularius Lamarck, N. praelucasi Douville, Assilina placentula(Deshayes), Dictyoconides flemingi, Trochammina sp, Textularia punjabensis Haque, Triloculina trigonula(Lamarck), Stainforthia dubia Haque, Stianforthia sp and Cibicides libatulus.*** **SHIMLA HILLS: Subathu Formation: Succession** ***B:B5 Biota: Nummulites sp. ex gr. N. tobeleri Gyphithyreus.*** **SHIMLA HILLS: Subathu Formation: Succession** ***B:B6 Biota: Assilina laxispira de La Harpe, A. cuvillieri Schaub, A. Placentula grande Schaub, A. papillata Nuttall, FAD of Nummulites beaumonti d' Archaic and Haime and bryozoan genus Steginoporella.*** **SHIMLA HILLS: Subathu Formation: Succession** ***C:C1 Biota: Assilina spira abrardi Schaub, A. hamzehi Mojab, Nummulites obesus, d'Archaic and Haime, N. lehneri Schaub, N. gallensis, Heim, and N. praediscorbinus Schaub, besides oysters.*** **SHIMLA HILLS: Subathu Formation: Succession** ***C:C2 Biota: A. cuvillieri Schaub and A. medanica Schaub, besides Nummulites discorbinus.*** **SHIMLA HILLS: Subathu Formation: Succession** ***C:C3 Biota: Pediastrum and chara Raskvella pepckii.***

Fig. 6 Paleogene Litho- Bio-Chrono- sequence stratigraphy and paleoenvironmental settings in the Himalayan Foreland Basin, India. Compiled by D. S. N. Raju, O. N. Bhargava and Birendra P. Singh

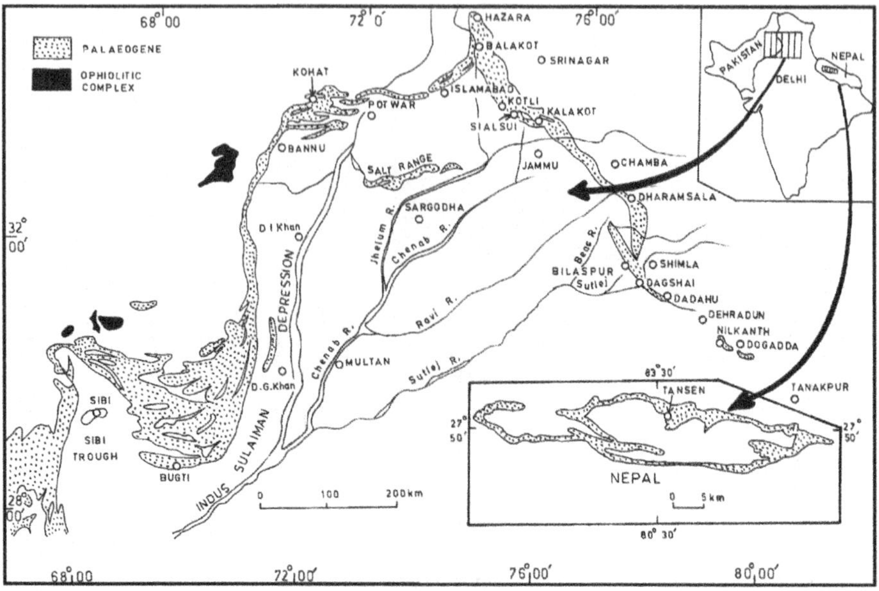

Fig. 7 Distribution of the Paleogene sediments of the Himalayan Foreland Basin. (After Bhatia and Bhargava 2006)

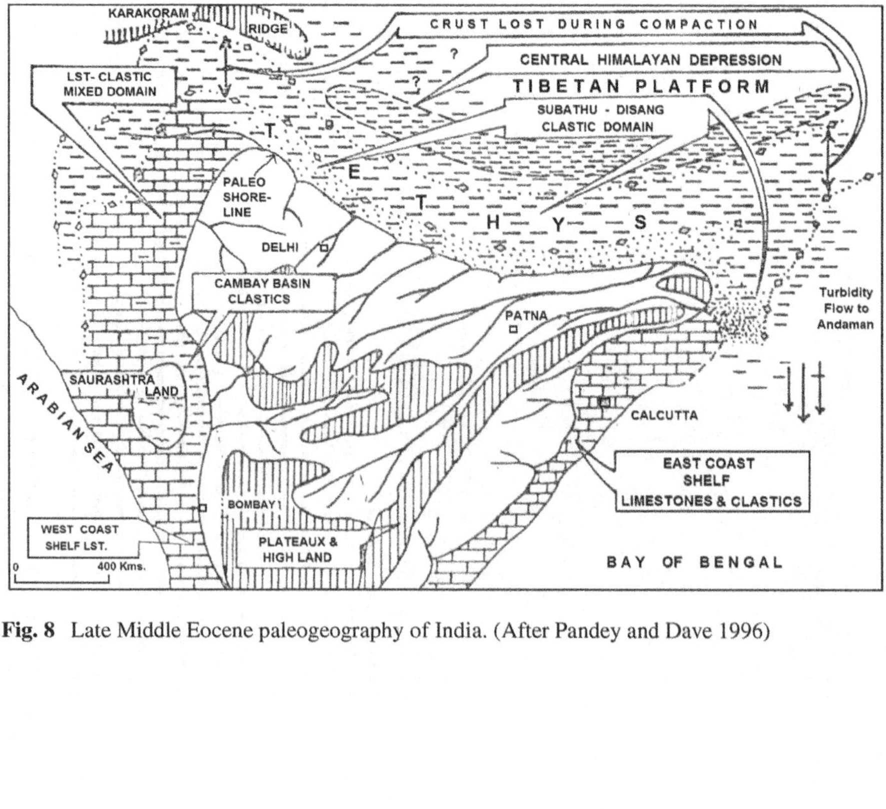

Fig. 8 Late Middle Eocene paleogeography of India. (After Pandey and Dave 1996)

Subathu sediments were deposited in open-sea and shallow marine environments. Mainly open marine conditions appear to have prevailed in the upper part of the sequence in the north western areas near Punch and Jammu as indicates by green shales and fossiliferous limestone. Most of the fossiliferous zones, contain mainly *Nummulites* which is restricted to tropical and subtropical waters and occurs abundantly in the shallower parts of the sublittoral (neritic) depths. In Triyath Nala traverse, a few significant ostracod genera have been recorded which are indicative of shallow normal marine condition.

Dagshai Formation: This formation is encountered only in a few deep wells in the basin. The base of the section is not penetrated in any of the wells.

HOLO: Dagshai; B **THICK**: The thick of the Dagshai is around 1000 m in Jammu area measured in relatively undisturbed section (Ranga Rao 1986). In Himachal Pradesh maximum thick is measured near Jwalamukhi area. In the subsurface maximum thick penetrated was 2300 m in Jwalamukhi–B, 1303 m Jwalamukhi-2, 1164 m in Balt\h-1 and 1028 m in Suruinsar-2. **LB**: The formation is underlain unconformable by Subathu. The mangnitude of hiatus is, however, not known definitely. **UB**: The Dagshai is conformably overlain by the Kasauli with a gradational contact. In the outcrop the boundary is commonly marked on the appearance of first laterally extensive sandstone body in the succession, above which the entire section is more arenaceous. **LITH**: In the northwestern part of the basin this formation is mainly argillaceous and contains thick sections of deep purple to chocolate coloured shale siltstone with subordinate sandstone and several calcrete horizons. In well Suruinsar-2 this section is composed of reddish brown, micaceous, calcareous siltstone and grey, fine grained, calcareous and occasionally micaceous sandstone. In Himachal Pradesh Dagshai is predominantly argillaceous with laterally impersistant sandstone. In the subsurface consists of reddish brown to chocolate brown, massive and occasionally calcareous shale and light grey to greenish grey, fine to very fine grained non calcareous sandstone. **BIOTA**: No diagnostic fossile of any significance. **AGE**: If the K/Ar dating of Khardung volcanics in Ladakh as 25–35 My (Sharma and Gupta 1983) are correct, Oligocene in the Himalayan region may be represented by volcanic activity with negligible sediment accumulation. Magnetic polarity stratigraphy across Udhampur Syncline suggests and age of 19.88 $\pm$1.23 Ma for the outcropping base of Upper Muree (=Kasauli) (Ranga Rao 1983). **ENVIRON**: The Dagshai represents diachronous deposition on the feather edge of the foreland basin. Depositional environments comprised semi-arid meander plain with sediment source from the rising but still distant proto-Himalayas to the north in which direction the Dharamshala foreland basin center was located.

Assam, Assam-Arakan Formation

Garo Hills oldest rocks are of Permian age after another hiatus of 108 My, the succession of Campanian to Miocene succession is known. The litho- bio- and chronostratigraphic succession of North East basins from Garo Hills to Upper Assam is shown in Fig. 9.

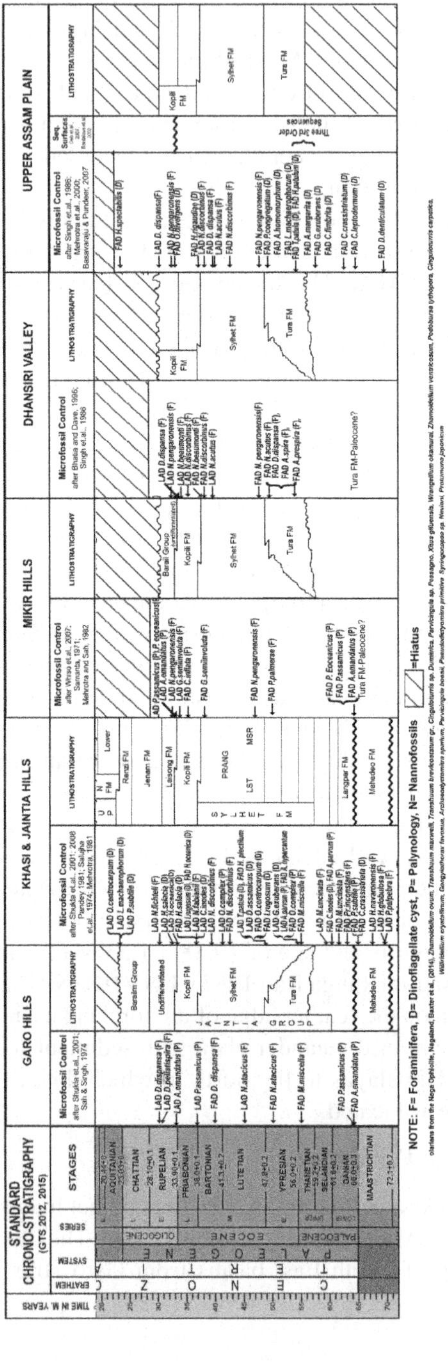

Fig. 9 Paleogene Litho- Bio- Chrono- Stratigraphic Succession of North East basins from Garo Hills to Upper Assam basins, India. Compiled by D. S. N. Raju

Generalised stratigraphy of Assam shelf showing formation names litho-facies and hydrocarbon occurrences are shown in (Fig. 10), Wandrey (2004b). The Paleogene formations are briefly described below.

Garo Hills

Nine formations and alluvium deposits mapped in the Garo Hills are described here.

Tura Formation (*Au*: Fox *in* Heron 1937)

HOLO: Around Tura Village in western Garo Hills. **LB**: Unconformably overlies the Mahadeo Formation. **UB**: Conformably overlain by thick limestones of the Sylhet Formation. **LITH**: A thin conglomerate band dominated by vein quartz and gneissic pebbles at the base and claystones/shales and occasional coal seams at the top in the type section. **THICK**: 244–500 m in the western Garo Hills. Ext: Typically exposed in west Garo Hills and continues to the Shillong Plateau, Mikir Hills, Kopili Valley, Dhansiri Valley and Upper Assam. **BIOTA**: Larger foraminifera viz. *Miscellanea* sp., *Nummulites* sp. and *Discocyclina* sp. **AGE**: Paleocene to Early Eocene. **ENVIRON**: Fluvial deltaic to shallow marine.

Sylhet Formation (*Au*: Evans 1932)

HOLO: Um Sohryngkew river section, south of Cherapunji. **LB**: Conformable and gradational contact with the Tura Formation. **UB**: Conformable and gradational contact with Kopili Formation. **LITH**: It consists mainly of massive limestone with intercalated sandstone and calcareous shales **THICK**: In Lakwa Well it is 130 m. Ext: The Sylhet Formation is exposed along the southern edge of the Shillong plateau from Tura in the west to the Kopili Valley in the east, it continues to the Mikir Hills in the northeast. The Sylhet Formation continues to the subsurface of Dhansiri Valley and Upper Assam and in the subthrust block of the Schuppen Belt, (Deshpande et al. 1993). **BIOTA**: Several species of larger foraminifera. **AGE**: Late Paleocene to Late Eocene. **ENVIRON**: Shallow shelf condition.

Kopili Formation (*Au*: Evans 1932)

HOLO: Khorungma section, in Kopili Valley. **LB**: Overlie the Sylhet Formation conformably with a gradational contact. **UB**: Conformably overlain by Barail Group. **LITH**: Dark grey, hard compact and splintery shales intercalated with thin limestone and sandstone. **THICK**: 500 m or more in the Garo Hills. **BIOTA**: Foraminifera fossils include *Nummulites fabianii*, *N. beaumonti*, *N. pengaronensis*, *Discocyclina* sp., *Hantkenina alabamensis*, *G. (T). cerroazulensis* **AGE**: Late Eocene. **ENVIRON**: Outer shelf to transitional, mainly lagoonal.

Barail Group (*Au*: Evans (1932) as Barail series)

LB: Conformable contact with the underlying Kopili Formation. **UB**: Unconformable contact with the overlying Angartoli Formation. **LITH**: Sandstones interbedded with shales. **AGE**: Late Eocene to Oligocene. **ENVIRON**: Brackish.

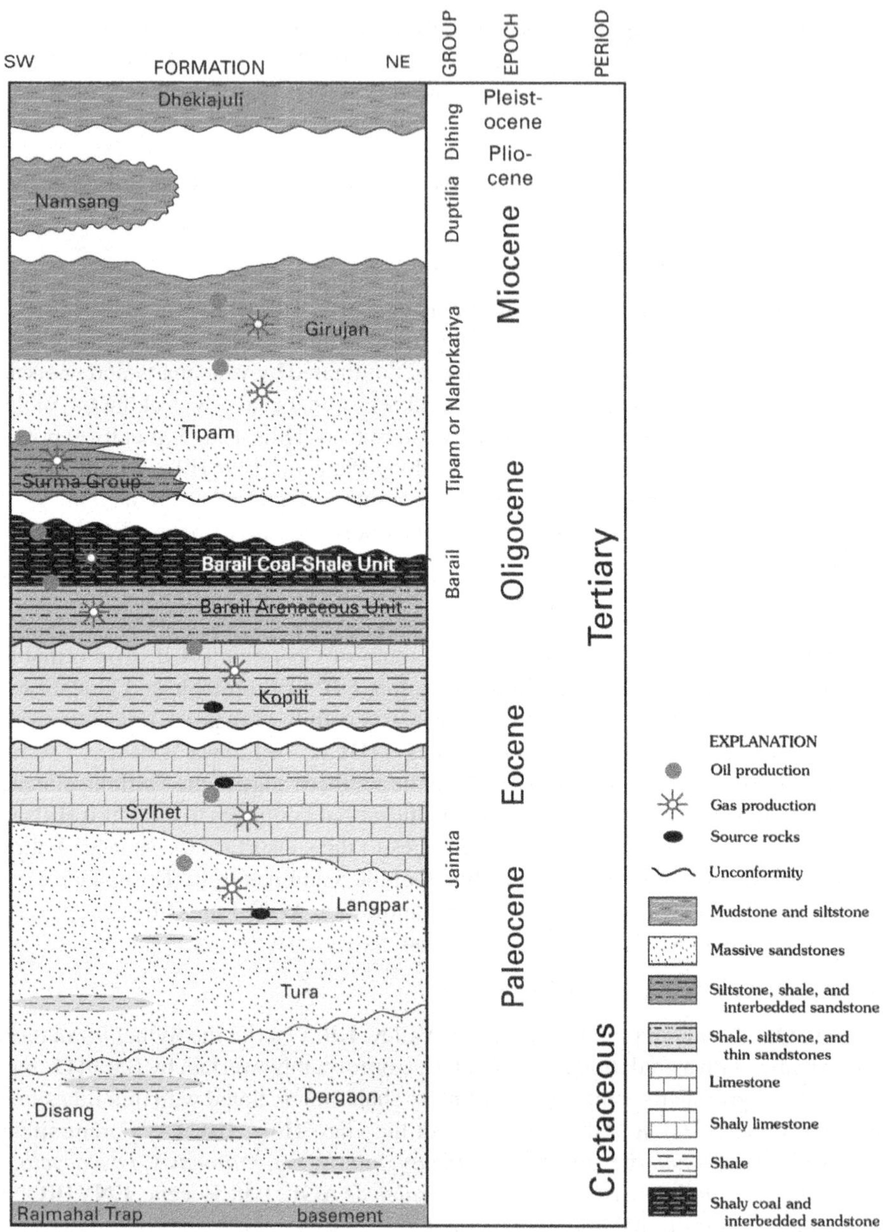

Fig. 10 Generalized stratigraphy of Assam Shelf (After Wandrey 2004b)

Khasi and Jaintia Hills

Oldest rocks are Sylhet rocks. They are overlain by the Mahadeo Formation. Hiatus of duration 13 My. The Paleogene formations and members are described below.

Langpar Formation (*Au*: Medlicott 1869)

HOLO: Langpar spur, south of Mawsmai in the Khasi Hills. **LB**: Conformably overlies the Mahadeo Formation. **UB**: Conformably overlying by the Theria Formation of the Jaintia Group. **LITHO**: It consists mainly of calcareous shales with sandy limestones and interbedded sands towards the top. **THICK**: A maximum thick of 95–151 m. **BIOTA**: *Guembelitria cretacea, Chiloguembelina* sp., *Subbotinid* sp., *Gglobigerina and Turborotalia eugubina* etc. from Um Sorhyngkew River section, Meghalaya. **AGE**: Maastrichtain to Danian age. The K/T boundary in the Um Sohhryngkew River section has been identified near the base of Langpar. **ENVIRON**: Iridium enrichment at Cretaceous/Tertiary boundary in Meghalaya, Bhandari et al. (1987). Marginal marine to open marine environment.

Therria Formation (*Au*: Medlicott 1869)

HOLO: Therriaghat section. **LB**: Gradational and comformable contact with Langpar Formation. **UB**: Overlain conformably by a massive Lakadong Limestone of Sylhet Formation. **LITH**: It consists of massive Lakadong limestone at the base followed by sandstones at the top **THICK**: A maximum of 95 m. **BIOTA**: Lower limestone member is poorly fossiliferous yielding algal remains and rare foraminifera. The coals in the upper sandstone member yield leaf impressions. **AGE**: Late Paleocene. **ENVIRON**: Shallow lagoon to fluvial.

Sylhet Formation

In the Khasi and Jaintia Hills, Sylhet Formation is divided into Lakadong Limestone, Lakadong Sandstone, Umlatodoh Limestone, Narpuh Sandstone and Prang Limestone members. Each member is described separately.

Lakadong Limstone Member (*Au*: Wilson and Metre 1953)

HOLO: Um Sohryngkew River. **LB**: It has a comformable lower contact with the Therria Formation. **UB**: Gradational, conformable contact with the Lakadong Sandstone Member. **LITH**: Pinkish grey massive hard, crystalline limestone. **THICK**: Thick varies between 220 and 250 m. **BIOTA**: *Miscellania miscella, Lokhartia haimei, Discocyclina ranikotensis, Fasciolites primaevis, Nummulites thalicus* and *Globorotalia pseudomenardii*. Paleocene calcareous algae viz. *Sporolithon* sp., *Mesophyllum* sp., *Lithoporella* sp., *Sporolithon* sp., *Amphiroa* sp., *Lithothamnion* sp., etc. from Khasi Hills Meghalaya. A new genus *Lakadongia* is described from the Thanetian (Paleocene) of Meghalaya, NE India by Matsumaru and Jauhri (2003). **AGE**: Based on the occurrence of typical Paleocene marker algae *Distichoplax biserialis,* the age of the limestone is considered as Thanetian (Late Paleocene); **ENVIRON**: Jauhri et al. (2006) identified two facies associations in the carbonate sequence: (i)

Facies A characterising the lower 50 m part and (ii) Facies B characterising the upper 50 m part. The facies A corresponds to the Thanetian sedimentation cycle in which deposition occurred in a relatively low-energy environment on a shallow subtidal ramp (protected lagoon). The Facies B correlates with the Ilerdian sedimentation cycle during which deposition occurred under low to moderate energy conditions on a relatively deeper mid-uppermost outer ramp, which allowed development of sediment-binding coralline algae and foraminifera such as discocyclinids and *Ranikothalia.*

Lakadong Sandstone Member (*Au*: Wilson and Metre 1953)

HOLO: Um Sohryngkew River. **LITH**: Dirty white, light yellow, fine to coarse grain, subangular to angular sandstone. **LB**: Gradational contact with underlying Lakadong Limestone. **UB**: Gradational contact with the Umlatodoh Limestone. **THICK**: In the type section it is 25 m. **BIOTA**: Leaf impressions. *Archaeolithothamnium* sp., *M. meghalayensis*, *Distichoplax biserialis*, *D. raoi*, *Amphiroa* sp. and *Jania occidentalis*. Jauhri et al. (2006), *Glomalveolina*, *Polystrata alba*, *Aberisphaera*, *Ranikothalia nuttalli*, *Nummulites* sp., *Discocyclina* sp. **AGE**: Late Paleocene to Early Eocene. **ENVIRON**: Marginal marine to deltaic conditions. Jauhri et al. (2006) suggested carbonate plateform.

Umlatodoh Limestone (*Au*: Wilson and Metre 1953)

HOLO: Um Sohryngkew River. **LB**: Gradational contact with Lakadong Limestone Member. **UB**: Gradational contact with Nurpuh Sandstone Member. **LITH**: Dark grey to pure bluish grey, crystalline and oolitic limestone. **THICK**: 55 m in the type section. **BIOTA**: reported *Lithophyllum*, *Lithothamnion*, *Coralline Ovulites*, *Halimeda* from South Jaintia Hills, Meghalaya. **AGE**: Eocene. **ENVIRON:** shallow, warm shelf environment of normal salinity with some minor sea-level changes within the transgressive phase.

Nurpuh Sandstone Member

HOLO: Um Sohryngkew River. **LB**: Gradational contact with the underlying Umlatodoh Member. **UB**: Gradational contact with overlying Prang Member. **LITH**: White to dirty white, medium to coarse grained quartzose sandstone. **THICK**: 20 m thick in the type section. **BIOTA**: *Orbitolites complanata* and *Nummulites atacicus.* **AGE**: Eocene. **ENVIRON**: Shallow marine to deltaic conditions.

Prang Member (*Au*: Wilson and Metre 1953)

HOLO: Um Sohryngkew River. **LB**: Gradational contact with the underlying Nurpuh Member. **UB**: Gradational contact with the overlying Kopili Formation. **Lith**: In the type section, it is represented by grey to greyish white crystalline and massive limestone. **Thick**: Thick varies between 130 and 300 m. **BIOTA**: *Lithoporella*, *Lithothamnion*, *Sporolithon*, *Spongites* and *Actinoporella* from South Jaintia Hills, Meghalaya. Record of *Discocyclina dispansa*, *D. omphalus*, *N. acutus*, *N. discorbinus*, *N. fabianii*, *Assilina granulosa*, *A. spira*, *Pellatispira* sp., *Truncorotaloides rohri*,

Hantkenina alabamensis and Turborotalia cerroazulensis are mentioned. **AGE**: Middle to Late Eocene. **ENVIRON**: A shallow water depth with more or less low to moderate energy condition.

Kopili Formation

See other details in Garo Hills.

LB: Conformable contact with the underlying Prang Member. **UB**: Conformable contact with the overlying Laisong Formation of Barail Group. **THICK**: 350–355 m.

Barail Group

Laisong Formation (*Au*: Laisong stage of Evans 1932; Emend: Laisong Formation of Ranga Rao 1983 and many later workers)

HOLO: Near Laisong Village along Jenam River, *Hypo*: Silchar–Haflong road. **LB**: Conformable contact with the underlying Kopili Formation. **UB**: Gradational contact with the Jenam Formation. **LITH**: In the reference section, it consists of dark grey, fine to medium grained, thin to thick bedded sandstones with thin shaly interbeds near the top. **THICK**: 185 m in the Larang River section in the Khasi-Jaintia Hills. **BIOTA**: Ranga Rao (1983) reported *Nummulites chavannesi*, *Operculina* sp. and *Biplanispira* sp. **AGE**: Late Eocene to Early Oligocene. **ENVIRON**: Deltaic with occasional marine incursion.

Jenam Formation (*Au*: Jenam stage of Evans 1932; Emend: Jenam Formation of Ranga Rao 1983).

HOLO: Jenam River section, *Hypo*: Silchar–Haflong road section **LB**: Gradational contact with the Laisong Formation. **UB**: Gradational contact with the overlying Renji Formation. **LITH**: Comprises of thin bedded, grey to brownish grey, fine to medium grained sandstone, dark grey to grey shale and sandy shale giving way to predominantly carbonaceous shale near the top. **THICK**: In the type section it is 1200 m and in the Prang River section it is 304 m. **BIOTA**: A few arenaceous foraminifera are recorded from Schuppen Belt. Palynoflora from Silchar–Haflong road section includes *Tricholpites reticulatus*, *Mayeripollis naharkotensis*, *Lygopodiumsporites*, *Biretisporites oligocenicus* (Kumar 1962). **AGE**: Early Oligocene. **ENVIRON**: Deltaic environment.

Renji Formation (Au: Renji Stage of Evans 1932; *Emend*: Renji Formation of Ranga Rao 1983)

HOLO: Renji Hills north of Badarpur. **LB**: Gradational contact with the Jenam Formation.

UB: Unconformably overlain by the Surma Group. **LITH**: An arenaceous unit with thin shale and sandy shale interbeds. **THICK**: In the type section, it is 700 m which increases to 950 m in the Jenam River section. Ext: It is well exposed in the Barail Range, Schuppen Belt, south eastern part of Shillong Plateau and eastern Manipur area. **BIOTA**: Palynological assemblages comprising of *Arecipites communis*, *Couperipollis rarispinous*, *Margocolporites* sp. etc. are recorded from the outcrops along Silcher-Halflong road section. **AGE**: Late Oligocene. **ENVIRON**: Deltaic environment.

Mikir Hills

The rock succession include Mikir traps of Early Cretacous to Pleistocene are known. The Paleogene formations are briefly mentioned below.

Tura Formation

See details in Garo Hills.

LB: Unconformably overlying the Mikir Trap. **THICK**: 20–30 m.

Sylhet Formation

See details in Garo Hills.

THICK: It is less than 100 m. **BIOTA**: *Nummulites atacicus, N. beaumonti, N. burdigalensis, N. discorbinus, N. garnieri, N. globulues, N. obtusus, Assilin aspira, Discocyclina javana, Bolivina* sp., *Bulimina pupoides, Cibicides lobatulus, Lagena striata, L. sulcata, Lagena* sp., *Nonionella* sp., *spiroloculina* sp., *Triloculina* sp., *Planorotalites palmerae, Pseudohastigerina* sp., Age: Latest part of Early Eocene to Late Eocene.

Kopili Formation

See details in Garo Hills.

THICK: 400 m. **BIOTA**: *Nummulits pengaronensis, Discocyclina omphalus, Assilina spira corrugatae, Globigerina gortanii, Cribrohantkenina inflata, Globigerapsis semiinvoluta* etc., Samanta (1971). **AGE**: Late Eocene. **ENVIRON**: The Mikir formation is essentially of continental to near shore deposits. The Garampani limestone and the lower part of the Kopilli Formation are open, marine shelf deposits. The upper part of the Kopilli Formation was probably deposited in brackish water conditions, Samanta (1971).

Barail Group

See details in Garo Hills.

Dhanasiri Valley

It contains about 7000 m thick Tertiary sediments the oldest rocks are non marine sediments the oldest rocks are nonmarine sediments of Early Cretaceous age. The formations known are mentioned below.

Tura Formation

See details in Garo Hills.

THICK: 20–30 m.

Sylhet Formation

See details in Garo Hills.

UB: Overlain by Charali Member of Kopili Formation.

Kopili Formation

See details in Garo Hills.

Kopili Formation in the subsurface of Dhansiri Valley and Upper Assam Plain is divisible into the lower Charali Member and the upper Amguri Member.

Charali Member (*Au*: Deshpande et al. 1993)

HOLO: Charali well #1. *ND*: Charali structure. **LB**: Conformably overlying massive limestone of Sylhet Formation. **UB**: Conformably overlain by the Amguri Member. **LITH**: In the type section, it is represented by grey to dark grey, fissile shales intercalated by thin limestone and sandstone. **THICK**: Varies from 135 to 460 m in the Charali well. Ext: Throughout the subsurface of Dhansiri Valley and Upper Assam. **BIOTA**: *N. discorbinus*, *N. pengaronensis*, *Discocylina dispansa*. **AGE**: Late Middle Eocene to Late Eocene. **ENVIRON**: Open marine to inner shelf to littoral with moderate clastic input.

Amguri Member (*Au*: Deshpande et al. 1993)

HOLO: Amguri well #2. **LB**: Conformably overlies the Charali Member. **UB**: Gradationally overlain by the Disangmukh Formation of the Barail Group. **LITH**: Grey hard, compact fissile shales **THICK**: 381 m in Amguri well 2. Ext: It is recognized in Upper Assam and Dhansiri Valley. **BIOTA**: *Nummulites* sp., *N. Pengaronensis*. **AGE**: Late Eocene. **ENVIRON**: Open marine to littoral

Disangmukh Formation (*Au*: Deshpande et al. 1993)

HOLO: Disangmukh well 1. **LB**: Conformable contact with the underlying Amguri Member of the Kopili Formation. **UB**: Gradational contact with the overlying massive sandstone of the Demulgaon Member. **LITH**: Interbedded sand-shale sequence. Occasionally thin coal streaks are also reported. The sandstones are dirty white, light grey to grey, fine to very fine, and moderately to fairly sorted. The shales are light grey to grey, brownish grey, moderately hard and compact, at places carbonaceous. **THICK**: Maximum thick varies from 270 to 310 m. **BIOTA**: *Nummulites* sp., *N. pengaronensis*. **AGE**: Late Eocene. **ENVIRON**: Distal delta front environment.

Demulgaon Formation (*Au*: Bhandari et al. 1973)

HOLO: Well Demulgaon 7. **LB**: Gradational contact with the underlying Disangmukh Formation. **UB**: Unconformable contact with overlying Bokabil Formation. **LITH**: In the type section it is mainly sandstones with subordinate shale and coal beds. **THICK**: In Dhansiri Valley, it is 100–120 m. **Biota**: The formation is generally unfossiliferous, rare presence of arenaceous foraminifers is known. **AGE**: Late Eocene to Oligocene. **ENVIRON**: Deltaic.

Khoraghat Sandstone

See details in Mikir Hills.
THICK: 390 m.

Upper Assam
Tura Formation

See details in Garo Hills.
LB: Unconformable contact with metamorphic basement complex. **UB**: Conformably overlain by thick limestones of Sylhet Formation with a conformable contact. **THICK**: 10–150 m.

Sylhet Formation

See details in Garo Hills.
THICK: 130 m.

Charali Member

See details in Dhansiri Valley.
THICK: 74–144 m.

Amguri Member

See details in Dhansiri Valley.
ENVIRON: Open marine to prodelta.

Disangmukh Formation

See details in Dhansiri Valley.
THICK: 260 m.

Demulgaon Formation

See details in Dhansiri Valley.
UB: Gradational contact with the overlying interbedded coal shale sequence of Rudrasagar Formation.

Rudrasagar Formation (*Au*: Bhandari et al. 1973)

HOLO: Well Rudrasagar #1. **LB**: Gradational contact with the underlying Demulgaon Formation. **UB**: Unconformable contact with the overlying undifferentiated Tipam Group. **LITH**: The sedimentary succession in the type section comprises essentially of coal and shale with minor sandstone. **THICK**: 570 m is recorded in the well Charaideo 1 in the Upper Assam subsurface **BIOTA**: A diverse assemblage of arenaceous foraminifera like *Haplophragmoides* sp., *Trochammina* sp., *Ammodiscus* sp., *Cyclammina* sp., *Lituola* sp. etc. is recorded. **AGE**: An Oligocene age is assigned to this formation. **ENVIRON**: The sediments were deposited under deltaic condition.

Schuppen Belt, Naga Hills Fold Belt and NE Manipur

The oldest rocks are ophiolitees of Late Albian to Maastrichtian, Santonian to Holocene sediments are known. Paleogene formations are included below. The litho-bio and chronostratigraphic successions from Manipur-Naga Hills to Mizoram is shown in Fig. 11.

Disang Group (*Au*: Mallet 1876)

HOLO: Disang (Dilli) river section, Hypo: Silchar-Halflong road section. **LB**: The lower contact is not exposed in most sections. **UB**: The upper contact with the Barial Group is gradational. **LITH**: The Disang Group of rocks comprises mainly shale, siltstone and fine-grained sandstone. It is divided into Lower and Upper Disang. The former consists of dark grey shales interbedded with thin bands of grey siltstone or fine-grained sandstone. In general, the frequency and thickness of siltstone/sandstone bands increases towards the top of this group (Upper Disang). The shales are argillaceous and carbonaceous in Lower Disang and more arenaceous in Upper Disang. **THICK**: It is 1500 in the type section and 3000 m in eastern Manipur. **BIOTA**: *Nummulites pengaronensis*, *N. discorbinus*, *Discocyclina dispansa*, *D. eamesi*, *Pellatispira madaraszi*, *P. inflata*, *Hantkenina alabamensis*, *Globorotalia cerroazulensis*, *Globigerina ampliapertura*, *Pseudohastigerina barbadoensis* etc., from near Heningkunglwa and Lotsu village, SW of Dimapur. Chungkham and Jafar (1998) reported foraminifera from the limestone beds within lower Disang exposed near Ukhrul contains *Abathomphalus mayaroensis*, *Rosita contusa*, *Dicarinella asymetrica*, *Gansserina gansseri*, *Globotruncana aegyptiaca*, *G. ventricosa*, *Globotruncanita angulata*, *G. conica*, *G. elevata*, *Margiotruncana sinuosa*, *Pseudoguembelina excolata*, *Globotruncana arca*, *G. stuarti* and *Heterohelix globulosa.* Lokho et al. (2004) reported *Hantkenina alabamensis*, *Hantkenina liebusi*, *Cribrohantkenina inflata*, *Pseudohastigerina naguewichiensis*, *P. micra*, *P. barbadoensis*, *Turborotalia cerroazulensis pomeroli*, *T. cerroazulensis cocoaensis*, *Globigerina* sp., *T. c. cerroazulensis*, *Chiloguembelina martini*, *C. cubensis*, *C.* cf. *tenuis*, *Chiloguembelina* sp., *Globigerinatheka semiinvoluta*, *Globigerina* sp., *Nummulites chavannessi*, *N. pengaronensis*, *Baggina cojimarensis*, *B. dominicana*, *B. dentata*, *B. subinaequalis*, *Cibicidoides* sp., *Cancris mauryae*, *Cyclammina* sp., *Lenticulina* sp., *Praebulimina* sp., *Osangularia* sp., *O. plummerae*, *Uvigerina continuosa*, *U. cf. eocaena*, *U. longa*, *U. cocoaensis*, *U. moravia*, *U. vicksburgensis*, *U. jacksonensis*, *U. glabrans* from Upper Disang Formation exposed in and around Pfutsero of southwest. A collection of fossil pteropods, including some unidentified species, provisionally referable to the families Limacinidae, Creseidae, and Clioidae (?) is reported from the late Middle Eocene–Late Eocene beds of the Upper Disang Formation exposed near the town of Pfutsero, Phek District, South Central Nagaland (Assam–Arakan Basin, northeastern India) Lokho and Kumar (2008). **AGE**: Lokho et al. (2004) suggest Late Eocene for the Upper Disang Formation of Nagaland. The exotic limestone blocks of Ukhrul are of Santonian to Maastrichtian. **ENVIRON**: Baruah et al. (1987) suggest that the fossiliferous sequence was deposited in a shallow marine environment ranging from mostly inner to middle shelf zones

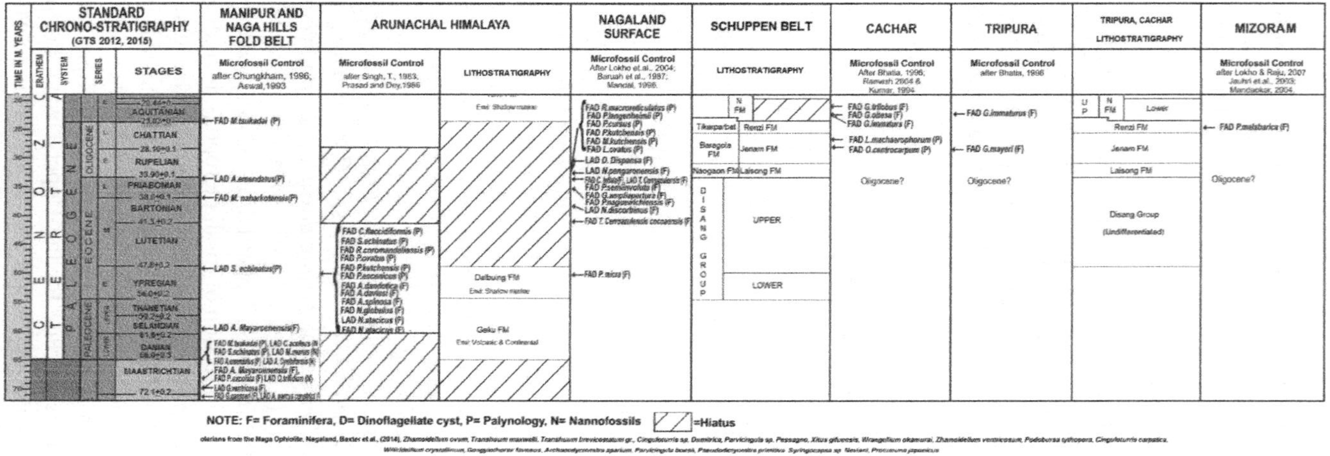

Fig. 11 Paleogene Litho- Bio- Chrono- Stratigraphic succession of North East basins from Manipur to Mizoram, India. Compiled by D. S. N. Raju

and also partly to outer shelf zone for a short period. Lokho et al. (2004) suggest lower part of upper bathyal set-up for the localities of Pfutsero 1, 2, Chobama and Leshemi in Phek District supported by dominant *Uvigerina* facies consisting *Uvigerina cocoaensis*, *U. continuosa*, *U.* cf. *eocaena*, *U. glabrans*, *U. jacksonensis*, *U. longa*, *U. Moravia* and *U. vicksburgensis*. The occurrence of pteropods in the Upper Disang Formation indicates deposition in an open marine basin above the aragonite compensation depth. The combined assemblages of pteropods and previously reported uvigerinid foraminifers from the Upper Disang Formation indicate a palaeobathymetry of ~500 m, i.e. upper bathyal zone, and a tropical–subtropical climate, Lokho and Kumar (2008).

Barail Group

See details in Khasi and Jaintia Hills.

BIOTA: Pteridophytic spores of *Cyathdites australis*, *C. minor and* angiospermous pollen of *Lakiapollis ovatus*, *Tricolporopilites robutus*. **AGE**: Late Eocene. **ENVIRON**: Brackish water, swampy environment of deposition.

Laisong Formation

See details in Khasi and Jaintia Hills.

LB: Conformably overlying the Disang Group. **THICK**: 1750+ m in the Schuppen Belt. **BIOTA**: Baruah et al. (1987) reported sporadic arenaceous foraminifera from Laisong Formation of Barail Group near Heningkunglwa and Lotsu village, SW of Dimapur, Nagaland. **ENVIRON**: Brackish Water environment.

Jenam Formation

See details in Khasi and Jaintia Hills.

THICK: 800 m in the Schuppen Belt. **LITH**: In the Schuppen Belt, it is predominantly an argillaceous sequence of dark grey, silty shale and mudstone with thin sand lenticles. Bioturbation, carbonaceous plant material, cross laminations and ripple marks are common.

Renji Formation

See details in Khasi and Jaintia Hills.

UB: Unconformable contact with middle Bhuban Formation. **THICK**: 850 m in the Schuppen Belt.

Tripura, Cachar, Mizoram and SW Manipur

Paleogene sediments are not known in Tripura and Mizoram. In Cachar Eocene and Younger sediments are known.

Disang Formation

See details in Schuppen Belt, Naga Hills Fold Belt.

LB: Metamorphic basement complex? **THICK**: Less than 1250 m. **BIOTA**: Kachhara et al. (2000) reported *Ostrea* sp., *Barbatia* sp., *Natica* sp., *Turritella* sp., *Assilina* sp., *Nummulites pengaronensis*, *Alveolina* sp. Tripathi and Satsangi (1982) reported *Ophiomorpha* sp. **AGE**: Upper age limit of the Disang Group in Manipur is Late Eocene, Kachhara et al. (2000). **ENVIRON**: Marine condition, Kachhara et al. (2000). Tripathi and Satsangi (1982) suggest shallow intertidal environment deposition.

Barail Group

See details in Khasi and Jaintia Hills.

BIOTA: *Chondrites* sp., *Skolithos* sp., *Thalassinoides* sp. **ENVIRON**: The depth of the sedimentary basin of Disang-Barial transition in the Imphal Valley and its surrounding areas may have been about 200 m.

Arunachal Pradesh

Proterozic and Paleogene sediments are known.

Yinkiong Group

This Group of rock is exposed around Yinkiong in the Siang Valley mapped by Jain (1974). It consists of two formations viz., Geku Formation and Dalbuing Formation.

Geku Formation (*Au*: Tripathi and Murti 1981; ND: Geku Village)

LB: Unconformably overlying the Abor Formation. **UB**: Gradational contact with overlying Dalbuing Formation. **LITH**: Volcanic and continental facies. **THICK**: 2774 m thick. **BIOTA**: Dicot plant fossils like *Apocynophyllum* sp., *Canavalia* sp., *Hicora* sp. and *Sophera* sp. **AGE**: Upper Paleocene. **ENVIRON**: Volcanic and continental facies.

Dalbuing Formation (*Au*: Singh 1984; *ND*: Dalbuing Village)

LB: Gradational contact with the Geku Formation. **UB**: Abor volcanic or the quartzite of the Tenga Formation tectonically lies over the formation along Roing Fault. **LITH**: Limestones, phyllites and shales. **BIOTA**: Foraminifera of *Nummulites atacicus*, *Assilina dandotica*, *A. daviesi*, *N. laminose*, *N. globulus*, *A. spinosa*, etc. Singh (1984). Prasad and Dey (1986) reported palynomorphs of *Dandotiaspora*, *Cyathidites*, *Malayaeaspora*, *Foveotriletes*, *Spinainaperturites* etc. **AGE**: Early Eocene. Prasad and Dey (1986) compared their palynomorphs assemblages with Early assemblages of Cauvery Basin, Venkatachala and Rawat (1972). **ENVIRON**: Shallow marine based on the foraminiferal assemblages. The Eocene basin was initially marked by fresh water regime which gradually turned into a marine basin.

Andaman Onshore

The oldest formation is of pre Cretaceous to probable Middle Cretaceous Port Meadow Formation. Baratang Formation of Late Cretaceous to Late Eocene overlies Port Meadow Formation. The Paleogene litho- bio- and chronostratigraphy and paleoenvironments settings of Andaman Basin are shown in Fig. 12.

Baratang Formation (*Au*: Boileau 1950)

HOLO: Gandhi ghat—Lorujig Anticline Traverse, Baratang Island; **LB** and **UB**: Both the upper and lower contacts of Baratang Formation with Port Blair Formation and Port Meadow Formation respectively are marked by distinct unconformities. For full details on lithology, biota, paleoenvironments reference is made to Pandey and Dave (1998)

Lithostratigraphy of Holostratotype: In the type section, the formation represents dominantly flysch sediments and consists mainly of claystones, shales and minor interbeds of sandstone and siltstone.

Lithostratigraphy of Hypostratotype: In the reference section of Dhani Khari Nala Traverse the formation consists of alternations of sandstone and shale. **THICK**: In the type section the thick of the formation is 458 m. *Subdivisions*: Ramachandra and Chatterjee (1963) subdivided the Baratang Formation into 2 broad lithological units. They recorded development of arenaceous units within Baratangs in Middle Andaman and suggested two broad subdivisions viz., the lower argillaceous unit and the upper arenaceous unit. No formally designated type sections were given.

Subsequently, Shrivastava et al. (1973) carried out detailed geological mapping of the Middle Andaman Island. They subdivided the Baratang Group into 10 formations. Later Shrivastava et al. (1974), proposed a formal subdivision of 'Baratang Group' into five 'formations' and the 10 formations proposed earlier were dropping as they were having limited geographical extent.

Karmatang Sandstone Member (*Au*: Shrivastava et al. 1973)

HOLO: In the outcrops of Karmatang Boliu in the eastern part of Middle Andaman; **LB**: The stratigraphic relationship of Karmatang Sandstone Member with the underlying chert/jasper, quartzitic beds of Port Meadow Formation remains mostly obscure.

Lithostratigraphy of Holostratotype: In the stratotype section, the Karmatang Sandstone Member consists of alternations of dark grey very fine to fine grained hard quartzitic standstone with dark grey slaty shale. The shale is more dominant in the lower part while sandstone dominates in the upper part of the section. At places sediments have undergone low grade metamorphism.

Lithostratigraphy of Hypostratotype: In the lower part of the section in Jhinga Nala section, the member dominantly comprises of sandy shale and shale. The shales are black to grey, slaty carbonaceous, ultrabasics of Saddle Hill Intrusives are also observed. **THICK**: Due to rapid alternations of competent and incompetent rock, acutely folded and faulted, the exact determination of thick is difficult. Based on detailed traverses along Jhinga Nala and Badamnala. Shrivastava et al. (1974) have calculated thick of 2750 ft (846 m) for this unit.

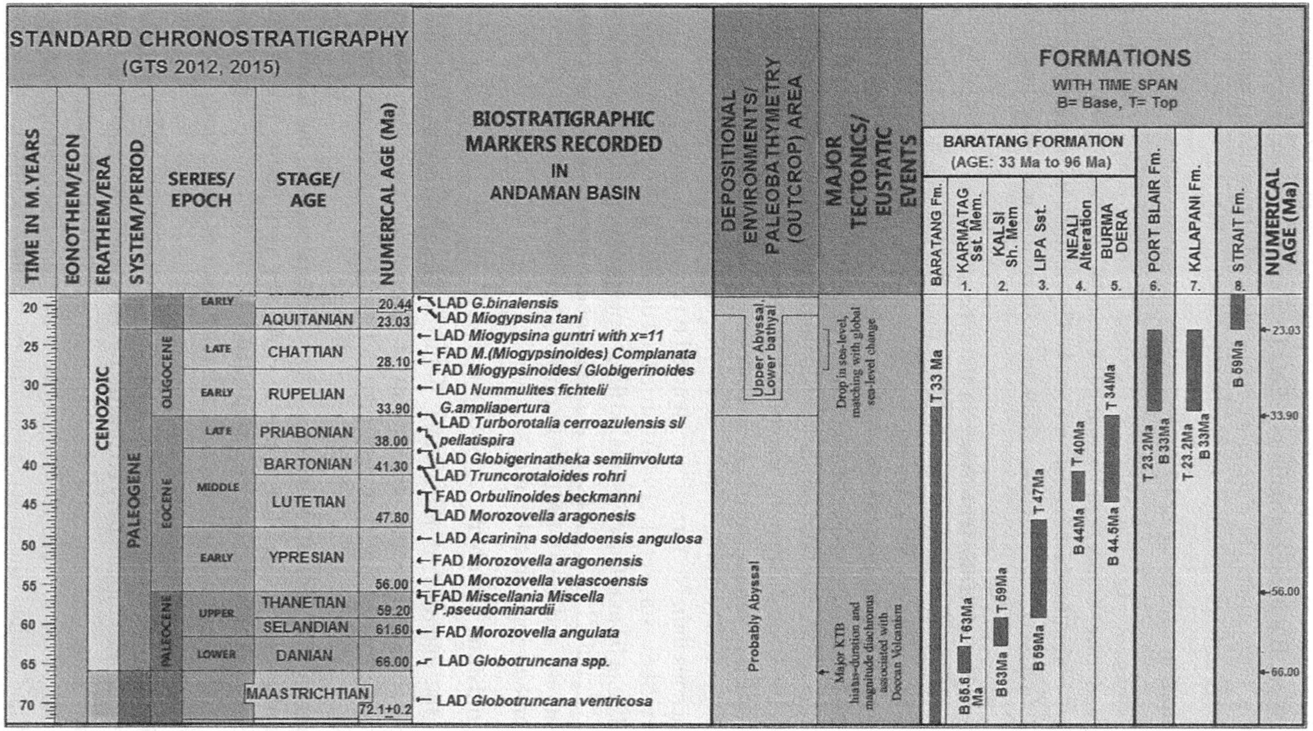

A BRIEF ON EACH FORMATION: 1.KARMATAG Sst Mem: THICKNESS:846m. LITHOLOGY: Sandstone with carbonaceous shale and minor limestone. 2. KALSI sh. Mem: THICKNESS: 260m, LITHOLOGY: Carbonaceous shale with claystones, intercalations of Sst. 3.LIPA Fm: THICKNESS: 452m. LITHOLOGY: Sst. conglomerate, minor shale. 4. NEALI ALTERATION: THICKNESS: 318m. LITHOLOGY: Alternation of Sst and Shale with minor Lst and bands of conglomerates. 5.BURMA DERA: THICKNESS: 410m, LITHOLOGY: Shale with thin beds of Sst, Siltstone and marble. 6. PORTBLAIR Fm: THICKNESS: >1000m. LITHOLOGY: Conglomerate and qirt beds. PALEOENVIRONMENT: Geosynclinal facies represents flysch deposition. 7.KALAPANI Fm: LITHOLOGY: Shale and clays with minor Sst and streaks of Lst. 8.STRAIT Fm: THICKNESS: 90m. LITHOLOGY: Sst sand LSt With ash layers, siliceous chalks. PALEOENVIRONMENT: Upper to middle bathyal.

Fig. 12 Paleogene Litho- Bio- Chrono- stratigraphy and paleoenvironmental settings in the Daman basin, India. Compiled by D. S. N. Raju

Kalsi Shale Member (*Au*: Shrivastava et al. 1972)

The name was given for thick sequence of dark grey, carbonaceous shale exposed on the western part of Middle Andaman by Shrivastava et al. (1972) after the Kalsi Bush Police camp no. 7.

HOLO: Kalsi Bush Police Camp. 7. **LB** and **UB**: The Kalsi Shale is exposed only on the western part of Middle Andaman. Its base is not exposed. Towards top, it is overlain by Lipa Sandstone, but the nature of their contact is not clear.

Lithostratigraphy of **Holostratotype**: In the type section the Kalsi Shale consists of alternations of shales with clay stones. The shales are grey to very dark grey, carbonaceous, pyritic and rich in organic matter. The claystones are white to cream colored and soft. At some places sandstone bands are also present.

THICK: The maximum thick of Kalsi Shale was estimated to be around 850 ft (259 m) in the northwestern part of Middle Andaman by Shrivastava et al. (1974).

Lipa Sandstone Member (*Au*: Shrivastava et al. 1973)

HOLO: Lipa Nala, Lipa North Karmatang Traverse, Northern part of Middle Andaman. **LB**: In the eastern part of the Middle Andamans, this unit rests unconformably over the Karmatang Sandstone. **UB**: It has a conformable contact with the overlying Neali Alternation.

Lithostratigraphy of Holostratotype: In the type section, the member dominantly consists of grey to buff coloured, medium to coarse grained, friable, occasionally massive, thick bedded sandstone and thin interbeds of conglomerates with pebbles of quarzite, chert/jasper etc. In the upper part, the alternation of dark grey, fine grained sandstone and grey, brown shale are seen. In the upper part of the sequence, sandstones are light greenish grey, medium to fine grained and moderately hard.

Lithostratigraphy of Hypostratotype: The lower part of the sequence in the Badam Nala traverse consists of grey to dirty white fine to medium grained sandstone with carbonaceous specks.

THICK: Shrivastava et al. (1974) calculated the thickness of Lipa Sandstone in the Badam Nala traverse to the order of 1450 ft (452 m) on the north and 1270 ft (387 m) along Kalsi Parnasala traverse in the south.

Neali Alteration Member (*Au*: Shrivastava et al. 1973)

HOLO: Neali Nala, Northwestern part of Middle Andamans (Neali-south Karmatang traverse); *Hypo*: Badam Nala Traverse, Middle Andaman. **LB**: In the lower part, the Neali Alternation shows conformable relationship with the underlying Lipa Sandstone. UB: The upper contact of the unit with Burmadera Member is disconformable, as evidenced by the presence of conglomerate and pebble beds at the contact of the two untis.

Lithostratigraphy of Holostratotype: In the Neali Nala traverse, the Neali Alternation Member consists of alternation and sandstone and shale with minor siltstone towards the top. Towards the base conglomerates with rounded pebbles of grey to brown sandstone are observed. **THICK**: Due to foldings and other structural complication the exact thick of this member is difficult to compute. However, Shrivastava et al. (1974) computed a thick varying from 268 to 345 m.

Burma Dera Member (*Au*: Shrivastava et al. 1973)

HOLO: Near Burma Dera Village (Karmatang Boliu Traverse), Middle Andaman; *Hypo*: Neali Nala Traverse, Middle Andaman. **LB**: The lower boundary of Burma Dera Member with the underlying Neali Alternation is disconformable as evidenced by presence of basal conglomerates, dip discordance and lithological changes. **UB**: The Burma Dera Member has an unconformable contact with the overlying Port Blair Formation representing a significant time gap.

Lithostratigraphy of Holostratotype: In the type section of Burma Dera Member, the sequence dominantly consists of shales with minor sandstone towards the base. The shales are greenish grey to grey with thin bands of siltstone and fine grained sandstone. At places few thin lenses of sandstone are observed in the shale. The sandstone are observed in the lower part and are fine grained.

Lithostratigraphy of Hypostratotype: In the Neali Nala Traverse, the member consists of dark grey to grey, fissile shale with fine grained sandstone and siltstone layers. Thin sandstone streaks are observed in the section. Towards the upper part sandstone are observed.

THICK: Shrivastava et al. (1974) have calculated the maximum thick of the unit to be around 1340 ft (400 m) along Badamnala traverse in the north, 130 ft (40 m) along Kalsi-6 Saguanala Traverse and Rangat Bakultala Traverse in the south. The thick is reduced to 1120 ft (341 m) along Kalsi-3 Pitchar nala due to folding and faulting.

Distribution: The Burma Dera Member has been extensively encountered in the core of a number of synclines in the Middle Andaman. **BIOTA**: Fauna of the Baratang Formation of Middle Andaman was examined by Pandey (1972) from Neali Nala and Chandbagh Nala traverses by Ramachandra. These expose, respectively, the upper and lower parts of the formation. The highest samples with profuse and diagnostic microfauna is about 350 m. below the observed top of the formation (contact with the Port Blair Formation not seen). The Stratigraphic Position of the lowest sample from Chandbagh Nala, above the Baratang base, is uncertain.

The sample from the upper Baratang contains a few specimens of *Nummulites fabianii*. This is the well known species of Late Eocene. The association of the *Globotruncana* forms of Late Cretaceous with *N. fabianii* of Late Eocene and differences in the infilling material in the two faunas clearly indicates the reworked nature of the *Globortruncana* assemblage. In the sample from the lower part of the formation also, differences in the infilling material of the *Globotruncana* assemblage even in the lower part of the Baratang Formation. However, in a locality sample from Middle Andaman Pandey (1972) came across an indigenous *Globotruncana Heterohelix* assemblage which is devoid of either arenaceous foraminifers or nereitic benthonics. Proportion of the planktonics in this sample, with the matrix, is very close to that of an ooze. The *Globotruncana* fauna recorded by Pandey (1972) may be referred to the *Globotruncana gansseri-A. mayaroensis* Zones of Bolli (1966). These two zones constitute the highest Cretaceous. It suggests that the lower horizons of the Baratang which yielded the indigenous *Globotruncana* assemblage may be placed in the Late

Cretaceous whereas the bulk of the Baratang Formation with reworked Cretaceous fauna may be assigned to a Tertiary age in the range of Paleocene to Late Eocene.

Diagnostic foraminifers are rather rare in the Baratang Formation. *Miscellanea miscella* D' Archiac & Haime, *Civillierina vallensis* Ruise De Gona, *Nummulites* sp., and *Discocyclina ramaroi* (?) Samanta were identified in thin sections of an allochthonous algal limestone. Paleocene-Lower Eocene planktonics identified by Kumar and Iyengar (unpub.) from the Baratang Formation include: *Globigerina triloculinoides* (Plummer), *Globorotalia pseudobulloides (*Plummer*)*, *G. uncinata (*Bolli) and *M. aragonensis Nuttal.* **AGE**: Based on the forminifera recorded from the Baratang Formation Late Cretaceous to Late Eocene age. **ENVIRON**: The lowest sample of the Baratang Formation, examined by Pandey (1972) is an ooze like material with *Globotruncana Heterohelix* assemblage and no benthonics. In this sample, terregenous material is marginally higher than that prescribed for an Ooze. Never the less, the ecological parameters of this rock may be inferred to be same as a 'Globigerina Ooze' and the Late Cretaceous basin floor may well be conjectured to be deeper than 3500 m.

For the greater part of the Tertiary Period, the indigenous Baratang fauna is constituted by above 90% holomarine agglutinated types-many of them of robust size. The assemblage also includes cold water genera like *Pelosina* and *Saccamina.* This arenaceous assemblage, in view of its composition showing cold water habitat in the tropical latitudes of Andaman, could referred to abyssal depths of low temperature. In the Peru-Chile Trench area, Bandy and Rudolfo (1964) observed that the faunal assemblages were persistently dominated (more than 90%) by arenaceous foraminifers at depths of 4000 m or more. The *Reophax-Asterorhiza* association is more characteristic of 4500–6000 m depth as per these authors. The Barantang assemblage recorded by Pandey (1972) is referable to the Reophax-Asterorhiza association indicating thereby depths of 4500 m and below during the Eocene part of the Baratang.

Lithological criteria suggest that bulk of the Baratang sediments are turbidite Graded bedding is common and wood fragments are also seen like the modern turbidites. There is a prgressive coarsening of the sediments towards north suggesting that the provenance of rock units was towards north, in Burma.

End of Cretaceous period and begining of the Paleocene was marked by a recognisable change in the Andaman Region. There was a noticeable uplift in the eastern shelf which contributed to some sediments with reworked *Globotruncana* fauna in the Geosyncline. Simultaneously the Geosyncline deepened further and there came a phase of igneous intrusion during Paleocene. This phase of ultrabasic intrusion, continued till Early Eocene in Burma where some Nummulitics have been also affected by this igneous activity (Pascoe 1964, p. 1411). However, fragments and pebbles of serpentine are commonly recorded from the Baratang sediments of North Andaman (Pascoe 1964, p. 1410), suggesting that much of the Baratang sedimentation has occurred after this intrusion.

Saddle Hill Intrusives (*Au*: Shrivastava et al. 1973)

HOLO: Saddle Hill Range, North Andaman. **LB & UB**: The ultrabasic rocks are found as intrusive in Baratang Formation. In Middle Andaman, they intrude all the five members of Baratang Formation viz. Karmatang Sandstone, Kalsi Shale, Lipa Sandstone, Neali Alternation and Burma Dera Members. The intensity of intrusions decreases in the younger sequences. Higher up in the Port Blair Formation, the intrusives are absent.

Description: The main rock type of Saddle Hill Formation are coarsely crystalline ultrabasic rock, serpentinized and rarely volcanic rock. The ultrabasic rocks comprise of dunite, peridotite, pyroxenite gabbro etc. and are dark greenish grey to black in appearance (Shrivastava et al. 1974).

AGE: Chhibbar (1934) has dated the ultrabasics to be of Cretaceous age. The fact that the ultrabasics intrude only the Baratang sediments and not the Port Blair's, suggest a Post Baratang Pre-Port Blair i.e. Pre-Oligocene age to these intrusives. **ENVIRON**: As already noted, the ultrabasics occurs as intrusives in the Baratang sediments.

Port Blair Formation (*Au*: Oldham 1885)

HOLO: Corbyn's Cove-South Point traverse near Port Blair town, south Andaman (between Gymkhana Ground to near Corbyn's Cove); *Hypo*: Jarwatikri-Uttara Macaurthiball Traverse, Middle Andaman. **LB**: The lower contact of this formation in unconformable with the underlying Baratang Formation. **UB**: The upper contact of this unit with Strait Formation is unconformable and is seen in Strait Island.

Lithostratigraphy of Holostratotype: In the type area of Port Blair, the formation consists of alternating sequence of sandstone and shales. In the basal part, the sequence comprises of conglomerate and grit beds.

Lithostratigraphy of Hypostratotype: In the reference section, the formation overlies the Burma Dera Member marked by a conglomerate. **THICK**: The maximum thick of Port Blair Formation was estimated to be +1100 m by Chandra et al. (1961). **BIOTA**: Tipper (1911) described *Nummulites atacicus* from this unit. *N. miarritzensis* and *Assilina granulosa* was also reported from sandstones of South Andaman. Gee reported *A. granulosa* from calcareous sandstone in northern part of Middle Andaman. **AGE**: Oligocene. **ENVIRON**: The presence of more than one band of conglomerate towards the base, suggest oscillation of shore line at the time of deposition of Port Blair Formation. The shelly conglomerates are regarded as modified subaequous talus rather than normal shore gravel (Pettijohn 1949). The smoothness of some of pebbles in conglomerates suggest that they were sufficiently worked by waves, whereas, the angular to subangular particles in greywackes suggest that they were not sufficiently moved by bottom traction and might have been brought to the side of deposition by turbidity currents. The better development of conglomerates in northern part of Middle Andaman suggest that during deposition the shore line lay towards the northern part of the present region. The littoral phase of earlier period gave place to deeper (epineritic) phase when sandstones and shales were deposited, Sandstones are well sorted suggesting slow rate of deposition. The shales were deposited during intermittent period of waning of bottom traction.

Kalapani Formation (*Au*: Pandey et al. 1992)

Remarks on nomenclature: The name has been proposed for the undifferentiated Pre-Neogene sequences of shales and clays encountered in the well drilled in Andaman basin. **HOLO**: Well An-42-1 between 520 and 3972 m; *Hypo*: Well An-1-1, from 2600 to 3740 m. **LB**: The formation unconformably overlies the basement of igneous rocks in An-42-1 and has a pronounced unconformity on the top. Where it is overlying Neogene Formation.

Lithostratigraphy of Holostratotype: In the stratotype of well An-42-1, the formation typically depicts flysch facies and consists of claystone, Shale and sandstone.

Lithostratigraphy of Hypostratotype: In the reference section in well An-1-A, the formation consists of light to dark grey, moderately hard highly calcareous shale and light grey to white, subangular to angular, feebly calcareous, fine to medium grained, poorly sorted sandstone. **THICK**: In the type section of AN-42-A, 3452 m of the formation have been encountered. On an average the thick of the formation ranges from 3452 to 100 m.

LV: As started above the Kalapani Formation has only been encountered in the subsurface of Andaman Basin. The formation mainly consists of shale, claystone and minor sandstone. The claystone is dominant facies and is interbedded with shale and minor sandstone. A few thin streaks of carbonaceous shale (HLO-A) and thin limestones have also been observed in the northern part of the basin.

Distribution: The formation is encountered in most of the wells drilled in Andaman basin including AN-42-A, HLO-A, An-1-A, AN-14-A and Oil India wells AOE-1 and AOE-2. **BIOTA**: In the Oil India wells AOE-1, AOE-2, the formation has yielded a rich suite of planktonic foraminifera of Late Cretaceous to Oligocene age. The diagnostic species include, *Morozovella Globorotalia velascoensis P. pseudomenardii, M. subbotinae, G. (Acarinina) soldadoensis angulosa, G. (A) pentacamerata, Asterocyclina* sp., *Globigerina ampliapertura, G. selli, G. gortanii, Lepidocyclina* sp *Asterigerina* sp., *G. opima opima, Catapsydrax dissimilis, G. tripartita, G. prasaepis, G. sastrili, G. opima nana, G. angulisuturalis.*

AGE: Cretaceous to Oligocene. **ENVRON**: The foraminiferal assemblage suggests an outer shelf to upper bathyal environment of deposition for major part of the sequence. The presence of pyritised foraminifera at places suggests suboxic to anoxic environment. The present of abundant reworked planktonic and smaller foraminifera suggest probable and smaller foraminifera suggest probable transportation of outer shelf sediments to bathyal environment by turbidity traction.

East Coast Basins

The bio-chrono and tectonostratigraphic frame work for lithofacies distribution and magnitude of hiatus in East Coast Basins (Cauvery, Krishna Godavari, Mahanadi, Bengal Basins) are shown in Fig. 24.

West Bengal Basin

The West Bengal Basin has a history of intracratonic rift basin. Permotriassic to recent sediments are known in the subsurface. Seven formations are briefly described.

The litho-bio- and chronostratigraphy and paleoenvironmental settings in the Bengal Basin is shown in Fig. 13.

Ghatal Formation (*Au*: Biswas 1963)

HOLO: Interval 2802–2971 m in well Ghatal-1; *Hypo*: Interval 3257–3592 m in well Karimpur-1. **LITH**: The holostratotype comprises of fine to very grained calcareous sandstones grey siltstone and oolites. The Ghatal Formation is subdivided into four members:

- (i) Karimpur Member: Comprising of fine grained sandstone alteration with buff or white limestone and shale.
- (ii) Mayapur Member is characterized by carbonaceous, calcareous shale and siltstone with occasional oolites, overlain by oolites.
- (iii) Amta Member is represented by thin grey shale sequence.
- (iv) Domjur Member is the youngest member of Ghatal Formation overlying Amta Member. It is characterized by fine grained sandstone, calcaceous siltstone and grey shale.

LB: Unconformable with Bolpur Formation. **UB**: Unconformable with Jalangi Formation. **THICK**: Ghatal Formation shows gradual increase in thick towards east. Maximum thick is observed in the well Karimpur-A, where it is around 335 m (Chandra et al. 1991) **BIOTA**: Ghatal Formation has yielded a few planktonics in Dhananjaypur-A, Domjur-A and Chandkuri-1 including *Turborotalia*, *trinidadensis*, *Globorotalia* (*T*) *inconstants*. **AGE**: Early Paleocene. **ENVIRON**: The formation is referred to the Early Paleocene accordingly. The formation represents a transgressive– regressive sequence. Its Karimpur and Mayapur member are transgressive deltaic deposit. Domjur Member is regressive represented in the well Domjur-1 by fine to very fine sandstones. Amta Member includes predelta shales of a regressive delta.

Jalangi Formation (*Au*: Biswas 1963)

HOLO: Interval 2407–3095 m in well Jalangi-1. **LITH**: The type section is represented dominantly by coarse to medium grained, white to brownish sandstone. Carbonaceous matter and red shales are occasionally present. Towards top, carbonaceous shale, coal and fine to medium grained sandstone are present. The Jalangi Formation is subdivided into lower, middle and upper. The Lower Jalangi is dominated by coarse to medium arkosic, light grey to dirty white sandstone. The Middle Jalangi is a coal shale unit with fine grained sandstone, while the Lower is characterized by coarse to medium grained glauconitic sandstone. **BIOTA**: Foraminifera in Jalangi are rather sparse. In the lower and middle Jalangi non-diagnostic agglutinated forms viz; *Trochammina*, *Ammobaculites*, and rare calcareous foraminifera occur. In well Domjur-1 the limestone sequence yielded very few *Nummulites*, *Fasciolites* and *Lockhartia*. Upper Jalangi yielded *Nummulites acutus*, *N. discorbinus* etc. **LB**: Unconformable with Ghatal Formation. **UB**: Unconformable with overlying Kalighat Formation. **THICK**: There is a general eastward increase in thick. Maximum thick of 640 m (+) has been met in Memari-1 well and still higher thick is inferred within the

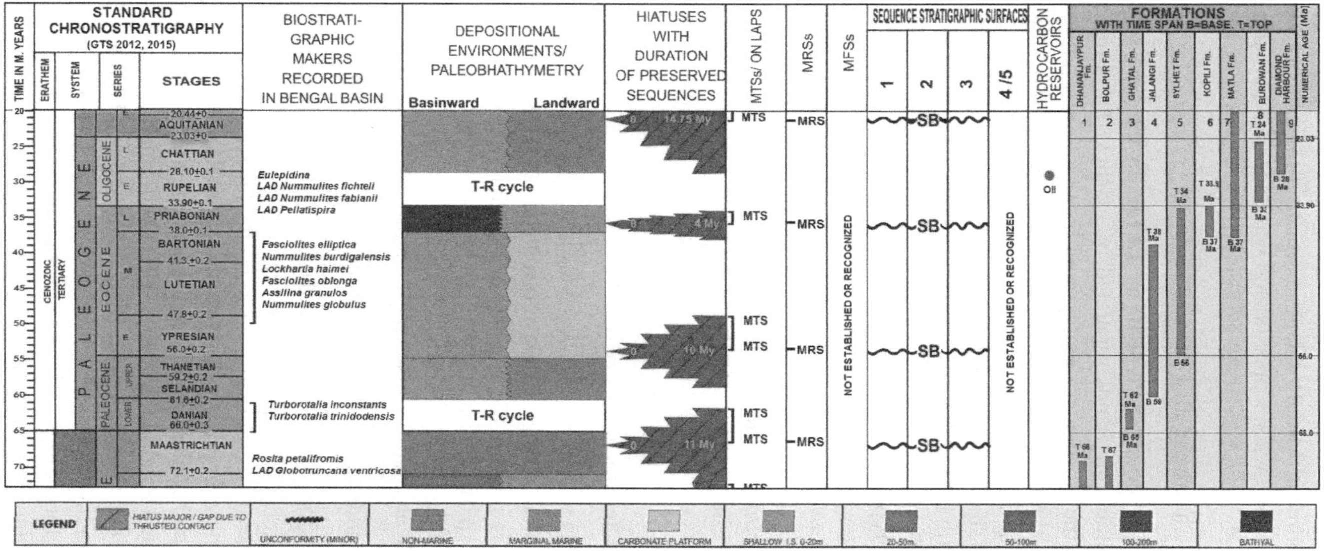

A BRIEF ON EACH FORMATION:

1) DHANANJAYAPUR Fm: THICKNESS: 94m-600m; LITHOLOGY: Shale and Silty Shale; PALEOENVIRONMENT: Paleobathymetry 70m-300m. 2) BOLPUR Fm; THICKNESS: 156m; LITHOLOGY: Coarse pebbly sandstone with red clay ; PALEOENVIRONMENT: Alluvial fan. 3) GHATAL Fm: THICKNESS: Max. 536m; LITHOLOGY: 4Members are recognized Sst alternating with Lst, Sh, Oolites and Sst; PALEOENVIRONMENT: Recognises a transgressive regressive cycle. 4) JALANGI Fm; THICKNESS: Max. 650m; LITHOLOGY: Sandstone, Carbonaceous shale and coal divided into 3 units; PALEOENVIRONMENT: Fan delta in inner part and marine in upper part. 5) SYLHET Fm; THICKNESS: 795m In type section; LITHOLOGY: Area referred at Khali ghat Fm, Lst bands represented clastics divided in to 3 units. 6) KOPILI Fm; THICKNESS: 424m; LITHOLOGY: Carbonaeous Shale; PALEOENVIRONMENT: Shallow inner neritic. 7) MATLA Fm; THICKNESS: 1500m-4000m ; LITHOLOGY: Shale clay with occasional glauconitic sand bands; PALEOENVIRONMENT: Marine. 8) BURDWAN Fm; LITHOLOGY: 2Members were recognized Sandstone, Shale, glauconitic sandstone; PALEOENVIRONMENT: Deltaic marine members represents transgressive regressive cycle. 9) DIAMOND HARBOUR Fm; THICKNESS: 500m; LITHOLOGY: Siltstone thin layers of Sst, upper part Coal; PALEOENVIRONMENT: Shallow marine and Coastal.

Fig. 13 Paleogene Litho- Bio- Chrono- sequence stratigraphy and paleoenvironmental settings in the West Bengal basin, India. Compiled by D. S. N. Raju

paleodepression between Memari and Ghatal and between Memari and Radha **AGE**: Upper Paleocene to Middle Eocene. **ENVIRON**: Lower–Middle Jalangi Formation represents fan delta environment. Middle Jalangi Formation indicates presence of marginal delta/lower delta plain facies. Upper Jalangi was deposited in marine environment.

Kalighat Formation (*Au*: Chandra et al. 1993)

HOLO: Domjur-1 well, interval 3845–4640 m. **LITH**: In the type section, it consists of a number of limestone bands separated by clastics. The limestone is grey or buff coloured and the clastics are represented by fine to medium grained, glauconitic sandstone and shale. The entire Kalighat is divisible into three units: lower middle and upper. The lower Kalighat is represented by alternation of sandy limestone and sandstone or shale. The middle Kalighat unit is characterized by thick compact limestone body with little or no clastics. The uppe Kalighat is characterized by occurrence of a coarse to fine grained sandstone topped by a limestone band. **THICK**: In the holostratotype it is 795 m thick **BIOTA**: The lower Kalighat is characterized by: *Nummulites globulus*, *Assilina granulosa*, *A. laminosa*, *A. daviesi*, *Fasciolites oblonga*, *Lockhartia hunti*, *L. conditi*, *L. haimei*. The middle Kalighat is characterized by *N. burdigalensis*, *N. globulus*, *N. dijodjokartae*, *Assilina spira*, *A. placentula*, *Fasciolites elliptica*. In the upper Kalighat *N. fabianii, Pellatispira glabra* and *P. madraszi* are characteristic. **ENVIRON**: Faunal assemblage recorded indicate shallow inner shelf condition with minor pulses of regression in an overall transgressive phase.

Hooghly Formation (*Au*: Chandra et al. 1993)

HOLO: Interval 5200–5324 m in Golf Green-1 well. **LITH**: In the type section lithology comprises of grey to dark grey calcareous shale. **THICK**: 124 m in the holostratotype. **BIOTA**: Foraminifera recorded are dominated by larger foraminifera *N. fabianii* and *Pellatispira* sp. Assemblage. **AGE**: An Upper Eocene age is suggested based on the above assemblage. It is equivalent to Kopili Formation. **ENVIRON**: Recorded foraminifera indicate shallow inner shelf conditions.

Matla Formation (*Au*: Chandra et al. 1993; Modified after Biswas 1963)

HOLO: Interval 1330–4560 m in well SME-5. Hypo: SHE-5 between 13.30 and 4560 m.

LITH: This unit is represented by dark grey, grey, fissile shale or clay, occasionally glauconitic. Sand bands, wherever present, are very thin. **BIOTA**: The Malta Formation is highly fossiliferous and comprises of forms represented in the Hooghly, Burdwan, Diamond, Harbour and Pandua Formation. **UB**: Conformable with Ranaghat of Debagram Formation. Unconformable with Bengal Alluvium. **THICK**: 1500 m. in 1 Chapur-1 and 4000 m in SMM-2 and Raghunod wells. **AGE**: This formation has inter tonguing with several formations and the age is ranging from Upper Eocene to Pliocene. **ENVIRON**: Marine environment.

Burdwan Formation (*Au*: Biswas 1963)

HOLO: Interval 1545–1675 m in well Burdwan-1. **LB**: Unconformable with Hooghly Formation. **UB**: Unconformable with Pandua Formation. **LITH**: The type section is characterized by the dominance of sandstone with varying grain size from coarse to fine and are occasionally feldspathic and lignitic. Few shale bands, ligntic and pyretic and microspherulites of Siderite are common. The Burdwan Formation is subdivided into (i) Gold Green Member comprising of fine to medium grained, white, dirty white sandstone, light to dark grey siltstone and ark grey shale. At the bottom a fossiliferous limestone band is also recorded. (ii) Memari Member comprises of grey to dark grey glarcconitic, carbonaceous shale, fine to medium grained, occasionally glanconitic sandstone and a thin limestone band in the lower part. **THICK**: 500 m in Golf Green area. **BIOTA**: In basin margin area Burdwan Formation is devoid of microfauna. Very few arenaceous foraminifera includes *Trochammina, Ammobaculites*. Oligocene form *Eulepidina* is found in a number of wells. In wells like Maju-1, Golf Green-1, Chandkuri-1, Lower Oligocene form *Nummulites fichteli* occurs along with *Lepidocyclina*. Oligocene planktonic forms like *Globigerina gortanii, G. tripartita, G. ouchitensis* also occur. Burdwan Formation is Early to Late Oligocene in age. **AGE**: Lower to Upper Oligocene. **ENVIRON**: The Golf Green Member comprises transgressive deltaic sand. Memari member not only includes transgressive but also regressive deltaic sand and shales.

Diamond Harbour Formation (*Au*: Mohan et al. 1981)

HOLO: Diamond Harbour-1, between 3905 and 4405 m. **THICK**: Maximum thick to the tune of 500 m is observed in type section i.e. in Diamond Harbour-1 well **LB**: Lower Boundary is conformable with Burdwan Formation. **UB**: Upper Boundary is unconformable with the overlying Pandua Formation. **LITH**: In type section, the formation comprises of grey to dark grey siltstones, thin layers of very fine grained or silty sandstones and shale. Upper part contains grey coaly shales and coal. Silty nature of Diamond Harbour Formation is observed only in type section; in other wells it is represented largely by a shale section. **BIOTA**: The formation contains arenaceous foraminiferal assemblage of *Trochammina, Haplophragmoides*, *Reuselola* alongwith *Ammonia* and *Bulimina*. **AGE**: The formation is assigned Upper Oligocene to Lower Miocene age. **ENVIRON**: Diamond Harbour Formation is inferred to be deposited under shallow marine and coastal environment.

Mahanadi Basin

Age of Mahanadi Basin is Early Cretaceous to recent. About 2078 m volcanic pipe with intratrappean overlies the Archaean Basement. Thin volcanic rocks equivalent to the Deccan trap in two wells. KTB represented by hiatus. Paleocene to recent sediments are encountered in several onshore wells. The Paleogene litho-bio-chrono-sequence and paleoenvironmental settings in the Mahanadi Basin is shown in Fig. 14.

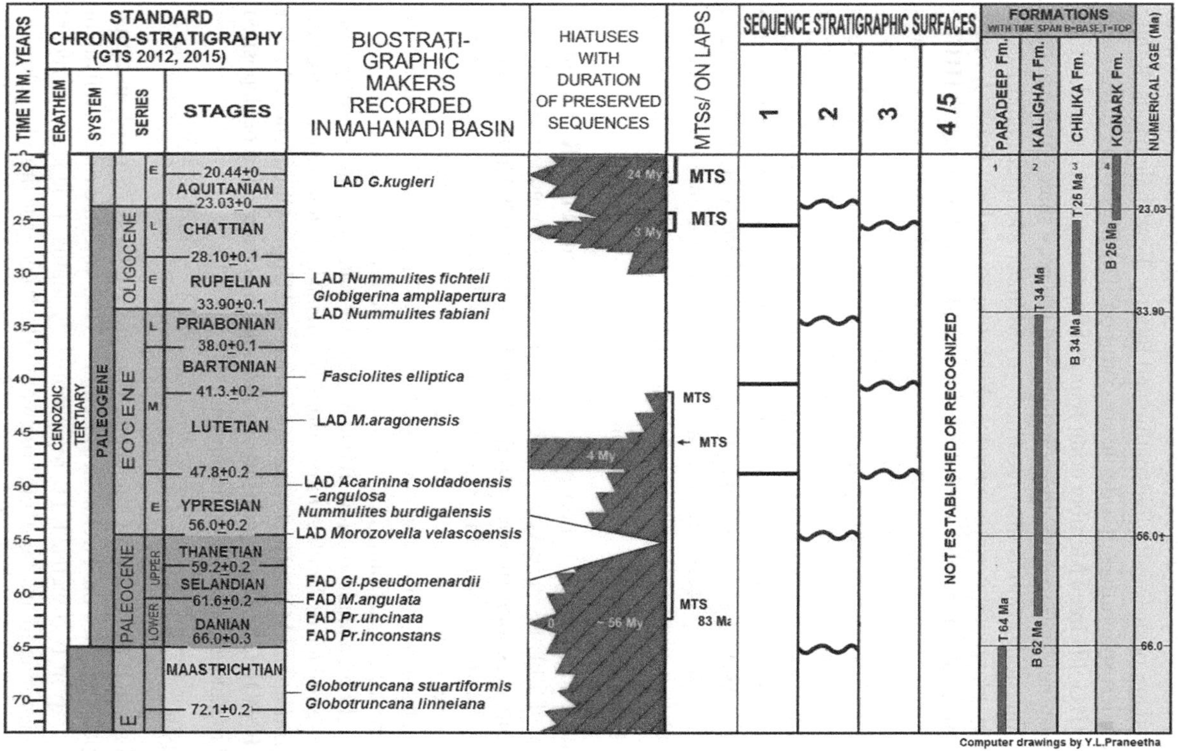

A BRIEF ON EACH FORMATION :

1. PARADEEP FM: THICKNESS: About 400m in graben areas, LITHOLOGY: Sandstone with minor shales/claystones, PALEOENVIRONMENT: Varies from outer neritic.
2. KALIGHAT FM: THICKNESS: 300m to 800m, LITHOLOGY: Limestone with sand and shale, PALEOENVIRONMENT: Stable marine shelf.
3. CHILKA FM: THICKNESS: About 630m, LITHOLOGY: Shale and some sand layers, PALEOENVIRONMENT: Shallow marine to outer neritic/ upper bathyal.
4. KONARK FM: THICKNESS: 375m to 1530m, LITHOLOGY: Shale with few layers of sand and limestone, PALEOENVIRONMENT: Outer neritic to bathyal in the offshore part.

Fig. 14 Paleogene Litho- Bio- Chrono- sequence stratigraphy and paleoenvironmental settings in the Mahanadi basin, India. Compiled by D. S. N. Raju

Kalighat Formation: (*Au*: Raju et al. 2009)

HOLO: 3298-2497 m in MND-2, *Hypo*: 2500–1782 m in MND-3. **LITH**: The bottom of this formation consists of a package of alternating sand and shale. Major part of the formation consists predominantly of calcareous interoperated with thin sandstone layers. The limestones are fossiliferous. Layers of calcareous shale occur in the upper part. **LB**: Coincides with K-T boundary, a major unconformity with base level fall and hiatus of 5–6 My equivalent Danian period. **UB**: The top is a pronounced erosional unconformity and subaerial exposures with major hiatus. **THICK**: 300–800 m **BIOTA**: The basal part contains *Daviesina langhami* and *Lockchartia haimei*. The Early Eocene interval yielded *Flosculina oblonga*, *Nummulites burdigalenis*, *N. globulus*, *N. praelucasi*, *Lockhartia hunti* and *L. conditi*. The Middle Eocene strata contain *Nummulites djodjokarte*. *M. discorbinus*, *Flosculina elliptica* and *Assilina spira*. The late Eocene strata yielded *Pellatispira inflata* and *Nummulites fabianii*. In the hypostratotype (MND-3) additional foraminifera including *Globorotalia uncinata* and *Lockhartia retiata* are found. **AGE**: Middle Paleocene to Late Eocene **ENVIRON**: Stable marine shelf. The lowest sands represent marginal marine condition.

Chilika Formation: (*Au*: Raju and Misra 2009)

HOLO: Interval from 3838 to 3204 m in MND-8 **LITH**: Predominantly shale with some sand layers

LB: Conformable in MND-8, **UB**: Unconformable **THICK**: About 630 m in the type section **BIOTA**: *Nummulites pengaroenensis* in lower part, *Eulepidina dialata* in the middle part, *Globigerina ampliapertura*, *G. selli*, *Sphaeroidina bullodes*, *Globigerinoides primordius* and *Cyclammina* sp., in the upper part of the type section. **AGE**: Early Oligocene to lower part of late Oligocene **ENVIRON**: Shallow marine conditions in the lower part while outer shelf/upper bathyal conditions in the upper part prevailed in the type section.

Krishna—Godavari Basin

The preserved and eroded records of sediments from Early Permian to recent are known. KTB is represented by volcanic an intratrappeans.The Paleogene litho- bio- and chronostratigraphy and paleoenvironment settings of Krishna Godavari (KG) Basin are shown in Fig. 15.

Cretacous Tertiary Boundary (KTB)

Numerous publications are available in KTB both from outcrops and subsurface. The most detailed study of the Deccan volcanism and mass extinction is by Keller et al. (2012) who described effects of the Deccan volcanism on stratigraphy and biologic contents in eleven deep wells from the KG Basin. In these wells, two phases of Deccan volcanism record the world's largest and longest lava mega-flows interstratified with the marine sediments in the KG Basin about 1500 km from the main Deccan volcanic province. The main phase-2 eruptions (~80% of total Deccan Traps) began in C29r and ended at or near the KTB, an interval that spans planktic

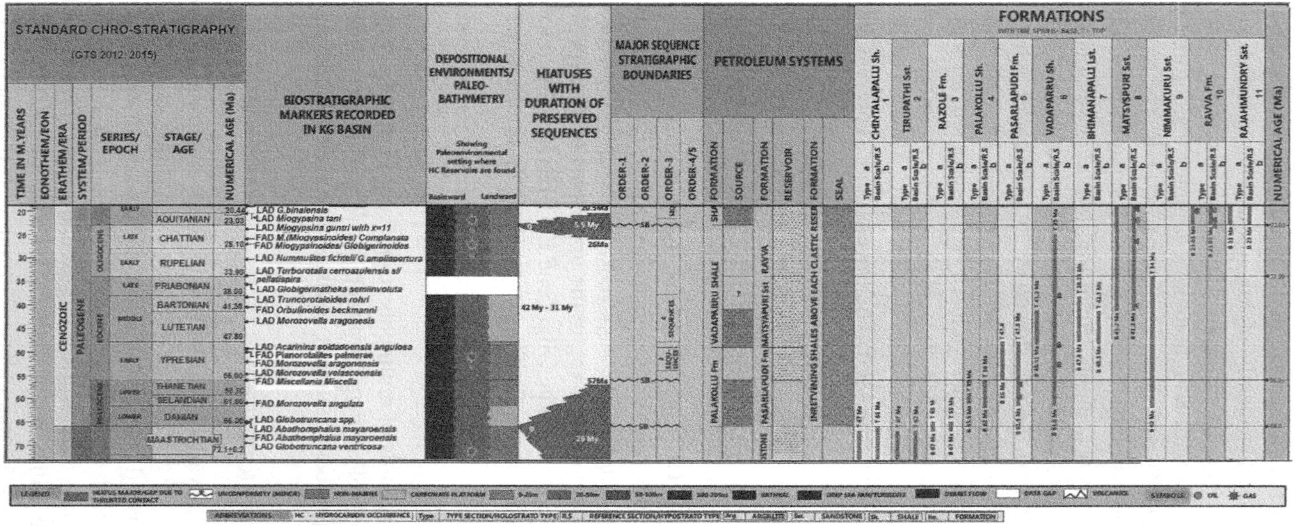

A BRIEF ON EACH FORMATION :

1) **CHINTALAPALLI Sh:**1.a THICKNESS:351m, LITHOLOGY: Clay,Sandstone, PALEOENVIRONMENT: Shallow marine 1.b. THICKNESS: Greater than 900m, LITHOLOGY: Sandstone with minor shale, PALEOENVIRONMENT: Shallow inner neritic to deeper than 70m.

2) **TIRUPATHI Sst:** 2.a. THICKNESS: 560m, LITHOLOGY: Shale with few sandstone layers in lower part. PALEOENVIRONMENT: Upper bathyl. 2.b. THICKNESS:700m to grater than 2000m. LITHOLOGY: Shale with thin beds of Sandstone. PALEOENVIRONMENT: Upper bathyal.

3) **RAZOLE:** 3.a. THICKNESS: 270m; LITHOLOGY: Basalts with inter trappeans, PALEOENVIRONMENT: Volcanics with marin incursion. 3.b. THICKNESS: 250m, LITHOLOGY: Sandstone, clay, laterised towards top, PALEOENVIRONMENT: Continental to marginal marine.

4) **PALAKOLLU Sh:** 4.a. THICKNESS: 445m, LITHOLOGY: Shale with occasionally silt bands, PALEOENVIRONMENT: Middle to outer neritic. 4.b. LITHOLOGY: Shale with occasionally thin beds of sandstone & silt stone, PALEOENVIRONMENT: Middle neritic to upper bathyal.

5) **PASARLAPUDI Fm:** 5.a. THICKNESS: 870m, LITHOLOGY: Alternative sandstone and shale with thin limestone bands. PALEOENVIRONMENT: Cyclical 0.50m bathymetry. 5.b. THICKNESS: 785m. LITHOLOGY: Sandstone with alternating shale beds, PALEOENVIRONMENT: Fluctuating 0 to >100m bathymetry. 6) **VADAPARRU Sh:** 6.a THICKNESS: 710m, LITHOLOGY: Shale with minor sandstone layers. PALEOENVIRONMENT: Lower 50- 100m, Upper : 0-20m bathymetry. 6.b. THICKNESS: > 2000m in offshore area, LITHOLOGY: Claystone, thin Sandstone& Siltstone, PALEOENVIRONMENT: Inner neritic to outer. 7) **BHIMANAPALLI Lst:** 7.a. THICKNESS: 565m, LITHOLOGY: Limestone with Sand, Shale alternative, PALEOENVIRONMENT: 20-50m bathymetry. 7.b. THICKNESS: 540m. LITHOLOGY: Limestone with beds of shale and sandstone with minor claystone and Limestone , PALEOENVIRONMENT: 0-20m bathymetry. 8) **MATSYPURI Sst:** 8.a THICKNESS: 692m, LITHOLOGY: Sandstone with minor clay and limestone. PALEOENVIRONMENT: 0-20m bathymetry. 8.b. THICKNESS: maximum 1200m, LITHOLOGY: Sandstone with few claystone beds. PALEOENVIRONMENT: Inner to middle neritic. 9) **NIMMAKURU Sst:** 9.a. THICKNESS: 225m, LITHOLOGY: Sandstone with occasionally pebbles and claystone. 9.b. THICKNESS: 510m, LITHOLOGY: Sandstone with occasionally pebbles and claystone. 10) **RAVVA Fm:** 10.a. THICKNESS: 1090m, LITHOLOGY: Thin sandstone alternatively with clay and siltstone, PALEOENVIRONMENT: Cyclical with bathymetry between 0 and 200m. 10.b. THICKNESS: 780m, LITHOLOGY: Sandstone and claystone alternation, PALEOENVIRONMENT: Middle bathyal. 11) **RAJAHMINDRY Sst:** 11.a. THICKNESS: 250m, LITHOLOGY: Sandstone, clay lateristressed towards top, PALEOENVIRONMENT: Continental to marginal marine 11.b. THICKNESS: 240-1090m, LITHOLOGY: Sandstone with clay beds, PALEOENVIRONMENT: Marginal marine to shallow neritic.

Fig. 15 Paleogene Litho- Bio- Chrono- sequence stratigraphy and paleoenvironmental settings in the Krishna Godavari basin, India. Compiled by D. S. N. Raju, In collaboration with Sudhir Shukla

foraminiferal zones CF1-CF2 and most of the nannofossil *Micula prinsii* zone, and is correlative with the rapid global warming and subsequent cooling near the end of the Maastrichtian. The mass extinction began in phase-2 preceding the first of four mega-flows. Planktic foraminifera suffered a 50% drop in species richness. The mass extinction was likely the consequence of rapid and massive volcanic CO_2 and SO_2 gas emissions. Deccan volcanism phase-3 began in the early Danian near the C29r/29n boundary. These data suggest that the catastrophic effects of phase-2 Deccan volcanism upon the Cretaceous planktic foraminifera were a function of both the rapid and massive volcanic eruptions and the highly specialzed faunal assemblages prone to extinction in a changing environment. Data from the KG Basin indicates that Deccan phase-2 alone could have caused the KTB mass extinction and that impacts may have secondary effects.

Razole Formation (*Au*: Rao 1990)

HOLO: Well Razole—1 **LB**: Conformably underlain by the Tirupati sandstone in West Godavari and Chintalapalli shale in parts of East Godavari sub-basins. **UB**: Conformably overlain by the Matsyapuri sandstone/Palakollu shale in East Godavari sub basin and unconformably by the Rajahmundry formation in Mandapeta area.

LV: Razole volcanics occur as outcrops. In West Godavari basin, these lava flows are encountered at a depth of 1000 m and Narasapur sub basin at a depth range of 3300 m. These volcanics are absent in Amalapuram, northeastern part of Narasapur and in offshore area.

Palakollu Shale (*Au*: Rao 1990)

HOLO: The interval from 2130 to 2575 m in deep well Palakollu-1. **LB**: This formation unconformably overlies the volcanics of the Razole Formation. **UB**: Conformably underlies the Pasalarapudi Formation. **LITH**: (Venkatarangan et al. 1993), In the type section the formation consists of dark grey fissile shales with occasional silt bands. Siltstones are grey to dark grey, hard, compact and micaceous. **THICK**: 445 m. in the type section. **BIOTA**: *Morozovella conicotruncana and M. aequa, M. angulata, M. velascoensis,* which suggest Late Paleocene (upper P3–P5) are recorded. The assemblage includes *Globorotalia (Turborotalia) inconstans, Eoglobigerina trivalis, Globorotalia (Acarinina) precursoria, and G. (T.) imitatea*. Between 2410 and 2575 m, between 2410 and 2575 m. *G. (T.) pseudobulloides* is recorded. The FAD/LDHO of *Cruciplacolithus primus* at depth 2560–65 m. and occurrence of *Cruciplacolithus tenuis* at depth 2315–20 m assign the sediments between 2315 and 2565 m to NP1 (upper part) *Markalius inversus nannoplankton* zone of Early Danian age. The occurrence of *Princius dimorphosus* in the sediments at 2445–50 m could further assign the sediments 2445–2515 m. to a combined NP2–NP4: *Cruciplacolithus tenius? Eolissolithus macellus nannoplankton* zones of Danian age.

The occurrence of *Toweius eminens, Heliolithus riedeli, Neochiastozygus* chiastus and *Princius bisulcus* between 2160 and 2445 m depth is significant. Based on the known ranges of the above species, a combined nannoplankton zones NP5–NP9: *Fasciculithus tympaniformis-Discoaster multiradiates* of Thanetian age is assigned to the sediments. **AGE**: Paleocene (Danian to Thanetian). **ENVIRON**: Middle to Outer shelf.

Pasarlapudi Formation (*Au*: Rao 1990)

HOLO: Well Palakollu—1 (depth interval, 1970–2840 m). **LB**: Conformably underlain by the Vadaparru shale. **UB**: Conformably overlain by the Bhimanapalli limestone. **THICK**: The thick of the formation in the well Pasarlapudi—1 is 870 m and in the well Razole—1 is 765 m. **LITH**: This formation consists mainly of alternations of sandstones and shales with thin limestone bands. **BIOTA**: Foraminifera: In the holostratotype, the sediments at 2850 m, just 10 m below the base of the formation yielded *Morozovella marginodentata/subbotina plexus* suggestive of earliest Lower Eocene. The entire succession from 2000 to 2850 m yielded rare larger foraminifera including, *Nummulites*, *Assilina*, *Discocyclina*, and *Fasciolites*. and *Miliolids*. **AGE**: The upper limit of the formation is within *Planorotalites palmarae* (Zone P9) of Lower Eocene. The lower limit is well within earliest Lower Eocene. **ENVIRON**: The entire succession of Pasarlapudi Formation, both in holo and hypostratotypes was deposited in a very shallow marine conditions in bathymetry between 0 and 50 m but often within 0-20 m.

Vadaparru Shale (*Au*: Rao 1990)

HOLO: Vadaparru–1 (depth interval, 1525–2235 m). *Hypo*: Bhimanapalii–1 (depth interval, 1780–2025 m). **LB**: Conformably underlain by the Razole Formation in onland area, not known in offshore due to lack of data. **UB**: Conformably overlain by the Matsyapuri and Ravva formations. **THICK**: The formation is 710 m thick in the well Vadaparru-1 and 245 m in the well Bhimanapalli-1. The formation thickens (more than 2000 m) in the offshore area. **LITH**: In the type section, this formation consists mainly of shales with minor sand layers. It is rich in organic matter and feebly calcareous with occasional presence of pyrite. **BIOTA**: Foraminifera: The hypostratotype in Bhimanapalli-1 gave a better control. In the lower part of the formation, *Morozovella acqua*, occurs at 1950 m and *M*. cf. *subbotinae*, is found at 1925 m, which suggest an age of upper P6 (lowest Lower Eocene). The upper part between 1800 and 1855 m yielded *Nummulites*, *Discocyclina*, and *Fasciolites*. The holostratotype in Vadaparru-1, yielded essentially larger foraminifera including *Nummulites* and *Discocyclina*. However, between 2000 and 2105 costate *Uvigerina* occurs. *Acarinina broedermani* is found at 2100 m. Nannoplanktics: The sediments between 1525 and 2235 m yielded low diversity and frequency of nannofossils. It is mostly dominated by Prinsiaceae and Braarudosphaeraceae members. *Reticulo-Fenestra dictyoda* is occurring almost throughout the section and based on the total known range of the species, a combined NP13–NP16: *Discoaster lodoensis–Discoaster tanii nodifer* nannoplankton zones of Late Ypresian-Lutetian age is inferred. The presence of *Braarudosphaerids* are very common in Eocene times and indicate a near shore depositional environment. **AGE**: Early to Middle Eocene. **ENVIRON**: In holostratotype the succession from 1525 to 1900 m was deposited under very shallow environment with a bathymetry between 0 and 2 m, but between 2000 and 2105 m a bathymetry of about 50–100 m prevailed.

Bhimanpalli Limestone (*Au*: Rao 1990)

HOLO: Well Bhimanapalli-1 **LB**: Conformably underlain by Pasarlapudi Formation or Vadaparru shale. **UB**: Conformably overlain by Matsyapuri sandstone. **THICK**: The thick of the Bhimanapalli Limestone in well Bhimanapalli-1 in 555 m and in Pasarlapudi-1 is 540 m. **LITH**: In Bhimanapalli-1 this formation consists mainly of highly fossiliferous limestone with thin sand–shale alternations in some areas. **BIOTA**: Foraminifera—The hypostratotype has yielded better control particularly in the lower part. The lower part and interval just below that from 1750 to 1989 m hqas yielded *Planorotaloides palmerae*, *Acarinina pseudotopilesis*, *Morozovella aragonensis*, *Nummulites*, *Fasciolites*, and *Assilina,* which clearly demonstrate, that the lower limit of the formation in the hypostratotype is within zone P9 of Lower Eocene. The interval between 1450 and 1700 m yielded Nummulimtes and few other larger foraminifera. At 1650 m *Lockhartia hunti pustulosa* has been recorded.

In the holostratotype, interval from 1230 to 1750 m, yielded larger foraminifera including *Nummulites*, *Fasciolites*, *Discocycline* and *Assilina*. Close to the upper limit of the formation at 1250 m in Bhimanapalli-1 *Truncorotaloides*, *Assilina* occur suggesting a Middle Eocene age. **ENVIRON**: The carbonates of the holostratotype were deposited under a bathymetry essentially between 20 and 50 m. In the hypostratotype, the sediments of this formation between 1750 and 1970 m were deposited under a moderate bathymetry between 50 and 80 m (Middle to Outer neritic).

Matsyapuri Sandstone (*Au*: Rao 1990)

HOLO: Well Matsyapuri-1 **LB**: Conformably underlain by Bhimanapalli Limestone. **UB**: Conformably overlain by Narasapur claystone. **THICK**: 540–1200 m. **LITH**: In the type section in well Matsyapuri-1 this formation consists mainly of sandstone with thin beds of claystone and limestone. The sandstones are dark grey to greenish, coarse to very coarse grained, moderately sorted, sub rounded with thin beds of dark grey carbonaceous claystone. Traces of pyrite at places with calcareous matrix occur. The claystone is brownish to dark grey, calcareous with traces of shell fragments; Reference section: The lithology of the formation in the well Bhimanapalli-1 is mainly sandstone with few claystone bands. **THICK**: The thick of the formation in Matsyapuri-1 is 692 m and in Bhimanapalli-1 540 m. The thick of the formation is maximum near GS-18-1 well where it is 1200 m. **BIOTA**: Foraminifera: Foraminifera are better represented in the hypstratotype. Important events from top to bottom include:

1. LAD of *Miogypsina cushmaniantillea*, at 790 m.
2. LAD of *M. (Lepidocyclina) excentrica* at 850 m.
3. LAD of *Nummulites fichteli* at 1040 m.
4. LAD of *Pellatispira at* 1070 m.

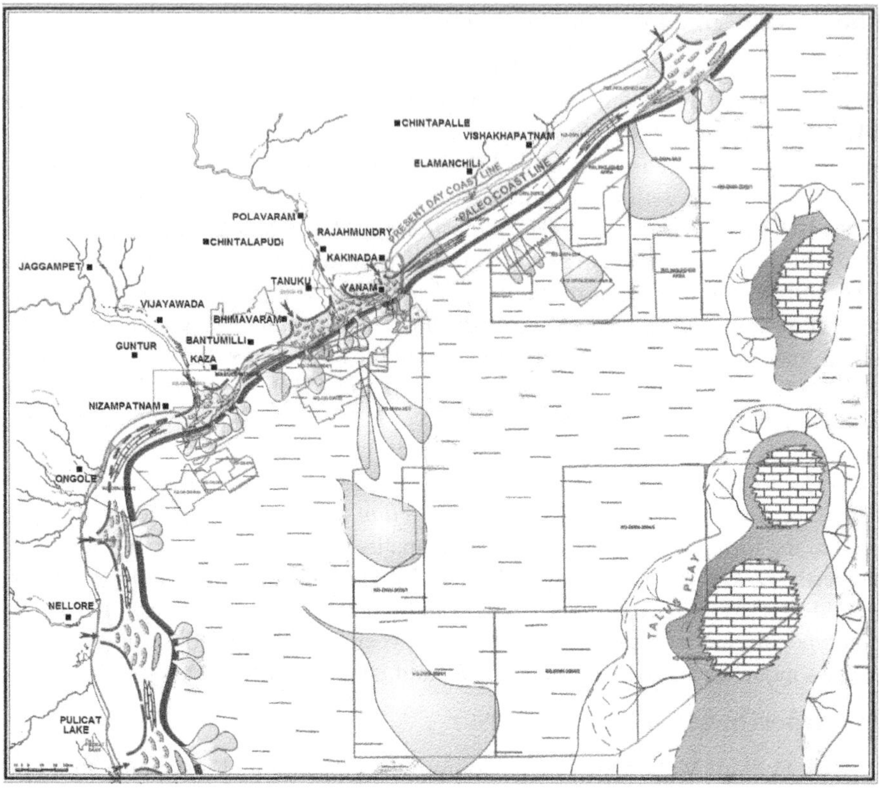

Fig. 16 Paleogeography of Eocene (Pasarlapudi/Vadaparru), Krishna Godavari Basin. (After Karuppuswamy 2013)

In between these events several larger foraminifera occur.

AGE: Middle Eocene—Middle Miocene **ENVIRON**: The succession is deposited under shallow marine condition with a bathymetry between 0 and 20 m, except at 850 m, where a bathymetry of 50 m might have been attained.

Paleogeography of Eocene (Pasarlapudi/Vadaparru Formations) show present coastline, paleo coastline shallow marine facies, shelf margin and also sub marine fan lobes in between shelf margin and 85° East Ridge. Carbonate deposits were predicted on the top of 85° East Ridge but they are yet to be proved (Fig. 16), Karuppuswamy (2013).

Paleogene-Neogene Boundary

The major events close to the Paleogene/Neogene boundary are: (1) drop in sea-level and hiatus.

Cauvery Basin

The oldest sediments recorded on the subsurface of Early Permian age. The formations and their biota are described below. The Paleogene litho- bio- and chronostratigraphy and paleoenvironments settings of the Cauvery Basin are shown in Fig. 17.

Kallamedu Sandstone (Fluvial Sandstone)

HOLO: Kallamedu **THICK**: 300 m **LB**: At places overlies Ottakovil sandstone and Kallankurchchi limestone, **UB**: Unconformably overlain by the Niniyur limestone (KT contact is observed in Periyakurichi mine) **BIOTA**: Nautiloid-*Hercoglossa danica* **AGE**: Danian. **ENVIRON**: Fluvial

Niniyur Formation (*Au*: Sundaram and Rao 1986)

HOLO: Niniyur village **LB**: Unconformably underlain by Kallamedu sandstone, **UB**: Unconformably overlain by Caddalore sandstone (Miocene) **THICK**: 65 m. **LITH**: Limestones interbedded with sandstones **BIOTA**: Nauitiloid–*Hercoglossa danica*, Foramnifera-*Thalmannita* sp., *Morozovella angulata*, Ichnofossils-*Ophiomorpha*, *Thalssinoides* **AGE**: Palaeocene (Danian) **ENVIRON**: Shallow marine.

Karasur Formation = Nerina = Niniyur (*Au*: Kossmat 1895)

HOLO: Ksrasur **LB**: Mettuveli Formation, **UB**: Manaveli Formation **BIOTA**: Planktic foraminifera **AGE**: Paleocene **ENVIRON**: Shallow shelf

Cauvery Basin Subsurface
Kamalapuram Formation: (*Au*: Venkatarengan et al. 1993)

HOLO: Well no. Kamalapuram-1 **LB**: Unconformably underlain by the Portonovo shale or Komarakshi shale. **UB**: Conformably overlain by the Karaikal shale. **THICK**: Thickness of the formation in type section is 680 m and in reference Sect. 840 m. Maximum thick recorded is 100 m (+) near Bhuvanagiri area. **LITH**: The formation is characterized by alternations of shaly sandstone and shale bands. Sandstone is greyish white to dirty white, sometimes greenish white, compact, hard to moderately hard, coarse, sometimes pebbly subangular to subrounded, well sorted, calcareous, at some places micaceous, pyritiferous and shaly. The shale is light grey to dark grey, compact, moderately hard, fissile. At some places, brownish limestone streaks are present along the fissile planes and encrustations of pyrite are also seen. **BIOTA**: *Globotruncana arca*, *Globorotalia pusilla*, *Morozovella velascoensis*, *M. subbotinae*, *M. spinulosa*, *M. aragonensis*. **AGE**: Maastrichtian-Middle Eocene. **ENVIRON**: Outer shelf to Upper Bathyal

Karaikal Shale Formation (*Au*: Rao 1972)

HOLO: Well no. Karaikal-1 **LB**: Conformably overlying the Kamalapuram Formation, at places unconformably underlain by Portonovo shale or Komarakshi shale. **UB**: Unconformably overlain by different formations. **THICK**: 1000 m in PY-3 structure. **LITH**: The formation is made of predominantly shale which are grey,

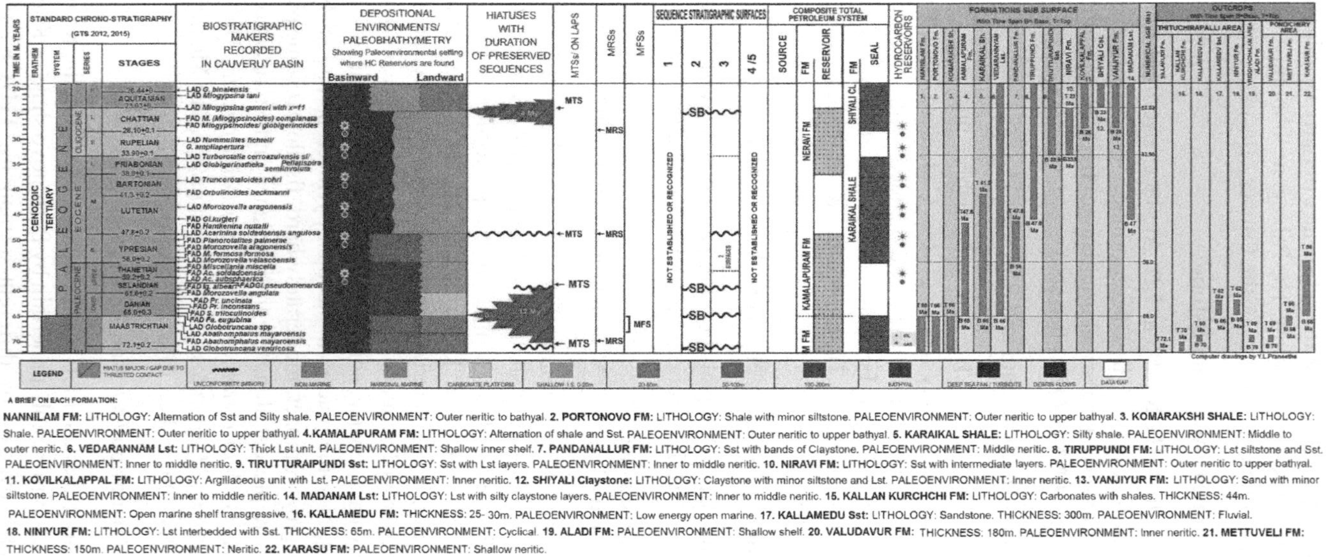

Fig. 17 Paleogene Litho- Bio- Chrono- sequence stratigraphy and paleoenvironmental settings in the Cauvery basin, India. Compiled by D. S. N. Raju, In collaboration with A. Nallapa Reddy, B. C. Jai Prakash and L. Chidambaram

silty, fissile and occasionally calcareous, micaceous, pyrite is seen as disseminated crystals; **BIOTA**: Diagnostic foraminifera: *Morozovella velascoensis*, *Planorotalites pseudomenardii*, *P. pusilla*, *Subbotina pseudobulloides*. **AGE**: Paleocene-Eocene **ENVIRON**: Middle to Outer shelf.

Pandanallur Formation (*Au*: Rao et al. 1971)

HOLO: Well no. Pandanallur-1 **LB**: Underlain by Karaikal shale, **UB**: Overlain by Tirupundi, Niravi, Madanam Limestone, Cuddalore formations. **LITH**: Pandanallur Formation consists mainly of clayey sandstone with bands of interbedded claystone. The sandstones are light grey to dirty white, fine to coarse grained, subangular to subrounded, feebly calcareous; silty with clayey matrix; pyrite and carbonaceous matter occur in traces. Claystones are dark grey to greenish grey, sticky, silty feebly calcareous. Pyrite limestone, chert and chalk also occur in traces. **THICK**: 100 m (+) in Bhuvanagiri area. **BIOTA**: Diagnostic foraminifera: *Nummulites fichteli*, *Globigerina angiporoides*, *Lepidocyclina* sp. **AGE**: Early Eocene **ENVIRON**: Middle shelf.

Thiruppundi Formation (*Au*: Rao et al. 1971)

HOLO: Well no. Thiruppundi-1 **LB**: Unconformably underlain by the Bhuvanagiri Formation. **UB**: Conformably overlain by the Nannilam Formation **THICK**: 613 m in type section. **LITH**: In the type section, the Tiruppundi Formation comprises limestone, siltstones and sandstones. The upper part of formation is marl, in the middle it is essentially a thick limestone and in the lower part, there is siltstone with thin bands of limestone. The sandstones consist of thin lenses of very fine sand occurring in between the limestone. In the reference section, most of the formation is massive limestone except the lower part which is claystone. **BIOTA**: Diagnostic foraminifera: *Turborotalia cerroazulensis*, *Turborotalia frontosa*, *Acarinina broedermanni*, *Orbulinoides beckmanni* **AGE**: Middle Eocene- Early Miocene (?). **ENVIRON**: Inner to Middle neritic.

Niravi Formation (*Au*: Rao et al. 1971)

HOLO: Well no. Karaikal-2 **LB**: Unconformably underlain by Tirupundi Formation/Karaikal shale. **UB**: Overlain by the Shiyali claystone/Kovilkalappal formations. **THICK**: 535 m **LITH**: In the type section, the formation comprises mainly of sandstone with minor limestone bands in between. **THICK**: The formation attain a maximum thick of 1000 m (+) in the Nagapattinam sub basin. **BIOTA**: Diagnostic foraminifera: *Globigerina angulisuturalis*, *G. binaensis*, *G. gortanii*, *G. ampliapertura*, *G. tripartita*, *Globorotalia opima* and costate *Uvigerina*, *Hoeglundina*, *Sphaeroidina*, *Gyroidina*, *Planulina*. **AGE**: Oligocene **ENVIRON**: Upper Bathyal.

Tiruttyraipundi Sandstone Formation (*Au*: Rao et al. 1971)

HOLO: Well no. Tiruttyraipundi-1 **LB**: Unconformably underlain by the Karaikal shale and Tiruppundi/Kamalapuram formations. **UB**: Overlain by the Madanam limestone and Tittacherri formations. **THICK**: 1000 (+) m in the Nagapattinam susbasin. **LITH**: In the type section, the formation comprises mainly of sandstone with minor limestone bands in between. The sandstones are medium to coarse grained, yellow, brown and grey in colour, at times pebbly and calcareous, subangular to subrounded. It is moderately to poorly sorted and is fossiliferous. Thin bands of light grey, compact siliceous limestone are seen at places. **BIOTA**: Diagnostic foraminifera: *Myogipsina complanata-formosaensis*, *Nummulites fichteli*, *Globigerina sastrii*, *Chiliguembelina cubensis*. **AGE**: Oligocene—Early Miocene **ENVIRON**: Inner to Middle shelf conditions.

Kovilkalappal Formation (*Au*: Venkatarengan et al. 1993)

HOLO: Well no. Kovilkalappal-1 **LB**: Unconformably underlain by the Bhuvanagiri Formation. **UB**: Conformably overlain by the Nannilam Formation. **THICK**: 955–1035 m. Thick of the formation in type section is 280 m where as it is 300 m in the reference section Extent: It occurs in Tanjore sub basin and Nagapattanam sub basin. **LITH**: The formation is an argillaceous rock unit with significant presence of limestone. Bands of sandstone are also observed. It is fossiliferous. It is calcareous and presence of pyrite is also noticed. **BIOTA**: *Pellatispira madaraszi*, *Discocyclina* sp, *Baculogypsina tetrahedra*, *Nummulites acutus*, *N. chavannesi*, *Fasciolites* sp, *Lockhartia* sp. **AGE**: Middle to Late Eocene. **ENVIRON**: Inner to Middle shelf.

Vanjiyur Sandstone Formation (*Au*: Rao et al. 1972)

HOLO: Well no. Karaikal-4 **LB**: Underlain by the Shiyali claystone. **UB**: Overlain by the Madanam Limestone and at places by Tittacheri formations. **THICK**: 65–155 m **LITH**: In the type section, this formation is predominantly a sandy rock unit with occasional occurrence of siltstone and calcareous matter. **BIOTA:** Diagnostic foraminifera: *M. (M.) complanata*, *M. formosensis*, *Globigerina angulisuturalis*, *Turborotalia opima*. **AGE**: Late Oligocene-Early Miocene **ENVIRON**: Inner to Middle shelf.

West Coast Basins

Distribution of lithofacies during the Paleogene in selected wells across Westren offshore is shown in Fig. 25.

Kerala–Konkan

The basin evolved the following breakup and separation of Sychelles and Madagascar from India Ca 110–90 Ma. In the subsurface 4 Basalt flows with Coniacian to Santonian to Danian were observed. Paleogene formations are briefly described below. The Paleogene litho- bio- and chronostratigraphy and paleoenvironments settings of Kerala-Konkan Basin is shown in Fig. 18.

Cochin Formation (*Au*: Mathur et al. 1993)

HOLO: Interval 3150–3942 in well CH-1-1; **LB**: Unconformably overlie the Deccan Traps, **UB**: Unconformably underlie the Kasargod Formation **LITH**: In the holostratotype the formation is divisible into four subunits; the lowest unit is predominately dirty white to grey, medium to coarse grained sandstones with thin grey to dark grey sandstones. The second unit consists of grey and dark grey siltstones with

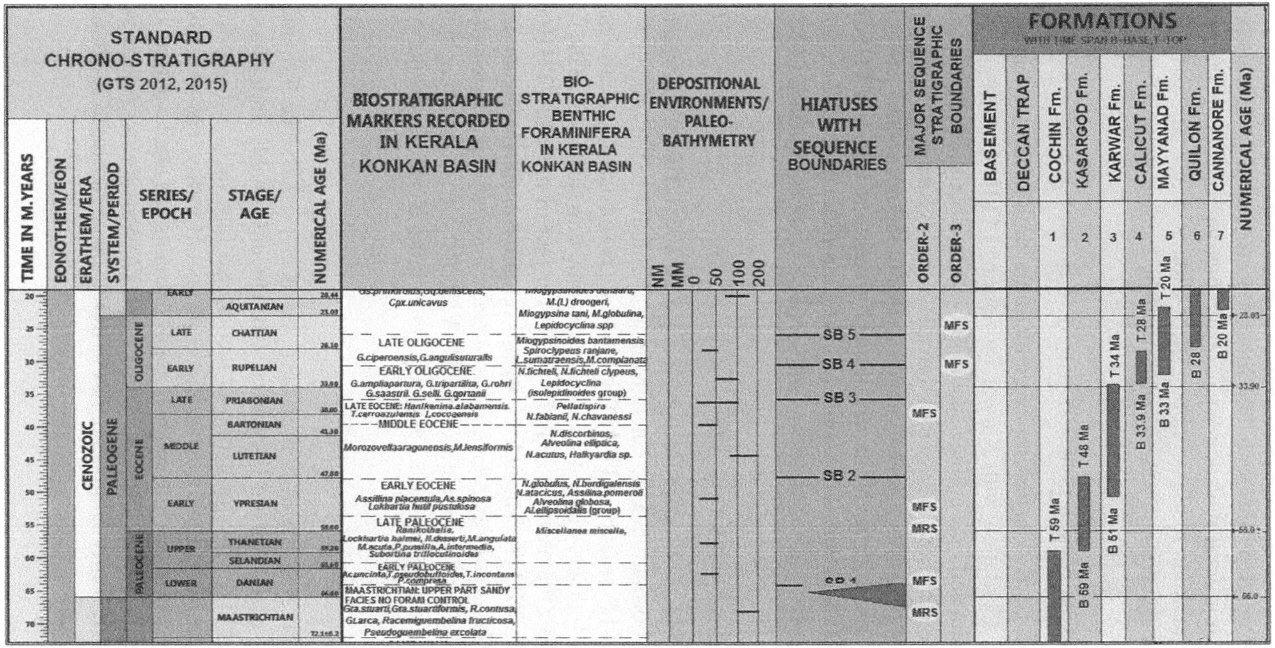

Suggested Faunal Abbrevation: EARLY PALEOCENE: *Glomo: Glomoalveolina, Ac: Acarinina, Pl: Planorotalites,*
LATE PALEOCENE- MIDDLE EOCENE: *N: Numulites, Ac: Acarinina, Mz: Morozovella, As: Assilina, Al: Alveolina*
LATE EOCENE: *Ps: Pellatispira, Gl: Globorotalia, T: Turborotalia*

BRIEF ON EACH FORMATION:

1. COCHIN Fm: THICKNESS:790m to 919m, LITHOLOGY:Sst.shale with thin Lst, PALEOENVIRONMENT: Deeper than 100m to middle neritic. 2. KASARGOD Fm: THICKNESS: 588m, LITHOLOGY: Sst with thin beds of Clay&Lst, PALEOENVIRONMENT: Inner neritic. 3. KARWAR Fm: THICKNESS: 87m, LITHOLOGY: Lst with thin shale beds, PALEOENVIRONMENT: Lagoonal to inner neritic. 4. CALICUT Fm: THICKNESS: 87m, LITHOLOGY: Lst.several intervals dolomitized, PALEOENVIRONMENT: Inner neritic to marginal marine. 5. MAYYANAD Fm: THICKNESS: 270m, LITHOLOGY: Gravelly sand interbedded with clays&lignite, PALEOENVIRONMENT: Shallow inner neritic-swamp 6. QUILON Fm: THICKNESS: 912m, LITHOLOGY: Limestone with thin shale/claystonebands, PALEOENVIRONMENT: Marginal marine to inner neritic. 7. CANNANORE Fm: THICKNESS: 227m, LITHOLOGY: Shale/Cst& Sst with thin Lst bands, PALEOENVIRONMENT: Fluctuating b/w mid.neritic to u.bathyal.

Fig. 18 Paleogene Litho- Bio- Chrono- sequence stratigraphy and paleoenvironmental settings in the Kerala–Konkan basin, India. Compiled by D. S. N. Raju, In collaboration with C.N.Ravindran

shales and thin limestone towards top. The top unit is dominantly shale with thin limestone bands. **THICK**: 919 m in K-1-1 and 790 m in CH-1-1. **BIOTA**: *Morozovella pseudobulloides*, *M.uncinata*, *Rosita contuse*, *Globotruncana venticosa*, *Dicarinella Concavata*. **AGE**: Santonian to Paleocene. This lithounits crosses the KTB boundary. **ENVIRON**: From foraminiferal evidences a paleobathymetry deeper than 100 m can be suggested for the Santonian to Masstrichtian succession. A hiatus across KTB is recognizable. Paleocene succession was probably deposited under middle shelf environment.

Kasargod Formation (*Au*: Mathur et al. 1993)

HOLO: Well KG-1-1 **LB**: Unconformable with either the Cochin formation or the Deccan traps. **UB**: Gradation with Karwar formation. **LITH**: Consists of predominately grey, medium and Fine grained sandstone with thin beds of grey tom dark grey clay and Limestone towards the top part. Inn CH-1-1, the Kasargod Formation consists of argillaceous limestone, sandstone and shales in the upper part .two thin basaltic flows were observed in the basal part of this formation in CH-1-1. **THICK**: 588 m in KG-1-1 **BIOTA**: *Nummulites globulus*, *Assilina placentula*, *Miscellania miscella*, *Ranikothalia* sp., and *Lockhartia hunti*. **AGE**: Late Palaeocene to early Eocene **ENVIRON**: Inner shelf.

Karwar Formation (*Au*: Mathur et al. 1993)

HOLO: Interval 1342–1429 m in deep well KK-OS–V-1-1 **LB**: Conformably overlies the Kasargod Formation. **UB**: Unconformable with either the Calicut Formation or the Mayyanod Formation **LITH**: The lithology is consistently a carbonate facies comprising grey-light grey limestone with occasional thin shale beds. This formation is present all over the basin, except the shallower part of the shell. **THICK**: 87 m in the type section. **BIOTA**: The index Larger foraminifera include *Nummulites fabianii*, *Nummulites burdigalensis*, *Assilina spinosa*, *Lituonella* sp., *Nummulites bagelensis* **AGE**: Early, middle to late Eocene **ENVIRON**: Varies from Lagoonal and restricted marginal marine towards basin margin to inner Neritic in deeper part of the basin.

Calicut Formation (*Au*: Mathur et al. 1993)

HOLO: Interval from 1342 to 1429 m in deep well CH-1-1 **LB**: The formation rests unconformably over the Karwar Formation. **UB**: Unconformably underlies the Cannare Formation **LITH**: This formation is recognized into include Limestone sandwitched between the Eocene and Lower Oligocene unconformity. The 87 m thick unit is characterized by greenish white and buff colored forams and algal wackestone. Several intervals are dolomitized. **THICK**: 87 m m **BIOTA**: *Nummulites fichteli* and *Nummulites vascos* **AGE**: Early Oligocene. **ENVIRON**: Inner Neritic to Shallower shelf and Marginal Marine.

Mayyanad Formation (*Au*: Desikachar 1976)

HOLO: Interval from 448 to 695 m in deep well K-1-1 **LB**: It rests unconformably over the Karagod Formation. **UB**: Unconformably overlie the Quilon Formation **LITH**: In the holostratotype this formation consists of well sorted coarser to medium grained, occasionally gravelly sands interbedded with dark clays and lignite. A few thin limestone bands are also present. The lower part of this formation laterally changes into carbonate facies of the Calicut formation beyond the shallow shelf. **THICK**: Maximum thick is 270 m. *Extent and LV*: The formation is exposed in detached outcrop and has also been met in bore hole nonconformably overlying Archean basement. Lower part of Mayyanad formation laterally changes into carbonate facies of Calicut foamation the shelf edge. **BIOTA**: *Nummulities fichteli*, *Nummulities vorcer*, Upper part is fossiliferous. **AGE**: Early Oligocene to probably early Miocene **ENVIRON**: Shallow inner-shelf to marginal marine and swampy environment.

Mumbai Offshore Basin

The oldest sediments named Panna Formation overlie the Deccan traps. The Paleogene Formations are briefly described below. The Paleogene litho- bio- chronostratigraphy and paleoenvironments settings of Mumbai Offshore Basin is shown in Fig. 19.

Panna Formation (*Au*: Chugh 1987)

HOLO: Well Bassein-1 **LB**: Nonconformably underlain by the Deccan Trap or the Crystalline basement, **UB**: Conformable and gradational with the overlying the Bassein Formation in most part. Unconformable in basin margin. **LITH**: Represented by conglomerate and sandstones at the basal part, overlain by a section of coal-shale alternation in the middle and succeeded by shale section at the top. The coal is lignite. The shale is laminated, carbonaceous and pyritiferous. Chakravorti et al. (Unpublished Report) have identified altogether 12 prominent lithological episodes, out of them 8 are fining-up cycles, 2 major coarsening-up cycles and 2 moderate cycles. There are 5 layers of coal and thin beds of open marine limestone in the Panna Formation. **THICK**: In the type section, the formation is 75 m. In the reference section the formation is 661 m. *LV*: The formation is spreads over the entire Mumbai offshore basin, excepting the paleo-high. Panna Formation has been informally divided into three units. The formation shows considerable variations. In the Tapti-Daman block, a good sand development is observed adjoining the Navasari low. A large part of this formation is represented by thick sandstone unit (Alluvial fan deposit) in the Panna East area. In south Ratanagiri, it is represented by a fairly thick trapwash sequence. **BIOTA**: Top of the Panna Formation coincides with top of the *Nummulites burdigalensis-Assilina spinosa* assemblage. Palynology played a major role in dating the strata of Panna Formation. Mehrotra et al. (2002) recognized three dinoflagellate bioevents in the Thanetian and 12 in Ypresian (within duration of 7.3 My). **ENVIRON**: Fluctuation (cycles) conditions varying from fluvial, lagoonal to shallow inner shelf prevailed. The cycles appear to comparable with 9–10 sequence recognized in the Ypresian of Europe (Hordenbol et al. 1998) **AGE**: Late Paleocene to Early Eocene.

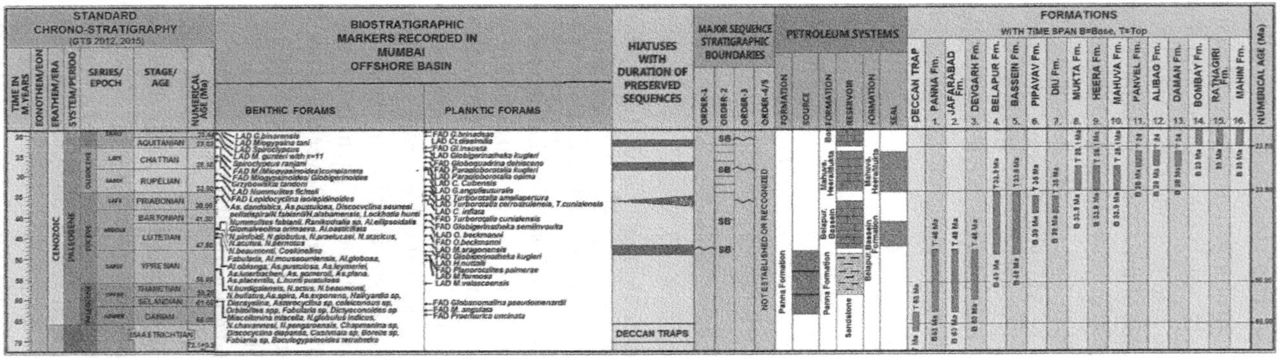

Suggested Faunal Abbrevation: CRETACEOUS: *Gta: Globotruncanita, Gt: Globotruncana, Ga: Gansserina, R: Rosita, P: Pseudotextularia, Mt: Marginotruncana, D: Dicarinella* EARLY PALEOCENE: *Glomo: Glomoalveolina, Ac: Acarinina, Pl: Planorotalites, T: Turborotalia*
LATE PALEOCENE- MIDDLE EOCENE: *N: Numulites, Ac: Acarinina, Mz: Morozovella, As: Assilina, Al: Alveolina* LATE EOCENE: *Ps: Pellatispira, Gl: Globorotalia, T: Turborotalia*
MIOCENE-PLIOCENE: *G: Globigerina, Gs: Globigerinoides, Gq: Globoquardrina, Gna: Globigerinatella, Gnta: Globigerinita, Cpx: Catapsydrax, P: Praeorbulina, Pu: Pullineatina*

A BRIEF ON EACH FORMATION:

1. **PANNA FM:** THICKNESS: 75m, LITHOLOGY: Conglomerate and Sst at basal part, coal shale alternate middle part, shale in towards top, PALEOENVIRONMENT: Cyclical fluvial, lagoonal to inner shelf 9 to 10 cycles are recognizable.
2. **JAFARABAD FM:** THICKNESS: 518m, LITHOLOGY: Shale and Limestone. 3. **DEVGARH FM:** THICKNESS: 513m, LITHOLOGY: Dolamitic fossiliferous Lst. 4. **BELAPUR FM:** THICKNESS: 42m, LITHOLOGY: Caclareous shale.
5. **BASSEIN FM:** THICKNESS: 305m, LITHOLOGY: Carbonate (Wackstone and packstone), PALEOENVIRONMENT: Restricted platform to open carbonate platform deeper towards shelf margin blocks. 6. **PIPAVAV FM:** THICKNESS: 94m, LITHOLOGY: Sandstone. 7. **DIU FM:** THICKNESS: 313m, LITHOLOGY: Silty Shale. 8. **MUKTA FM:** THICKNESS: 47m, LITHOLOGY: Massive Lst, Shale & marl stone. PALEOENVIRONMENT: Inner neritic. 9. **HEERA FM:** THICKNESS: 98m, LITHOLOGY: Shale, Lst alteration.
10. **MAHUVA FM:** THICKNESS: 314m, LITHOLOGY: Calcareous Shale. 11. **PANVEL FM:** THICKNESS: 262m, LITHOLOGY: Argellaceous Lst. 12. **ALIBAG FM:** THICKNESS: 84m, LITHOLOGY: Shale with a few Lst bands. PALEOENVIRONMENT: Shallow inner neritic.
13. **DAMAN FM:** THICKNESS: 370m, LITHOLOGY: Sandstone. 14. **BOMBAY FM:** THICKNESS: 378m, LITHOLOGY: Lst alternation with thin shale layers. PALEOENVIRONMENT: Cyclical shale deposited under marginal marine. 15. **RATNAGIRI FM:** THICKNESS: 373m, LITHOLOGY: Lst with Sh.thin layers. PALEOENVIRONMENT: Shallow Shelf Lagoon to open shelf. 16. **MAHIM FM:** THICKNESS: 75m, LITHOLOGY: Fossiliferous shale.

Fig. 19 Paleogene Litho- Bio- Chrono- sequence stratigraphy, paleoenvironmental settings and petroleum systems in the Mumbai Offshore basin, India. Compiled by D. S. N. Raju, In collaboration with C. N. Ravindran, Alok Dave and Sudhir Shukla

Bassein Formation (*Au*: Basu et al. 1982)

HOLO: Well BS-1, interval 1765 to 2070 m; **LB**: Unconformable with the Panna Formation. In the shelf margin block and Mahim graben, it overlies the shale of Belapur Formation. Over Ratanagiri and DCS area, it unconformably overlies the Devgarh Formation. **UB**: A regional unconformity (H-3B sequence boundary) separates the Bassein Formation form the overlying Mukta Formation. Over most of the shelf area, base of 15–20 m thick high gamma marker (clay) define its upper boundary. **LITH**: Bassein Formation is characterized by a monotonous, buff to light brown, highly porous wackstone in the upper part and hard packstone towards the base. Occasionally echinodermal wackstone constitute the middle part. **THICK**: The formation is 305 m thick in the type section. Its thick varies drastically, from less than 100 m over the Paleohigh to more than 700 min the western and southern parts of Bombay High-DCS block. **BIOTA**: In the Panna-Bassein area, the top part of the formation coincides with the top of *Coskinolina-Fasciolites* assemblage zone. In a few wells, typical upper Eocene forams like *Pellatispira* sp., *Nummulites fabiani* etc. occur in the uppermost part of the Bassein Formation. **ENVIRON**: A wide range of environments like restricted platform (Shelf lagoon) with isolated shales in the Bassin area to open carbonate platform in Bombay High and Ratanagiri area, and finally deeper water carbonates over most of the shelf margin blocks. **AGE**: Middle to Upper Eocene.

Mukta Formation (*Au*: Mathur et al. 1983 introduced it as a member; Zutshi et al. 1993 raised it to the rank of Formation)

HOLO: Well B-57-11 **LITH**: In the type area, the formation consists of a massive limestone unit underlain by a 15–20 m thick zone of radioactive shales and marlstone. The limestone is fossiliferous. The shale is fossiliferous and carbonaceous. In well SS-1, it shales out. **THICK**: 47 m thick in the type section **BIOTA**: Foraminiferal assemblage including *Nummulites fitchteli*. *LV*: The formation extends over the entire Bombay offshore Basin but shales out in Tapti-Dman block. EMW: Oil and gas.

Alibagh Formation (*Au*: Basu et al. 1982; Panvel Formation of Zutshi et al. 1993 is considered junior synonym of Alibagh Formation.)

HOLO: Well B-38-1 **LB**: The Heera Formation is unconformably overlain by dominantly shaly facies of Alibagh Formation, **UB**: Unconformably overlain by the massive limestone of the Ratanagiri Formation. Over most of the Panna-Bassein block, limestone of the Bombay Formation overlies the Alibagh Formation. **LITH**: In the type section, the formation consists of predominately of shale with a few thin limestone bends. The shale is occasionally pyrtic and fossiliferous. The limestone is fossiliferous and micritic. **THICK**: In the type section, the formation is 84 m thick. In B-30-1, it is 278 m. **BIOTA**: The formation is characterized by *Miogypsinoides complanata* and *Spiroclypeus*. **ENVIRON**: Deposited in shallow inner shelf environment **AGE**: Upper Oligocene.

Paleogeographic map of Mumbai high and western shelf shows from east to west clay/shale facies of fluvial to transitional environments, carbonate facies and shelf margin facies further seaward (Fig. 20a), Wandrey (2004b).

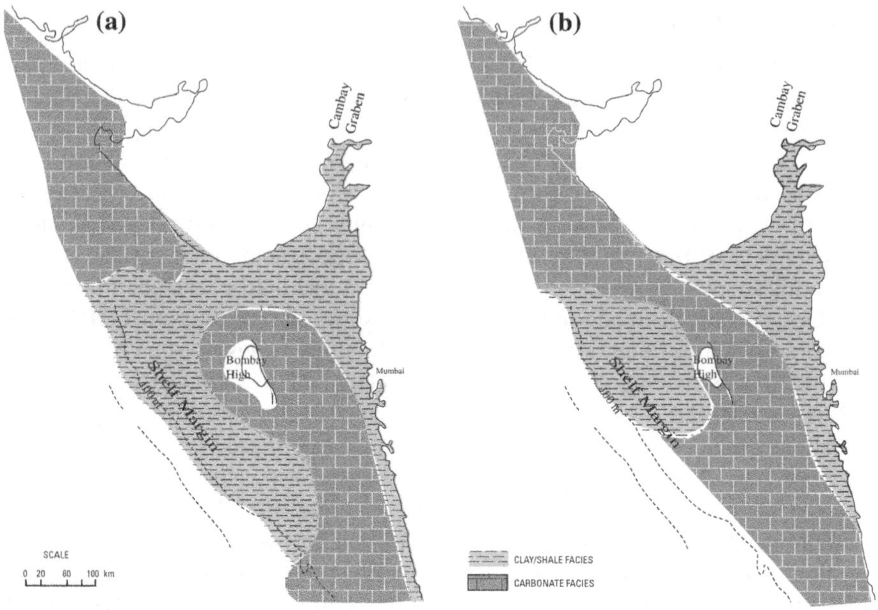

Fig. 20 Generalized Paleocene (left) and Oligocene (right) paleogeographic maps of Bombay-High and western shelf area. (After Wandrey 2004b)

A paleogeographic map of Oligocene (Fig. 20b) shows clay/shale facies of marginal marine along eastern part and wide carbonate facies and shelf margin clay/shale facies towards the west. There was no deposition from Paleocene to Middle Oligocene of Mumbai High.

Cambay Basin

The oldest sediments dated by palynofossils indicate Maastrichtian age and fluvial environment. The latest Cretaceous Early Paleocene is a time of the Deccan volcanism. Paleocene to Recent sediments were encountered in subsurface. The Paleogene litho- bio- and chronostratigraphy and paleoenvironments settings of Cambay Basin is shown in Fig. 21.

Deccan Trap

Hypo: Well Cambay-11, interval 2482–3487.75 m. **LB**: Unconformable with the Mesozoic, **UB**: Olpad Formation **LITH**: It comprises of vesicular and non-vesicular dark grey to greenish grey, melanocretic basalts associated with andesite, trachyte etc. About 14 lava flows are recognized separated by intertrapeans beds. The intertrapeans beds are made of variegated shales, chocolate brown with greenish grey specks and pockets pf calcite. **THICK**: 1500 m. *LV*: Present in Cambay Basin and absent in on land part in further sites. **BIOTA**: No micro fauna is reported from the intertrapeans **ENVIRON**: Effusive volcanic activity **Age**: An upper Cretaceous to Lower Paleocene age is suggested to the Deccan Trap.

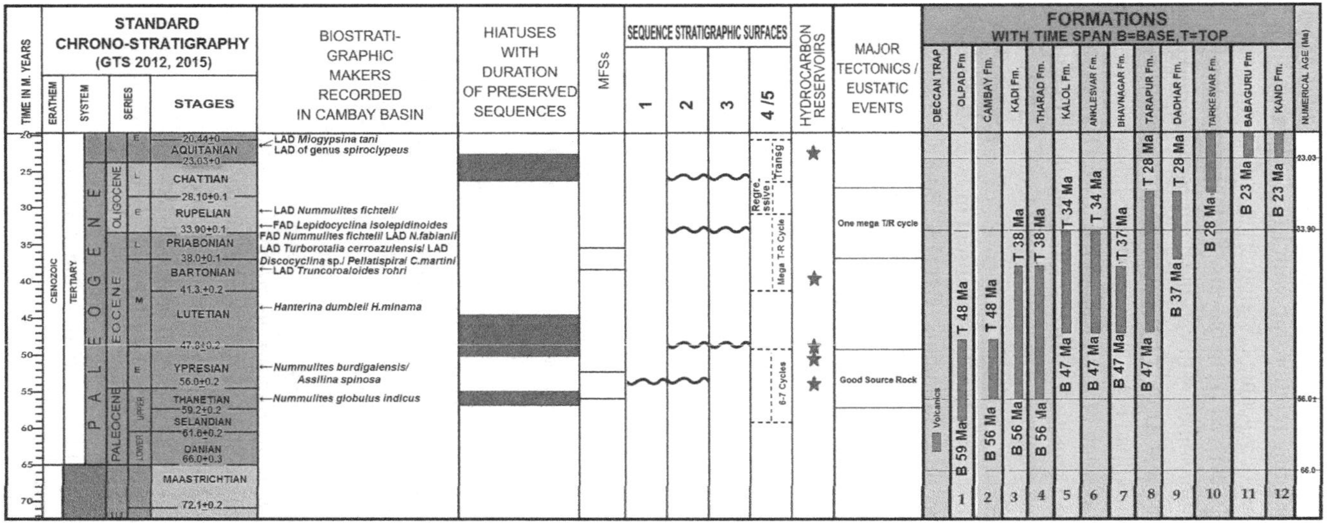

A BRIEF ON EACH FORMATION:

1. OLPAD FM: THICKNESS: 1773m; LITHOLOGY: Volcanic conglomerates, Sst, Silt, Claystone, 3 lithosources were recognized; PALEOENVIRONMENT: Fresh water to brakish, shallow marine in upper part. 2. CAMBAY FM: THICKNESS: 1500-3000m; LITHOLOGY: Dark grey to black shale with bands of sand, siltstone; PALEOENVIRONMENT: Cyclical, braking to shallow inner neritic, 6 to 8n cycles are recognized. 3. KADI FM: THICKNESS: 471-871m; LITHOLOGY: Alterations of sandstone, siltstone, coal; PALEOENVIRONMENT: Continental to littoral paralic. 4. THARAD FM: THICKNESS: 585-647m; LITHOLOGY: Sandstone, shale and thick coal beds; PALEOENVIRONMENT: Deltaic. 5. KALOL FM: THICKNESS: 200-150m; LITHOLOGY: Alterations of Sandstone, siltstone and carbonaceous shale; PALEOENVIRONMENT: Deltaic marginal marine, inner neritic transgressive – regressive cycle. 6. ANKLESVAR FM: THICKNESS: 390-602m; LITHOLOGY: Altering shale, sandstone and carbonaceous shale, arenaceous limestone in upper part; PALEOENVIRONMENT: Marine neritic to prodelta. 7. BHAVNAGAR FM: THICKNESS: 121m; LITHOLOGY: Shale, claystone and ferruginous sandy claystone; PALEOENVIRONMENT: Near shore fluctuating energy conditions. 8. TARAPUR FM: THICKNESS: 59-125m; LITHOLOGY: Silty and sandy shale; PALEOENVIRONMENT: Inner neritic to upper bathyal. 9. DADHAR FM: THICKNESS: 211m in type section; LITHOLOGY: Soft sandstone with shale alterations; PALEOENVIRONMENT: Shallow neritic. 10. TARKESVAR FM: THICKNESS: 50-125m; LITHOLOGY: Claystone, minor argillaceous sandstone, streaks of coal; PALEOENVIRONMENT: Represents regressive phase; 11. BABAGURU FM: THICKNESS: 180-580m; LITHOLOGY: Ferruginous sandstone, conglomerate and clays; PALEOENVIRONMENT: Fluvial to shallow marine. 12. KAND FM: THICKNESS: 150-305m; LITHOLOGY: Soft claystone with occasional of sandstone; PALEOENVIRONMENT: Fluctuating marginal marine to inner neritic.

Fig. 21 Paleogene Litho- Bio- Chrono- sequence stratigraphy, paleoenvironmental settings and hydrocarbon occurrence in the Cambay basin, India. Compiled by D. S. N. Raju, In collaboration with Sudhir Shukla

Olpad Formation (*Au*: Chandra and Chowdary 1969)

HOLO: Olpad well-1A, interval 1468–2443 m (+); **LB**: Unconformable with the Deccan Traps, **UB**: Intertonguing relationship with Lower Cambay shale **LITH**: In the type section it comprises of volcanic conglomerate, sandstone, silt, shale, claystone, and clays exclusively derived from basalts. Sudhakar and Basu (1973) have divided the Olpad into two lithosomes with varying degrees of intermixture: Grey and mottled claystone lithosome and Trap conglomerate-Sandstone lithosome. Pandey et al. (1993) have divided the Olpad Formation into three lithosomes; the upper sandstone lithosome, middle claystone lithosome, and the lower Trap lithosome. In one exploratory well the Olpad Formation is subdivided into three members: (i) Trapwash Member, (ii) claystone Member (iii) Nawagam Member **THICK**: Maximum thick 1778 m in Jambusar-P-I.

LV: Mostly developed in the Cambay Basin, absent in over Meshana host. Shows distinct vertical variation as three lithounits. **BIOTA**: Poorly fossiliferous and the fauna (after Pandey et al. 1993) recorded are thin walled fresh water Gastropods, bivalves shells and chitinous plates. Palynofossils are also recorded in the Olpad Member. **ENVIRON**: The formation was deposited under fresh water to slightly brackish water coastal environment, shallow marine towards upper part. **AGE**: Paleocene to Lower Eocene age has been assigned based on palynofossils and indirect stratigraphic methods (Pandey et al. 1993).

Cambay Shale (*Au*: Zubkov et al. 1966; Chandra and Chowdary 1969; Bhandari and Mathur 1968).

The Paleogene litho- bio- and chronostratigraphy and paleoenvironment settings of Cambay Basin is shown in Fig. 16.

HOLO: Cambay well No. 11, interval 1950–2382 m; **LITH**: In type section this formation is comprised of dark grey to black, fissile, frequently laminated and bituminous shale with occasional bands of sands and siltstones. **THICK**: in type section it is 432 m, in Kalol-263 it is 1153 m *LV*: thick increases in north in Linch-Warson-Sobhasan up to 1500–3000 m **BIOTA**: Cambay Shale, in general is poorly fossiliferous. Important foraminifera recorded from upper part of the Cambay Shale include: *Nummulites burdigalensis*, *Operculinoides* sp., and rare occurrences of *Assilina granulosa* and *Assilina* cf. *A. spira* and ostracodes *i.e. Alocopocythere longilinea*, *A. abstracta*, *Gyrocythere grandilavis*, *Butonia boldi*, *Paracypris* sp., and *Neocyprideis suratensis*, *Phylcetenophora meridionalis* and *Acanthocythereis vastanensis.* The younger Cambay Shale is rich in nonmarine ostracodes. They include *Canada cambayensis*, *Cythereidella gujaratensis*, *Cythereidella govindanii*, *Theriosynoeum*? *Danielopoli*, *Frambocythere colinii*, and *Metacypris bhatiai*. *Globorotalia* sp., *Chiloguembelina* sp., arenaceous foraminifera from the upper part of the formation. **ENVIRON**: The Cambay Shale generally was deposited in fluctuating environment varying from brackish to shallow inner neritic with intermittent fresh water conditions. **AGE**: Lower Eocene.

Kadi Formation (*Au*: Chandra and Chowdary 1969)

HOLO: Well south Kadi-4, interval 1345–1816 m; **LB**: Conformable with younger Cambay shale, **UB**: Unconformable with overlying Kalol Formation which is marked by the base of seismic "B" mark. **LITH**: In type section it is characterized by alternations of sandstone, siltstone and coal with layers of shale, shale is dark grey, sideritic and carbonaceous. Coal is black moderately hard, subvitreous to dull luster. **THICK**: 871 m in UA and WA-1 wells. In the type well South Kadi-4 the thick is 471 m.

LV: Restricted in Ahmedabad-Mehsana Block. For details please see Pandey and Dave (1998) **BIOTA**: The formation is mostly unfossiliferous. However, a few ostracods represented by *Alocopocythere abstracta*, *Buntonia boldi*, broken shell fragments, gastropods shells, dinoflagellates and pollen species are recorded from the Nandasan well-D. Important palynofossils: *Proxaperites hammenii*, *Psilodiporites hammenii*, *Tricolpites densioranatus*, *Marginipollis concinnus* and *Spinizonocolpites echinatus* etc. **ENVIRON**: The sediments in this formation deposited under continental to paralic to littoral environment. **AGE**: A Lower Eocene to Middle Eocene age is assigned to this formation based on palynofossils evidence.

Tharad Formation (*Au*: Roychoudhary et al. 1972)

HOLO: Well Tharad-1, interval 961–1546 m; **LB**: Unconformable with the Olpad Formation, **UB**: Conformable with Tarapur Formation **LITH**: The type section comprises of sandstone, shale and thick coal beds. Sandstone is grey, pink, coarse to medium grained. The formation in Sanchaor-1 is comprises of sand, shale and coal alteration. **THICK**: Maximum thick in the type section is 585 m, which increases in Sanchaor area to 647 m. *LV*: The thick generally reduced towards western and eastern margins. **BIOTA**: The formation is poorly fossiliferous, at places solitary forms of bryozoa and shell fragments along with sideritic globular bodies have been recorded (Sharma et al. 1990, unpublished report on well Tharad-P). **ENVIRON**: The Tharad Formation was deposited under deltaic regime (Sharma et al. 1990). Presence of phytoplankton and brownish grey claystone suggestive of fluctuating water conditions **AGE**: A Lower to Middle Eocene age is assigned based on palynofossils evidences.

Kalol Formation (*Au*: Zubkov et al. 1966)

HOLO: Well Kalol-19, interval 1310–1595 m; **LB**: Unconformable with Kadi Formation, **UB**: Conformable with the Tarapur Formation **LITH**: The type section comprises alternations of fine to medium grained sandstone, siltstone, grey to dark grey, silty and carbonaceous shale, sideritic claystone and coal. The Kalol Formation is subdivided into three units (Mehrotra and Ramakrishna 1980) **THICK**: Maximum thick 250 m, varies from 100 to 250 m. **BIOTA**: The formation is poor in microfauna; Sudhakar and Basu (1973) have reported *Truncorotaloides rohri*, *Nummulites* sp., and *Discocyclina* from the upper part. *Chilguembelina martini* is also reported from the upper part. *Discocyclina* sp., *Pellatispira* sp., and *Nummulites acutus* are also reported (Pandey et al. 1993). The characteristic micro-floral assemblage includes *Polycolpites flavatus*, *Polycolpites granulatus*, *Psilodiporites hammenii*, *Proxapertites cursus*, *Palmaepollenites* sp., and *Homotryblium* sp. **ENVIRON**: The formation was deposited under alternating regressive and transgressive marine environment. Shallow marine inner shelf, marginal marine and deltaic environments. **AGE**: Middle to Upper Eocene age is assigned to the Kalol Formation

Anklesvar Formation (*Au*: Rao 1969; Chandra and Chowdary 1969; Sudhakar and Basu 1973)

HOLO: Well Anklesvar-1, interval 800–1190 m; **LB**: Unconformably underlain by Cambay shale, **UB**: The Anklesvar Formation is conformably overlain by the Dadhar Formation. **LITH**: In type section the lower part is represented by an alternating sequence of shale, sandstone and claystone. The upper part of the formation in its type section consists of claystone, shale, sub-greywacke, sand, carbonaceous shale and arenaceous limestone. **THICK**: In type section the formation has thick of 390 m. In Gandhar-1, the thick is 602 m. **BIOTA**: Sudhakar and Basu (1973) reported arenaceous foraminifera such as *Haplophragmoides*, *Trochammina* sp., *Nummulites* sp., *Discocyclina* sp., *Trucorotaloides* and *Chiloguembelina* sp., *Nummulites acutus*, *N. striatus*, *N. maculatus*, *N. beaumonti* and *Pseudohastigerina micra*, *Proxapertites cursus*, *Polycolpites flavatus*, *Ploycolpites granulatus*, *Palmaepollenites ovatus*, *Dicolpopollis* sp., *Psilodiporites hammeni.*, *Glaphyrocysta exuberans* etc. **ENVIRON**: Open marine neritic environment in some parts of the basin, sandstone beds represents delta front sands of the proto-Narmada delta developed in the South Cambay Basin. **AGE**: Middle to Upper Eocene age.

Bhavanagar Formation (*Au*: Pandey et al. 1993)

HOLO: Well Gulf-1, interval 1891–2012 m; **LB**: Unconformable with the Olpad Formation; **UB**: Unconformable with the Tarapur Formation **LITH**: The type section includes grey shale, sandstone and ferruginous sandy claystone. The shales are intercalated with sideritic mudstone and sandstone mudstone and sandstone. The sandstone is hard, coarse to very coarse grained. **THICK**: 121 m in Gulf-1. **BIOTA**: The formation is poorly fossiliferous, includes: palynofossils *Polybrevicolporites cephalus*, *Plygalacidllites clarus*, *Pellicieroiporites* sp., and *P. langenheimii.* **ENVIRON**: Lower part of the formation is deposited in low energy environment; upper part is interpreted to deposit in high energy environment affected by fluvial conditions. In overall upper part indicate deposited in near shore fluctuating energy conditions. **AGE**: A middle Eocene age is assigned based on palynofossils.

Tarapur Formation (*Au*: Zubkov et al. 1966)

HOLO: Well Cambay-11 interval 1390–1745 m; **LB**: Conformable with the Kalol Formation in Tarapur-Cambay and Mehsana-Ahmedabad Blocks. **UB**: Unconformable with the Babaguru Formation. **LITH**: In the type section, Cambay well-11, the formation is composed of greenish grey, at places, dark grey, poorly fissile, silty and sandy shales, with occasional presence of sideritic bands in the lower part and argillaceous sandstone, siltstone, calcareous, grey shales in the upper part. In Limbodra-1, it is composed of grey sandy shales, oolitic and chamostic claystone and sand/silts are developed. **THICK**: In type section it is 355 m in Limbodra-1 its thick is 59 m *LV*: Mainly shale with minor sandstone **BIOTA**: The larger foraminifera are represented by *Nummulites acutus*, *N. djodjokartae*, *N. maculates*, *N. fabianii*, *N. beaumonti*, *N. stamineus*, *N. uranensis*, *Discocyclina*, *Dispansa*, *Discocyclina* sp., *Propocyclina* sp., *Asterocyclina*, *Dictyoconoides cooki*, and *Chilogumbelina maetini*. Planktic foraminifera include *Globigerina eocaena*, *G. yeguaensis*, *G. linaparta*, *Turborotalia cerroazulensis*, *Globigerina*. **ENVIRON**: Based on microfauna a shallow-inner neritic—middle neritic-outer neritic at places bathymetry greater than 350 m is suggested **AGE**: Middle Eocene-Upper Eocene-Early Oligocene.

Dadhar Formation (*Au*: Rao 1969)

HOLO: Well Dadhar-1, interval 2265–2381 m; **LB**: Unconformable with Tarkesvar Formation

LITH: The type section consists of thin soft sandstone with finely bedded shale alterations **THICK**: 211 m in type section *LV*: Mainly shale with minor sandstone **BIOTA**: *Baculogypsinoides tetraedra*, *N. beaumonti*, *Nummulites fabianii*, *Pellastispira* sp., *Discocyclina* sp., *N. fichteli*, *N. retiatus*, *Operculna* sp., *Rotalia* sp., and Gastropods. **ENVIRON**: Shallow marine neritic **AGE**: Upper Eocene-Lower Oligocene.

Tarkesvar Formation (*Au*: Sudhakar and Basu 1973)

HOLO: Outcrop in a canal section 1.5 km south of Tarkesvar village, Well Broach-B, interval 2687–2833 m; **LB**: Unconformable with the Dadhar Formation and the Tarapur shale; **UB**: Conformable with Babaguru Formation. **LITH**: In the type section lithology comprises of variegated claystone, mottled claystone, minor argillaceous sandstone and ferro-dolomitic sphercules, occasionally thin streaks of coal are encountered. In the reference section the formation consists of alterations of variegated soft claystone and coarse to medium grained sandstone. **THICK**: 50–125 m variables *LV*: More sand content in north of Narmada **BIOTA**: The formation is poorly fossiliferous, rare occurrence of *Quinqueloculina* sp., *Ammonia* sp., arenaceous foraminifera, gastropods and ostracode are recorded. **ENVIRON**: The formation represents a basin wide regressive phase and was deposited in a series of shallow, inland locked basins where the sediments are frequently subjected to sub aerial exposure. **AGE**: Upper Oligocene and Lower Miocene (Sudhakar and Basu 1973); Lower Miocene (Pandey et al. 1993).

In Cambay Basin inner neritic facies occur in the southern part of Cambay Basin (south of Ahmadabad) during Early Eocene as shown in paleogeographic map (Fig. 22a). During Late Eocene to Early Oligocene neritic deep inner neritic to outer neritic facies are characterized by *Uvigerina* and *Bollimina* facies. In the area south of Mehsana along the centre part of the basin. Inner neritic shallow inner neritic facies occur both along the eastern northern western part of the basins as shown in Fig. 22b. (Pandey and Srivastava 1996).

Kutch-Saurasthra Basin

Kutch had a history of Mesozoic to Quaternary. The Paleogene formations are described below. The Paleogene litho- bio- and chronostratigraphy and paleoenvironments settings of Kutch Basin is shown in Figs. 23, 24 and 25.

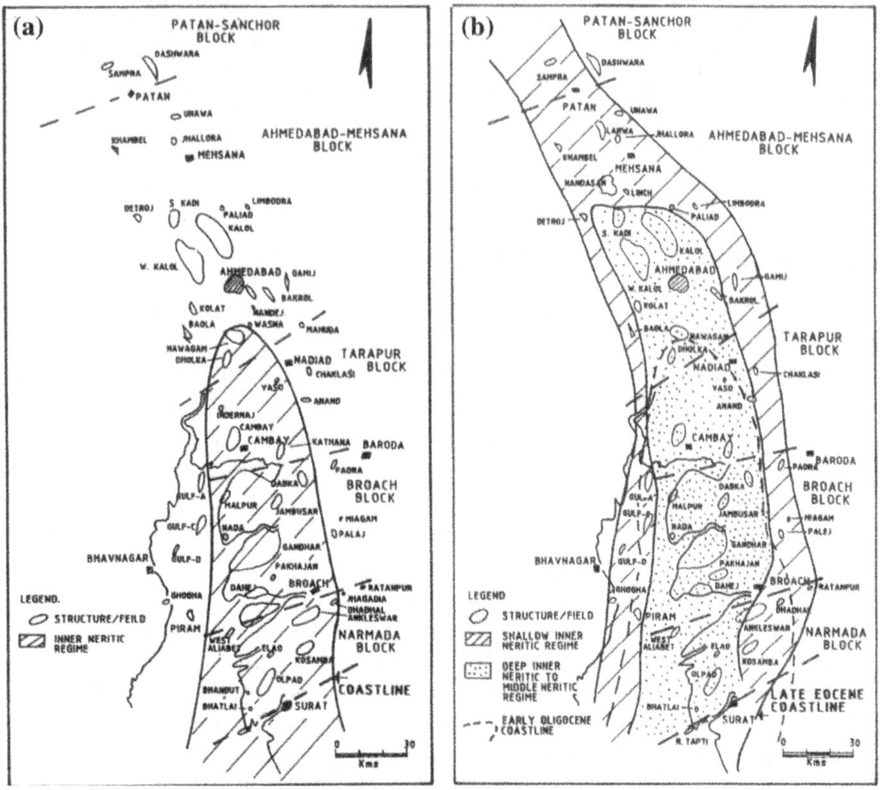

Fig. 22 **a** Early Eocene paleogeography of Cambay Basin. **b** Late Eocene and Early Oligocene paleogeography of Cambay Basin

Deccan Trap (*Au*: Anonymous)

HOLO: Deccan plateau (Maharasthra state); **LB**: Disconformable with the Bhuj Formation; **UB**: Unconformable with the Matanomadh Formation **LITH**: Alternating flows of columnar and a Mygdaloidal basalt with local intertrapeans beds **THICK**: Kutch onshore thick varies from 125 to 340 m, in offshore it varies from 145 to 1520 m **BIOTA**: *Physa* cf. *prinsepii*, *Paludina* sp., *Lymnea* sp., *Eucrypsis Illiocypris*, *Cypris*, *Paracypria*, *Candona* and *Eucandona* (Biswas 1983) **AGE**: Masstrichtian to Danian **ENVIRON**: Sub aerial in Kutch onshore and in the offshore it is sub aerial to subaquous in environment.

Matanomadh Formation (*Au*: Biwas and Raju 1973)

HOLO: Type section is exposed in Bhuj-Lakhpat road section, east of Matanamadh village and in Madhwali Nadi section to south of the village **LB**: Lower boundary sharply defined by red laterite or trap-pebbles conglomerate above Deccan traps; **UB**: The upper contact is marked by lignite bands in the type locality, elsewhere the

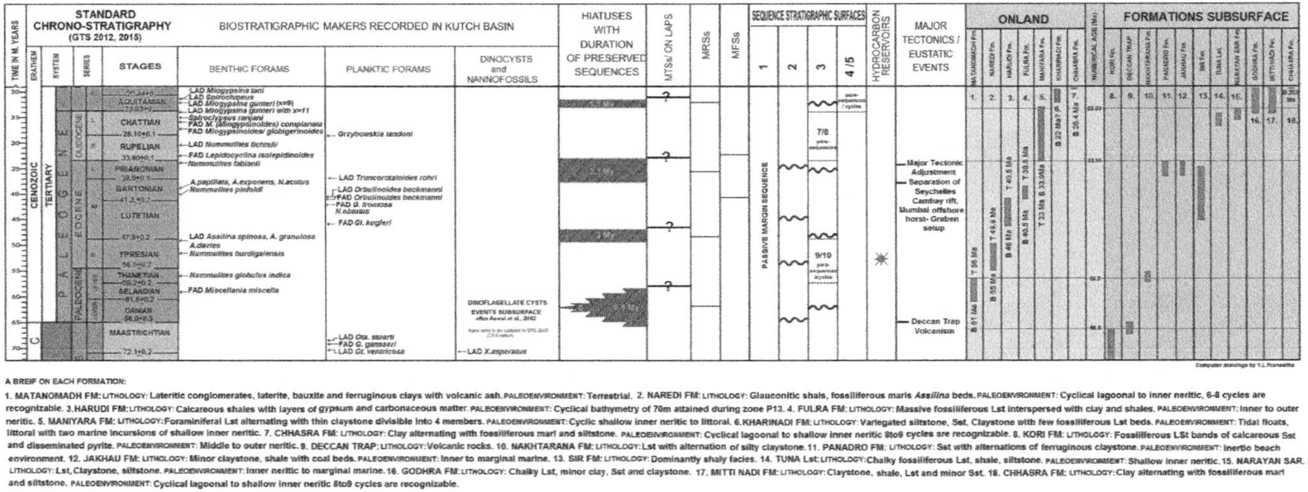

Fig. 23 Paleogene Litho- Bio- Chrono- sequence stratigraphy, and hydrocarbon occurrence in the Kutch basin, India. Compiled by D. S. N. Raju, In collaboration with C. N. Ravindran and Sudhir Shukla

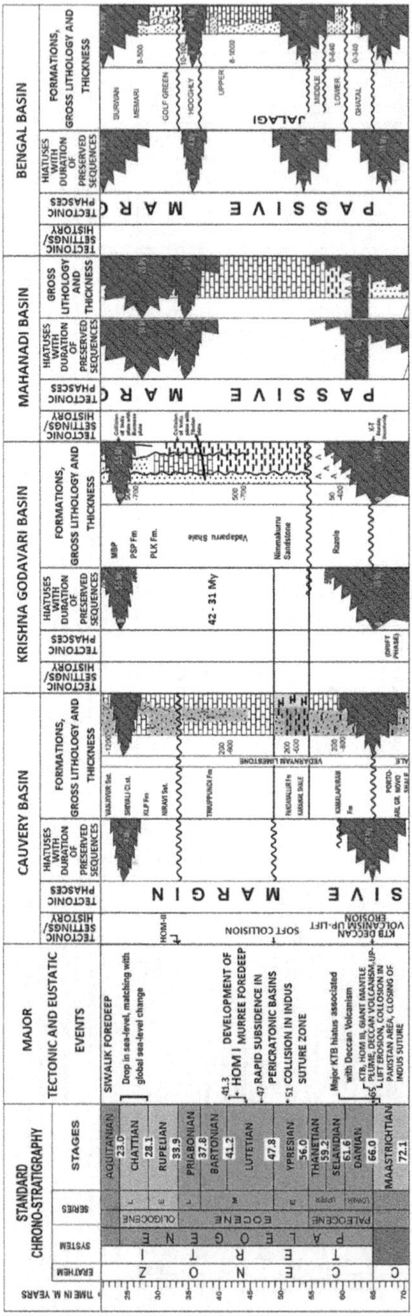

Fig. 24 Bio-chrono and tectonostratigraphic framework for East Coast basins of India. Compiled by D. S. N. Raju

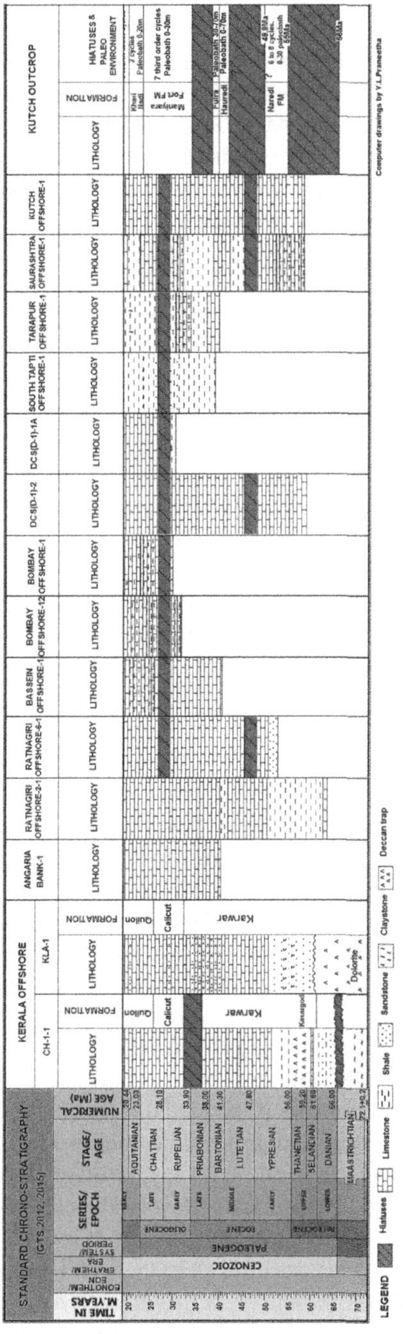

Fig. 25 Distribution of lithofacies during the paleogene in selected wells across western offshore, India. Compiled by D. S. N. Raju and C. N. Ravindran. *Source of well data* Pandey and Dave (1998)

contact is disconformable **LITH**: The Lithological section is variable and consists of brightly colored clastic and volcanic material, laterite to clays in the lower part of the formation. The common rock types are red laterite, bauxite, lateritic trap-pebble conglomerate, bentonitic clays, ferruginous clays tuffaceous sandstone and occasional layers of lignite. **THICK**: 49 m in type section **BIOTA**: Locally rich in dicot leaf impressions and pollen/spores **AGE**: Pollen/spores probably indicate Palaeocene age **ENVIRON**: The vocano-clastic sediments were deposited during vaning phase of the Deccan volcanism. Volcanic debris were reworked and deposited along with clastic in fluvial and lacustrine environment.

Nakhtarana Formation (*Au*: Zutshi et al. 1993)

HOLO: Well GK-1-1, interval 1900–2025 m **LB**: Unconformable with the Deccan trap in the Kutch shelf, excepting the well KI-1A where it is unconformably underlain by the Kori Formation. **UB**: Unconformable with the Jakhau Formation. **LITH**: Three members are identified: Nakhtarana limestone member: consists of white buff, fossiliferous compact limestone along with alternations of siltstone; Nora member: claystone lithounits with minor basalts and fine grained sandstone; Anjar member: dominantly sandstone lithounits and comprises shale-sandstone alterations in lower part. **THICK**: In type section it is 152 m; over the Kutch thick varies from 6 to 410 m. **BIOTA**: *Nummulites-Lockhartia* biofacies, *Nummulites ranikothalia*, *Discocyclina*, *P. pseudomenardii*; *M. miscella.* **AGE**: Upper Paleocene **ENVIRON**: Inner shelf environment, lower part of the formation in KD-1 and GKH-1 wells is barren of fauna and deposited in non-marine to marginal marine environment.

Naredi Formation (*Au*: Biswas and Raju 1971)

HOLO: Type section is exposed in the cliff along the Kakdi Nadi, south of Naredi village and partly (only upper part) along Guvar stram NNW of Naredi village **LB**: In the type locality, the formation directly overlies the Deccan Trap Formation as observed in the Nala 700 m west of Naredi. In other places a well marked disconformity separates this formation from the Matanomadh Formation. **UB**: The upper contact with the Harudi Formation is marked by red ferruginous and erosional surface. The surface occurs above the barren lateritic shales of the ferruginous claystone member of Naredi Formation **LITH**: Three distinct members are identified: (i) Gypseous shale member: It is about 24 m thick and consists of grey, brown, glauconitic sandstone and splintery shales with thin layers of gypsum with sideritic concretions with occasionally contains fossils of *Nautilus*. (ii) *Assilina* limestone member: it is 6 m thick and consists of bedded dirty white limestone and alloys grey marlite with *Assilina* (iii) Ferruginous claystone member: It is about 50 m thick and consists of grey and brown claystone with layers of gypsum and red ferruginous laminae. The lower and middle part locally develop black shale facies, pyretic shales and lignite beds, **THICK**: 40 m in the type section. It is over 100 m thick in the sub outcrops in Babia syncline (Panadra mines). In the well Lakhpat-1, the thick is 93 m. **BIOTA**: *Nummulites globulus nanni*, *N. burdigalensis*, *Assilina spinose.* **AGE**: Late Paleocene to Early Eocene. Age of the Naredi Formation was controversial for more than 5 decades. Considerable progress has been made in the last 5 years. The Naredi

Formation overlies last (NOT CLEAR) the Deccan trap. Keller et al. (2013) dated the lower part of Naredi section as of Early Eocene (Ypresian) age (SBZ8, E4) based on larger foraminifera and rare planktic foraminifera in assemblages. The top of the section is of late early Eocene (SBZ11, E6) age. The lower assemblage (SBZ8, E4) is indicative of brackish to normal marine inner shelf environment; the top assemblage (SBZ11, E6) is the most diverse and indicates an inner shelf to marginal marine environment. Two phases of marine transgression were identified. Sediments of the transgression/regression cycles were derived from physical and chemical-weathering processes of basaltic rocks as indicated by the different geochemical proxies. Carbon isotope analyses of bivalve shells and organic matter reveal a negative excursion that is correlative with the global Early Eocene excursion. The presence of fish bones, fish teeth and organic matter can be related to the Early Eocene climatic optimum. Clay mineral data from the Naredi Formation indicate variably hot humid to arid climate conditions. A negative carbon isotopic excursion detected in bivalves, fish bones and teeth and organic matter could be related to the Early Eocene Climatic Optimum 2. Khozyem et al. (2013), Anwar et al. (2013) carried out strontium Isotope study and is of Early Eocene age spanning from 50.58 Ma. **ENVIRON**: The paleobathymetry ranges from non-marine to 45 m.

Jakhau Formation (*Au*: Zutshi et al. 1993)

HOLO: Interval 1900–2025 in well GK-1-1 **LB**: Unconformably underlain by Nakhtarana Formation; **UB**: Unconformably overlain by the Fulara limestone in Kutch offshore, and marked by sharp fall in resistivity and gamma logs **LITH**: consists of dominantly argillaceous foraminifera packed limestone which is partially dolomitised and occasionally pyretic. The formation has coal bands at the top. **THICK**: Maximum thick in well KD-1 is 255 m; the thick increases to 485 m in GKS-1 **BIOTA**: *N. globulus*, *A.spinosa*, *G. subotinae*, *N. globulus*, **AGE**: Lower Eocene (Ypresian) age **ENVIRON**: Shallow warm water, inner shelf environment is suggested in the eastern part of the Kutch offshore.

Pandara Formation (*Au*: Zutshi et al. 1993)

HOLO: Interval 663–774 m in well GK-29A-1. **LB**: Unconformable with the Nora Member of the Nakhtarana Formation in the well GK-29A-1; **UB**: Unconformable with the Jakhau Formation, the contact is distinct between argillaceous limestone of the Jakhau Formation and dominantly sandy facies of the Pandara Formation **LITH**: The formation consists dominantly of sandstone facies in the upper part and the alteration of sandstone, feeurginous shale and slity siltstone in the lower part. **THICK**: 111 m in GK-29A-1 well **BIOTA**: *Nummunlites-Dicocyclina-Assilina* biofacies **AGE**: Lower Eocene **ENVIRON**: This formation got deposited in near-shore beach environment.

Harudi Formation (*Au*: Biswas 1971)

HOLO: Exposed in an escarpment to the west of Harudi village, the section is continuously exposed over a short distance of 300 m **LB**: Lower contact is disconformable and is fixed at the top of the laterite bed at the top of the Naredi formation; **UB**: Upper contact is conformable and is placed at the base of lowest massive foraminiferal limestone bed containing characteristic saddled to undulated *Discocyclina* **LITH**: Consists of splintery shale with yellow limonitic partings in the lower part and calcareous claystone and siltstone with occasional layers of gypsum and carbonaceous shale in the upper part. Two to three Coquina beds occur near the base. Occasionally concretionary and oolitic fossiliferous limestone bands are present in the lower part. Two to two and half feet thick ferruginous, gypseous clayey marlite with large diameter (1–2 cm) *Nummulites obtusus* is characteristic marker bed within the formation. The Nummulites is easily recognised by the doubly convex tablet shape and ornamented surface with curved sutures resembling that of a thumb impression. The bed is extensive. **THICK**: 14 m thick in the exposures **BIOTA**: *Truncorotaloides topilensis*, *Orbulinoides beckmanni*, *Turborotalia frontosa*, *Nummulites obtusus*, *N. acutus*, *Brarrudosphaera biglowi*, *Discoaster barbadiensis*, *Bolies*, *Xomcus*, *Porocidaris* and *Cidaris.* **AGE**: Lutetian to lower part of Bartonian. **ENVIRON**: Littoral to lagoonal in the lower part and inner to lower part of middle shelf. Cyclical, transgressive phase of deposition. A sharp raise of paleobathymetry by an order of 40/50 m immediately above the *obtusus* beds.

Fulra Limestone (*Au*: Biswas and Raju 1973; Samanta 1989; Related names: *Nummulites* Group of Wynne 1872; Kiddar series of Tewari 1957; Sengupta 1964; Poddar 1959).

HOLO: The type section is best exposed in southern flank of Babia Hill, about 17.7 km SW of Fulra village. The upper part is also well exposed in the nala to the south of Fulara **LB**: The lower contact is marked at the base of the massive limestone overlying the Harudi Formation. In the subsurface the lower boundary is marked by the sharp change in resistivity from serrated character of the underlying Jakhau Formation and 15–20 m thick clastic bed in the wells SP-1-1 and GK-1-1, GK-33A. **UB**: Unconformable but locally disconformable showing cut and fill structures and bioturbation. In the Kutch offshore, the Fulara limestone is unconformably overlain by the Tuna limestone and the boundary coincide with last occurrence of *Pellatispira* **THICK**: Maximum thick of 60 m exposed in the Berawali stream section. It is only 23 m thick in the type section. Thick of Fulara limestone is more than 400 m in the well GK-1-1 and GK-S-1 and GK-36-1. **BIOTA**: Two zones namely *Orbulinoides beckmanni* zone in the lower part and *Truncorotaloides rohri* zone in the upper part. The larger foraminifera belong to *Discocyclina sowerbii* zone in the lower part and *Fasciolites ellipticus* zone in the upper part. **AGE**: Late Middle Eocene age **ENVIRON**: A low energy, clear waters, probably outer middle-shelf environment.

Sir Formation (*Au*: Zutshi et al. 1993)

HOLO: Interval 3869–3966 m + in well Kutch-A **LB**: Not encountered; **UB**: Conformable with Fulara limestone. **LITH**: Dominantly shale/claystone and also contains minor limestone. Shale/claystone is occasionally carbonaceous and pyretic and feebly calcareous. **THICK**: 97 m in Kutch-A **BIOTA**: Characterised by *Truncorotaloides rohri*, and *Turborotalia cerrozulensis* **AGE**: Middle to Upper Eocene **ENVIRON**: Middle to outer neritic environment.

Tuna Formation (*Au*: Zutshi et al. 1993)

HOLO: Interval 1225–1285 m in well GK-1-1 **LB**: Unconformably with the Fulara limestone. This boundary typically is marked by gamma and resistivity breaks on logs. This boundary also marked by last occurrences of *Pellatispira* zone in the Kutch shelf wells and *Nummulites discorbis-Discocylina* assmeblaege zone top in wells GKH-1 and Sutri-1 and top of *Turborotalia cerruzoulensis* top in other wells of Kutch; **UB**: Unconformable with the Narayan sharovar Formation. It can be marked on the logs by increase in gamma and reduction in resistivity in the well KD-1. **LITH**: Consists of limestone with minor shale and siltstone. **THICK**: 60 m in GK-1-1, the thick in Kutch offshore ranges from 15 to 115 m. Towards Saurasthra its thick increases further. **BIOTA**: *N. fichteli., Globgerina tapuriensis* **AGE**: Rupelian (Early Oligocene) **ENVIRON**: varies from inner shelf to outer shelf.

Maniyara Fort Formation b (*Au*: Biswas 1971; related names *Nummulites* Group, Wynne (1872); Nari Series, Nagappa 1959; Chatterjee and Mathur 1966).

HOLO: Type section is continuously exposed along Bermoti River between Maniyar fort and Bermoti village from a locality 1.6 km NNE of Bermoti to a locality 450 m SE of Bermoti village **LB**: Disconformable. Ochre and sudden appearance of glauconitic pellets make it easy to recognise in the field; **UB**: persistent blue clay at the base of overlying formation and the top of limestone of Bermoti member distinguish unconformable contact.

LITH: Four members:

1. Basal member is 4 m thick and consists of alternating beds of foraminiferal, glauconitic, brownish to yellowish siltstone and calcareous gypseous claystone, the glauconite gets locally concentrated in the cut and fills of the lower disconformable surface.
2. The lumpy clay member is about 5 m thick and consists of cement colour to brownish calcareous lumpy claystone, occasionally containing thin limestone and marlite beds.
3. The coral limestone member is about 10 m thick and consists of dirty white nodular limestone, alternating with calcareous claystone in the lower part. The upper part comprises grey to dirty white massive limestone with abundant coral which frequently forming bioherms. The limestone is glauconite, biomicritic and biosperitic.
4. The Bermoti member is about 10 m thick and is best developed in the stream SE of Bermoti and also NNE of Wuaior. It is also well developed and top of the Maniyarfort Hill. The lower part is best exposed in Wuaior and consists of rusty brown, friable, glauconitic argillaceous sandstone with pseudoolites. The upper part is composed thinly bedded very hard grey to yellowish foraminiferal limestone with interbeds of silty marlite full of *Spiroclypes*.

THICK: In type section it is about 30 m **BIOTA**: Five zones/biochrons are recognised within this formation. In the lower part biochron RAM-1 defined as an interval from FAD of *Nummulites fichteli* to FAD of *Lepidocyclina isolepidinoides* is recognised. Dr. Jafar and others recognised nannoplanktons assemblage referable to zone

NN23 was recorded from zone. RAM-11 defined as interval from FAD of *L. isolepidinoides* and LAD of *N. fitcheli* is recognised in the coral limestone member. *Nummulites clypeus* is common in this zone.

Three zones/biochrons namely WA-I is defined from LAD of *N. fichtlei* to FAD of *Miogypsinidae*; WA-II defined as an interval from FAD of *M. bermudezi* to FAD of *M. (M.) complanata*; and WA-III defined as an interval from FAD of *M. (Miogypsinoides) complanata* to a level within a range of *M. (M) gunteri* are recognised. Two subzones/subchrons were recognised in WA-II and WA-III. **AGE**: Oligocene **ENVIRON**: Seven to eight cycles/parasequences are recognised in the outcrops.

Raju (2009) recognized 7 third order stratigraphic cycles/parasequences (4 in Ramanian Stage and ¾ in the Waiorian Stage). They are close to European data.

Narayansarovar Formation (*Au*: Zutshi et al. 1993)

HOLO: Well KD-1, interval 695–742 m **LB**: Unconformable with the Tuna limestone. It can be marked on the log by increase in gamma and reduction in resistivity; **UB**: Unconformable with the Godhra Formation. It can be marked on the log by a sharp break on gamma and resistivity logs **LITH**: consists of dirty white, greyish, brownish, hard, massive, compact and fossiliferous limestone, greenish grey to grey, feebly calcareous and fossiliferous claystone and colourless fine grained, sub-rounded sandstone. **THICK**: 47 m in well KD-1. there is general increase in thick towards west and NW. **BIOTA**: *M. compalanata*, *G. tandoni*, *S. ranjanae*, *G. gortani*, *Lepidocyclina*, *Spiroclypeus* **AGE**: Chatian (Upper Oligocene) **ENVIRON**: Over the Kutch shelf and GKH area was inner shelf, whereas it was outer shelf in the western part of the Kutch offshore.

Oxygen isotopic measurement of larger benthic foraminifers from western India suggest that the Paleogene temperature varied between 22 and 32 °C, deteriorated in late Middle Eocene (corresponding to planktic zones P13–P14) when the temperature dropped by 6 °C. With progressive cooling through the Late Eocene, the temperature reached 22 °C during Early Oligocene times. The cooling trend set in the Middle Eocene seems to have been terminated towards the end of the Paleogene whent the temperature rose to 25 °C.

Saurasthra Basin

Deccan Traps/Laterites

Jafrabad Formation (*Au*: Zutshi et al. 1993)

HOLO: interval 3884–4502 m in well SS-1; *Hypo*: Interval 1648–1883 m in well OS-2-G-1

LB: Unconformably overlies Deccan Traps/Laterites; **UB**: Unconformable with overlying the Belapur Formation. **LITH**: It consists of alternating sequence of limestone/shale with few quartzite sandstone layers in lower parts. **THICK**: in OS-2-G-1 is 165 m, its thick increase towards Diu Arch **BIOTA**: *N. burdigaliensis-Lokhartia hunti* occur in the upper part. The lower part of the formation is poorly fossiliferous. **AGE**: Upper Paleocene to lower Eocene. **ENVIRON**: Shallow marine

Belapur Formation (*Au*: Zutshi et al. 1993)

HOLO: Interval 3024–3428 m in well B-12-1; *Hypo*: Interval 1587–1700 m in well OS-2-G-1. **LB**: Unconformable with Jafrabad Formation; **UB**: Conformable with overlying Diu Formation in Bombay Basin **LITH**: Calcareous shale/claystone with a limestone marker bed at the bottom. **THICK**: 424 m thick in type section, in well OS-2-G-1 it is 61 m. **BIOTA**: *Fasciolites-Lokhartia alvelata-Nummulites diokartae, N. burdigalensis-Lokhartia hunti, Pustulosa, O. beckmanni, Discorbinus-Discocyclina, Globigerinatheka kugleri-S. frontosa-howeri* **AGE**: Middle Eocene.

Fulra Limestone (*Au*: Biswas and Raju 1971)

HOLO: See in Kutch Basin **LB**: Conformable with the Belapur Formation; **UB**: Unconformable with the Tuna limestone. **LITH**: In well OS-2-G-1 limestone is white to medium grade. There is abundant carbonaceous matter. **THICK**: 60 m in Kutch, it increases towards west in Kutch offshore and completely sails out over Diu Arch where it is called Belapur and Diu formations **BIOTA**: *Nummiulites fabianii, Pellatispira* sp., **AGE**: Upper Eocene age **ENVIRON**: Shallow marine lower energy clear water environment. 6 shallowing upward cylces are recogniziable within the major 3rd order cycles of approximately 2 My.

Tuna Limestone (*Au*: Zutshi et al. 1993)

HOLO: Discussed in Kutch; *Hypo*: Interval 1360–1483 m in well OS-2-G-1 **LB**: Unconformable with the Fulra limestone; **UB**: Unconformable with the Narayan Sarovar Formation **LITH**: In the well OS-2-G-1, consists of limestone with clays, carbonaceous matter. Towards Diu Arch limestone facies of this formation changing into dominantly shaly facies. **THICK**: In well OS-2-G-1 it is 123 m thick **BIOTA**: *Nummulites fichteli* **AGE**: Rupelian (Lower Oligocene) **ENVIRON**: Shallow inner shelf to littoral environment.

Narayan Sarovar Formation (*Au*: Zutshi et al. 1993)

HOLO: See Kutch Basin; *Hypo*: 1360–1555 m in well OS-2-G-1 **LB**: Unconformable with the Tuna Limestone; **UB**: unconformable with the Nabibanda Formation **LITH**: It consists of limestone (grainstone/packstone, with abundant carbonaceous material). In the northern part of Suarasthara Basin there is reduction in argillaceous matter and increase in carbonaceous matter within this formation. **THICK**: 205 m thick in well OS-2-G-1, toward Diu Arch its thick increase up to 378 m in the well S-2-1 **BIOTA**: *M.(M.) complanata* and *S. ranjanae* in the well OS-2-G-1. **AGE**: Chattian (Upper Oligocene) **ENVIRON**: Inner-shelf environment suggested by faunal and Lithological data.

Western Rajasthan Basin

Permo-Triassic to Eocene sediments are known. Three formations are briefly described below.

Jaisalmer Basin

Sanu Formation (*Au*: Dasgupta et al. 1973)

HOLO: Sanu village with type section exposed in the Pariwar Hill located west of village Pariwar; *Hypo*: Interval 79–1008 m in well Kharatar and interval 1340–2020 m in well Lang-2. **LITH**: The formation is composed of reddish unconsolidated, current bedded, glauconitic sandstone. The formation is composed in the reference well Kharatar-1 is divisible into (a) Mohammad Dhani Member comprising medium to coarse sandstone. The basal part of the sandstone is argillaceous with layers of chocolate brown clay. Lignite streaks and grey shale also characterize the upper part of the unit (b) Kharatar Member comprises greenish brey fine to coarse grained argillaceous feebly clacarious sandstone. It is overlain by argillaceous, light greenish grey, glauconitic limestone intercalated with marl. The upper part comprises dark grey to black fissile shale and sandstone interbeds. The top of the member is marked by chocolate brown silty shale/clay. The formation in the outcrop is devoid of microfauna. In the subsurface, limestone and marl in the upper part is rich in foraminifera and ostracoda. **BIOTA**: The characteristic foraminifera include, *Planororalities pseudomenardii*, *Morozovella velascoensis*, *P. pusilla pusilla*, *P. chapmani*, *P. compressa*, *M. angulata*, *Discocyclina seunesi*, *D. ranikotensis*, *Assilina dandotica* and *Miscellanea miscella*. The characteristic ostracodes include: *Acanthocythereis retispinata*, *Brachycythere rajasthanensis*, *Hermanites crecens*, *Protobuntonia ghotaruensis*, *Macrocypris guhai*, *Acanthocythereis procapsus*, *Cytherelloides jaisalmerensis*, *Phalcocytehre improcera*, *Anommatocytehre indica* and *Echinocythereis contexta*. **ENVIRON**: The basal part of Sanu Formation is inferred to have been deposited under continental environment with little marine influence. The limestone with glauconitic sandstone with intercalated shale in the upper part has yielded planktics land benthonics, indicative of middle shelf conditions. **AGE**: The faunal assemblage recorded suggest Late Paleocene age. The lower unfossiliferous part could not be dated however based on its stratigraphic position it is inferred to be of Paleocene age.

Khuiala Formation (*Au*: Narayanan 1959)

HOLO: Te-Takkar scarp located near west of Mohhammad Ki-Dhani. Hypo; Internal 511–791 m in well Kharatar-1 **THICK**: About 25 m **LITH**: The type section comprises of gypseous clay/shale followed by a grayish white to saffron shale consisting of *Assilina*. This is overlain by dirty white, thin bedded argillaceous limestone, yellow to brown hard, shell limestone, white to yellow chalk, occasionally sandy and at the top is pinkish yellow, thick bedded, massive, hard, compact, crystalline and fossiliferous limestone. Based on distinct lithological variation, Singh (1984) has subdivided the Khuiala Formation into two members. Te-Takar Limestone Member and Khinsar Shale Member. The Khuiala Formation is fossiliferous both in outcrop and subcrop sections. **BIOTA**: (Microfauna): include: *Nummulites burdigalensis*, *Assilina daviesi*, *A. daviesi var. pustulosa*, *Orbitolites complanatus*, *Praeindicola bikanerensis and P. sigali*, *A. granulosa*, *N. praediscorbinus*.

In the subsurface sections planktic foraminiferal assemblage comprises *Globotrotalia triangularis*, *Acarinina acarinata*, *A. soldadoensis angulosa.* Ostracodes:

Important ostracodes recorded are: *Alocopocythere abstracta*, *Hornibrookella rajasthanensis*, *Paragrenocythere reticulospinosa*, *Phalcocythere sentosa*, *Hermanites goeli*, *Schizocythere bikanerensis*, *H. avadhesi*, *Gyrocythere parvicarinata*, *Paijenborchella narediensis*, *Paijenborchellina indica*, *Orthonotocythere kutchensis*, *Anommatocytehre laqueta*, *Schizocythere spinosa*, *Stigmatocythere obliqua* and *Alocopocythere longilinea*. **ENVIRON**: Based on the recorded foraminifera in the outcrop and subsurface, a bathymetry ranging from shallow inner shelf to more than 100 m is inferred for this formation. **AGE**: The faunal assemblage recorded indicated in Early Eocene age. In the lower part (Te-Takkar Member) planktics belonging to *P. pseudomenardii* Zone were recorded by Sigal et al. (ibid) from most of well sections except Kharatar well where *M. velacoensis* occurs at the top. Based on these evidences, the outcrop section is of definite younger age (Early Eocene) while in the subsurface the Khuiala Formation ranges from Late Paleocene to Early Eocene.

Bandh Formation (*Au*: Narayanan 1959)

HOLO: Exposure south of Bandah Village, Hypo: In two locations. Batrewala Tibba and Shumarwali Talai, and well Kharatar-1, (interval 302–511 m). **THICK**: About 30 m; **LITH**: In the type area it consists predominantly of limestone. Yellow calcareous siltstone, greenish marl, bentonitic greenish clays thin beds of calcite at the base. The middle and upper parts comprise of shales and limestone occasionally laterites. **BIOTA**: Foraminiferal assemblage recorded by Sigal et al. (1971) and Singh (1984) includes: *Nummulites acutus*, *N. maculatus*, *N. beaumonti*, *N. pengaronensis*, *N. cuivelleri*, *Discocyclina javana var. indica*, *D. dispansa*, *D. sowerby*, *Fasciolites ellipitica*, *Assilina spira*, *A. papillata*, *A. subpapillata*, *Dictyoconoides cooki*. Others includes *Globigerinatheka kugleri*, *Truncoroloides topilensis* and *Turborotalia cerroazulensis*. Singh (1984) has recorded a different assemblage from the top most part of this formation in the Kharatar area. They include, *N. beaumonti* and *Heterostegina* sp. Sharma and Bhandari (1986) have recorded *Nummulites fabianii*, *N. pengaronensis*, *Pellatispira* sp., *Baculogypsinoides* sp. and *Chiloguembelina* sp. from this horizone. Ostracodes: Characteristic ostracodes recorded are: *Alocopocythere transversa* morp.A, *Anommatocythere confirmata*, *Echinocythereis (S.) sahnii*, *Gyrocythere exaggerate*, *Cytherelloidea khartarensis*, *Alocopocythere transversa* morph-A, *Stigmatocythere lumaria* morph.B and *Echinocythereis (S.) sparsa*. **ENVIRON**: Foraminifera from the limestone greenish grey clay and glauconitic marl in association with planktics suggest inner to middle shelf conditions. The upper part of Bandah Formation deposited in shallow inner neritic. **AGE**: Based on the faunal evidences a Middle to Late Eocene is suggested.

Bikkaner Nagaur Basin

Late Proterozoic-Cambrian oldest sediments after major hiatus. There is a major hiatus (120 My) covering the Late Cambrian to the Carboniferrous between the Permo Carboniferrous marine are known. Four formations of Paleogene are discussed.

Palana Formation (*Au*: Srivastava and Srinivasan 1962)

HOLO: It is present in quarry section at Palana (interval 460–500 m). **LITH**: In the type area the lithology is characterized by the association of nummulitic limestone, Fullers earth, gray to variegated and carbonaceous shale, medium to coarse grained sandstone and lignite bands. **AGE**: Based on palynofloral evidences. This formation is dated as Paleocene.

Marh Formation (*Au*: Dhar and Khar 1977)

HOLO: Rocks of this formation are exposed at Marh village and around Kolayat. **THICK**: Based on borehole data at Gariyala and other places, it is about 270 m thick (in subsurface). **LITH**: The formation is characterized by soft, friable pebble sandstone composed of pebbles of chert and quartz, ferruginous sandstone and red clays with rare presence of hematite bands. **ENVIRON**: The lower part of the formation was deposited under shallow marine environment, while the upper part of the formation, based on lithology and presence of leaf impression and cross bedding is suggested to have been deposited under fluvial conditions.

AGE: The lower part of the formation has yielded foraminifera and ostracoda of Early Eocene age (Kulshreshtha et al. 1989). The upper arenaceous sequence is devoid of fauna.

Jogira Formation (*Au*: Srivastava and Srinivasan 1962)

HOLO: The type section is located half a mite Joira Talav **THICK**: About 170 m. **LITH**: In the type locality the lithology is represented by alternation of marl, fuller's earth and foraminiferal limestone. **BIOTA**: Forminifera are recorded in this formation and the characteristic ones are: *Morozovella formosa formosa*, *M. subbotinae*, *Pseudohastigerina wilcoxensis*, *Discocyclina sella*, *Fasciolites elliptica*, *Truncorotaloides rohri*, *T. collactea*, *Morozovella lehneri*, *Globigerina eocaena* and *Chilogenumbelina cubensis*. **ENVIRON**: Based on the foraminferal assemblage a middle shelf (or deeper) conditions is suggested during the depositions of Jogira Formation. **AGE**: The age of this formation ranges from Lower to Middle Eocene.

Barmer Formation (*Au*: Dasgupta et al. 1973)

HOLO: Lunu Hill section **THICK**: It is about 26 m. Lithologically it is divided in two members, viz. Barmer Hill Member and Mandal Scarp member. **LITH**: The Barmer Hill Member comprises of siltstone, bedded chert, intraformational conglomerate, sandy siltstone and sandstone. **ENVIRON**: In the Barmer Hill section (Reference Section) a continental/Fluviatile environment is suggested while in the North, Kota Hill section shallow marine conditions prevailed (Dasgupta 1974). **AGE**: A Paleocene age, is suggested based on plant fossils.

Barmer Basin

Five formations in the span of Paleocene to Early Eocene are briefy described below.

Fatehgarh Formation (*Au*: Dasgupta 1974)

HOLO: Fault scarp south of fatehgarh village. **LITH**: The formation is divided into two members. The lower Vinjori Member comprises variegated silty sandstone and quartzite. The upper Sajit Member contains coquina bed having gastropods and bivalves. The upper boundary with Barmer Foundation is conformable. **THICK**: Maximum thick in type area is about 140 m. **BIOTA**: The formation is devoid of fauna. However, it has been correlated with non-marine sandstone of Mohmad ki Dhaini Member of Sanu Formation of Jaisalmer Basin. **ENVIRON**: The lower portion deposited in near shore while upper part laid under shallow inner neritic. **AGE**: Paleocene.

Barmer Formation (*Au*: Blanford 1876; refined by Dasgupta 1974)

Holo: Lunu Hill Section. **LITH**: The formation comprises of arenaceous unit. It has been divided into two members viz. the lower Barmer Hill member siltstone and leaf impression. The upper Maidai scarp comprises a current bedded, white to pink colour sandstone. The upper boundary is unconformable with overlying the Akli Formation while lower contact is conformable with Fatehgarh Formation. **BIOTA**: The formation is devoid of fauna. **ENVIRON**: Continental to fluvial **Age**: Occurrence of plant fossils suggest a probable Paleocene age.

Akli Formation (*Au*: Siddiqui and Bahi 1965)

HOLO: Quarry section at Hathi Singh Ki Dhani. **LITH**: This formation is divided into two members. The Thumbli Member comprises essentially sandstone with lignite and bituminous layers occasional gypseous and marl bads. The lower contact with the Barmer Formation is unconformable while upper contact is disconformable with the Mata ji ka Dungar Formation. Maximum thick of this formation is about 280 m. **BIOTA**: No microfauna. **ENVIRON**: Paralic to desiccating conditions **AGE**: On the basis of stratigraphic position Paleocene to Early Eocene age has been assigned.

Mata Ji Ka Dungar Formation (*Au*: Dasgupta et al. 1973)

HOLO: Outcrops at Mataji ka Dungar. **LITH**: it is comprising dominantly of ferruginous coarse grained sandstone. At base, a clay bed is seen. The lower contact, of the formation with the underlying Akli Formation is gradational. However, in the eastern most part around Jhak village the formation directly overlies the basement. The contact of the formation is unconformable with the overlying Kapurdi Formation. **THICK**: The maximum thick of this formation is about 180 m. **BIOTA**: The formation is poorly fossiliferous except for the lower part where indeterminate angiospermic leaf impressions are seen. **ENVIRON**: Continental

AGE: Based on stratigraphic position an Early Eocene age has been assigned.

Kapurdi Formation (*Au*: Shrivastava and Srinivasan 1963)

HOLO: Pond section near Rajputon ki Dhani. **LITH**: It comprises fuller's earth with carbonaceous streaks and gypseous clay in the lower part while the upper part consists of dull white argillaceous limestone and bioclastic marl. **LB**: The lower boundary seems to be unconformable with the underlying Mata ji ka Dungar Formation. **UB**: While the upper contact is marked by a pronounced unconformity with the overlying Uttarlai Formation. **THICK**: The maximum thick of this formation in quarry section is about 12–15 m. **BIOTA**: Lakhanpal and Bose (1951) have reported plant fossils, fish, crabs, shrimps, etc. and Dasgupta (1975) have reported *Assilina daviesi*, *A. granulose* and *N. atacicus* (mostly juvenile). **ENVIRON**: Near shore to shallow inner neritic. **AGE**: The microfauna recorded at the bioclastic at the top of the fuller's earth suggest Early Eocene age.

Tectono-stratigraphy

TS Unit ix: Duration: Paleocene to middle Eocene.

It is equivalent to super sequence xi of Ravi Shankar et al. (2005, Table 3) and lower deccan synthem of Pandey (1986). According to Ravi Shankar et al. (2005, P. 31) the sedimentation of the super sequence xi initiated with a marine transgressions—the Subathu sea. The sedimentation terminated in the Middle Eocene in the first phase of the Himalayan orogeny (HOM-1 ca 46–41.3 Ma), when the sea regressed.

The tectonic setting along West coast, East coast and North-East basins was stable platform between Deccan volcanism and HOM-1. In Cambay and Mumbai offshore lower part of this unit represents synrift facies. There was a carbonate build up represented by Bassein Formation (Mumbai offshore), Fulra Limestone (Kutch) and sylhet Limestone (Meghalaya, Bengal basin, Mahanandhi Basin and Bhimannapalli Limestone. The major source rocks along Western basins appear to be associated with Paleocene- Eocene maximum.

TS Unit- X: Duration: Late Eocene to Oligocene.

It is equivalent to the upper Deccan Synthem. Towards the upper part of this unit a major hiatus was detected and dated in west coast and East coast basins. In cambay basin this interval is represented a major transgressive cycle. In several basins of India, a drop in Sea-level and again in the Late Oligocene is well documented.

Acknowledgements The author is greatful to Dr. O. N. Bhargava for suggesting to prepare a paper on the Paleogene, critical review of the manuscript and useful suggestions. He is thankful to Dr. Satish C. Tripathi for accepting the paper for publication. Miss Y. L. Praneetha for assistance in preparing the figures and typing of the text. Thanks to My wife Mrs. D. Satyavati (Neelu) for help at home where the paper was prepared.

References

Ali JR, Jolley DW (1996) Chronostratigraphic framework for the Thanetian and lower Ypersian deposits of SE England. Geol Soc Spec Publ 101:129–144

Anwar D, Choudhary AK, Saraswathi PK (2013) Strotium isotope stratigraphy of the Naredi Formation, Kutch. Geol Soc India, Spec Publ 1:197–202, Bangalore

Ardunio G (1759) Lettera seconda sopra varie osservazioni fatti in diversiparti del territorio di Vicenza, ed altrove, appartenenti alla teoria terrestre, ed alla mineralogia. Venezia

Babu VRRM (2001) Plate tectonic history of the Indian plate, Nellore-Khammam Schist Belt. Indian Academy of Geosciences, Hyderabad, 183 p

Bandy OL, Rodolfo KS (1964, October) Distribution of foraminifera and sediments, Peru-Chile Trench area. In: Deep sea research and oceanographic abstracts. Elsevier Publications, no 5, pp 817–837
Baruah RM, Singh NP, Rao DC (1987) Foraminiferal biostratigraphy of Disang and Barail groups of a part of Nagaland. In: Proceedings of the national seminar on tertiary Orogeny in indian subcontinent, UBS Publishers' Distributors, pp. 305–327
Basu DN, Banerjee A, Tamhane DM (1982) Facies distribution and petroleum geology of Bombay Offshore basin, India. Jour Pet Geol 5:57–75
Bhandari LL, Mathur RB (1968) Strandline sand—Eocene prospects in Northern Cambay Basin. Second Indian Petroleum Conference, Baroda
Bhandari LL, Fuloria RC, Sastri VV (1973) Stratigraphy of Assam valley India. Bull AAPG 57(s):642–656
Bhandari N, Shukla PN, Pandey J (1987) Iridium enrichment at Cretaceous/tertiary boundary in Meghalaya. Curr Sci 56(19):1003–1005
Bhatia SB, Bhargava ON (2006) Biochronological continuity of the Paleogene sediments of the Himalayan foreland basin: Paleontological and other evidences. J Asian Earth Sci 26(5):477–487
Bhatia SB, Bhargava ON, Singh Birendra P, Bagi Harmeet (2013) Sequencestratigraphic framework of the Paleogene succession of th Himalayan Foreland Basin: A case study from the Shimla Hills. J Paleontol Soc India 58(1):21–38
Biswas B (1963) Results of exploration for petroleum in the western part of Bengal basin, India. In: Proceedings of the 2nd symposium on the development of petroleum resources ECAFE Mins Res De Vol. Ser. No.18, pt. 1, pp 241–250
Biswas SK (1971) Note on the geology of Kutch. Q J Geol, Min Metall Soc India 43(4):223–235
Biswas SK (1980) Mesozoic rock stratigraphy of Kutch, Gujrat. Q J Geol, Min Metall Soc India 49:1–52 (for 1977)
Biswas SK, Deshpande SV (1983) Geology and hydrocarbon prospects of Kutch, Saurashtra and Narmada Basins. Petroleum Asia J 6(11):1–126
Biswas SK, Raju DSN (1971) Note on the rock stratigraphic classification of the Tertiary sediments of Kutch. Q J Geol, Min Metall Soc India. 43(3):177–180
Biswas SK, Raju DSN (1973) The rock-stratigraphic classification on the Teritiary sediments Kutch. ONGC Bull 10(1 & 2):37–46
Blanford WT (1876) On the geology of Sind. Ibid. Recs 9:8–22
Boileau VH (1950) Preliminary report on a geological reconnaissance in the Andaman Island. Progr Rep Geol Surv, India
Bolli HM (1966) Zonation of Createaceous to Pliocene marine sediments based on planktonic foraminifera. Boletin informative de la association venezolana de geologia, Mineriay. Petroleo. 9:3–32
Costa LL, Manum SB (1988) Dinoflagellate cysts: the description of the international zonation of the paleogene(D1-D15) and the Miocene (D16-D20). Geol Jahrb A100:321–300
Chandra PK, Chowdary LR (1969) Stratigraphy of the Cambay basin. ONGC Bull 6(2):37–50
Chandra PK, Banerjee A, Chatterjee PK (1961) Progress report of the work done in part of Andaman. Oil Nat Gas Comm Direct Geol 1–24, pls 1–8. (Dehradun)
Chandra M et al (1991) Exploration of Miocene sandstone reservoirs in the deep basin. West Bengal. Unpublished ONGC Report, Bodra-Port Canning area
Chandra M, Mallik S, Das D, Thakur RK, Sinha MK (1993) Lithostratigraphy of Petroliferous Basins. West Bengal Basin, KDMIPE ONGC Publication, Document IX, pp 1–216
Chatterjee AK, Mathur UB (1966) Note on the Nari Series of Kutch, Gujarat. Bull Geol Soc India 3(1):9–11
Chhibbar HC (1934) The geology of Burma. H.C. Milton and cott, London
Chugh ML et al (1987) Paleogene clastic sediments in the Bombay Offshore Basin and their hydrocarbon prospects: an unpublished ONGC report

Chungkham P, Jafar SA (1998) Late cretaceous integrated coccolith globotruncanid biostratigraphy of pelagic limestones from accretionary prism of Manipur, northeastern india. Micropalaeontology 44(1):69–83
Dasgupta SK (1974) Stratigraphy of western Rajasthan shelf: proceedings IV Colloquim. Indian Micropalaeontology and Stratigraphy, Dehrdun, India, pp 219–233
Dasgupta SK (1975) A revision of the Mesozoic-Tertiary stratigraphy of the Jaisalmer Basin, Rajasthan. Indian J Earth Sci 2(1):77–94
Dasgupta SK, Dhar CL, Metha VK (1973) Progress report in westeren Rajasthan shelf, ONGC report (Unpublished)
De Lapparent A (1883) Traite de geologie. Savy, paris, p 1280
Deshpande SV, Goel SM, Bhandari A, Barua RM, Deshpande J, Kumar A, Rana KS, Chitrao AM, Giridhar M, Chaudhuri D, Kale AS, Phor L (1993) Lithostratigraphy of Petroliferous basins, Document-X, Assam Arakan basin, 2 volumes. ONGC Publication, KDMIPE, pp 1–122
Desikachar SV (1976) Geology and the hydrocarbons prospects of the kerala west coast basin. Paper presented in the workshop on coastal sedimentaries of India south of 18°N, ONGC, Madras
Dhar CL, Khar BM (1977) Report on the geology of Bap and adjoining areas. ONGC report, Western Rajasthan (unpublished)
Dumont A (1849) Rapport sur les travaux de la belgigue. Bulletin de l academie royales des sciences des lettres de la belgique 16:351–373
Evans P (1932) Teritary succession in Assam. Trans Min Geol Inst India 27:155
Gaetani M, Garzanti E (1991) Multicyclic history of the Northern India Continental Margin (North western Himalaya). Am Assoc Petrol Geol Bull 75(9):1427–1446
Gaetani M, Casnedi R, Fois E, Garzanti E, Jadoul F, Nicora A, Tintori A (1986) Stratigraphy of the Tethys Himalaya in Zanskar, Ladakh. Riv Ital Paleontol Stratigr 91(4):443–478
Gopendra Kumar R, Maithy PK (2005) Paleogeographic evaluation of India and its implications for hydrocarbon resources: a new approach. Assoc Petrol Geol, Spec Publ 1:28–40
Govindan A (2013) Larger foraminiferal biostratigraphy of early paleogene sections in India. Geol Soc India, Spec Publ 1:24–45
Gradstein FM, Ogg JG, Schmitz MD, Ogg GM (2012) Geologic time scale 2012, vol 1 & 2. Elsevier Publications, pp 1–1144
Gradstein FM, Ogg JG, Smith A (2004) Geologic time scale 2004. Cambridge press, pp 1–589
Grover R, Hansa R, Kotnala SK, Murthy MS, Boruah AC (2004) Palynological marine index (PMI) studies in Early Eocene section of older Cambay shale and Kadi Foramation of wells from Worosan-Mewod_jotana areas of Mehsana Block, Cambay Basin. Abstract Volume, 2nd APG conference and exhibition, Khajuraho, India, pp 110–111
Haq BU, Hardenbol J, PR Vail (1987) Chronology of fluctuating sea levels since the Triassic. Science 235, 1156–1167
Hardenbol J, Thierry J, Farley MB, Jacquin T, de Graciansky PC, Vail PR (1998) Mesozoic and Cenozoic sequence chronostratigraphic framework of European Basins. In: de Graciansky PC, Hardenbol J, Jacquin T, Vail PR (eds) Mesozoic and Cenozoic sequence stratigraphy of European Basins. SEPM Special Publication 60, pp 3–14
Haug E (1908–1911) Les Periodes geologiques. In: Traite de Geologie, vol II. Armand Collin, Paris. In 3 parts 1766–1776
Head MJ, Gibbard P, Salvador A (2008) The tertiary: aproposal for its formal definition
Helimann-Clausen C (1988) The Danish sub-basin: paleogene dinoflagellates. Neues jahrbuch fur geologie and palaontologie abhandlungen A101:339–343
Heron AM (1937) General report of the geological survey of India for the year 1936. Record Geol Survey India 72:85–90
Jain SL (1974) The first codocanth from India. J Palaeontol 48:49–62
Jauhri AK, Mandaokar BD, Mehrotra RC, Tiwari RP, Singh AP (2003) Corals and foraminifera from the Miocene (upper Bhuban Formation) of Mizoram India. J Paleontol Soc India 98
Jauhri AK, Mishra PK, Kishore S, Singh K (2006) Larger foraminiferal and calcareous algae facies in the Lakadong Formation of the South Shillong Plateau, NE India. J Paleontol Soc India 51:51–61

Jenkins DG, Luterbacher HP (1992) Paleogene stages and their boundaries: Introductory remarks. Neus Jahrbuch Fur Geologie und Paleontologie Abhandlungen 186(1–5):135–138

Juyal KP, Mathur NS (1990) Ostracoda from the Kakara and Subathu formations of Himachal Pradesh and Garhwal Himalaya. J Himalayan Geol 1:209–223

Kachhara RP, Soibam I, Jamir NM (2000) Upper age limit of the Disang Group in Manipur. In: Proceeding Indian colloquium of micropalaeontology and stratigraphy. ONGC Bull 37(2):215–218

Karuppuswamy G (2013) Play perspectives and emerging scenarios: K-G Basin. Bul Oil and Nat Gas Comm 48(2):18–56

Keller G, Armstrong H, Courtillot V, Harper D, Joachimski M, Kerr A, MacLeod N, Napier W, Palfy J, Wignall P (2012) Volcanism, impacts and mass extinctions (long version). Geological society of London, London

Keller G, Khozyem H, Adatte T, Spangenberg J (2013) Biostratigraphy and foramiferal paleoecologym of the Early Eocene Naredi Formation, SW Kutch, India. Geol Soc India, Spec Publ 1:183–196

Khozyem H, Adatte T, Keller G, Spangenberg JE, Saravanan N, Bajpai S (2013) Paleoclimate and Paleoenvironment of the Naredi Formation (Early Eocene), Kutch, Gujarat, India. Geol Soc India, Spec Publ, India

Kossmat F (1895) Importance of the Cretaceous rocks of South India in estimating the geographical conditions during later Cretaceous times. Records 28:39–54

Kulshreshta SK, Singh RY, Sobesh AV (1989) Stratigraphy of the lower tertiary sediments in Bikaner, Western Rajasthan. In: Proceedings of the XII Indian colloquium on micropalaeontology & stratigraphy, pp 193–202

Kumar A (1962) Palynology of the teritary sediments exposed along Silchar-Haflong road section, unpublished ONGC report

Lakhanpal RN, Bose MN (1951) Occurrence of foraminifera of the Guttiferae family in the Fuller's earth at kapurdih. Curr Sci, Jodhpur, District Rajasthan, pp 135–136

Lokho K, Kumar K (2008) Fossil pteropods (Thecosomata, holoplanktonic Mollusca) from the Eocene of Assam–Arakan basin, northeastern India. Curr Sci 647–652

Lokho Kapesa, Venkatachalapathy R, Raju DSN (2004) Uvigerinids and associated foraminifera: their value as direct evidence for shelf and deep marine paleoenvironments during Upper Disang of Nagaland, Eastern Himalaya and its implications in hydrocarbon Exploration India. J Petrol Geol 13(1):79–76

Mallet FR (1876) On the coal fields of Naga hills bordering the Lakhimpur and Sibsagar districts. Mem Geol Surv India 1292:14–94

Mathur et al (1983) Age and paleoecology of the Basal Clastics in Bombay High. Unpublished Report, ONGC, Heera-Panna and Ratnagiri Area

Mathur RB, Sreekantswamy HN, Ananthkrishna K, Srivastava UC, Biswas SK, Krishna A, Bhdsale JS, Rathod GD, Karwal S (1993) Lithostratigraphy of Indian petroliferous basins, document VI, Kerala-Konkan Basin, KDMIPE, ONGC, Publ, pp 1–33

Matsumaru K, Jauhri AK (2003) Lakadongia, a new orbitoidal foraminiferal genus from the Thanetian (Paleocene) of Meghalaya, Micropaleontology 49(3):277–292

Medlicott HB (1864) On the geological structure and relation of the southern portion of the Himalayan Range between the rivers Ganges and Ravee. Mem Geol Surv India 3

Medlicott HB (1869) Geological sketch of the Shillong Plateau. Rec Geol Surv India 2(1):1–24

Medlicott HB, Blanford WT (1879) A manual of geology of India, Chiefly compiled from the observations of the Geological Survey of India, vol II, pp 445–817

Mehrotra RB, Ramakrishna V (1980) A relook in the stratigraphy and hyrocarbon occurrences of North Cambay Basin with special reference to Kadi formation. Unpublished Report, ONGC, Ahmedabad

Mehrotra NC, Venkatachala BS, Swamy SN, Kapoor PN (2002) Palynology in hydrocarbon exploration—The Indian scenario, Part-1. Mem Geol Soc India, no 48, 162 p

Miller Kenneth G, Kominz Michelle A, Browning James V, Wright James D, Mountain Gregory S, Katz Miriam E (2005) The phanerozoic record of global sea-level change. Science 310(5752):1293–1298

Mohan M et al (1981) Biostragraphy of the Teritiary succession met in well diamond Harbour-1 and its contribution to the Paleoecology and Geological History of Bengal Basin, Unpublished ONGC Report

Nagappa Y (1959) Foraminiferal biostratigraphy of the Cretaceous Eocene Succession from India, Pakistan & Burma regions. Micropalaeontology 5:145–192

Narayanan K (1959) Progress report on the geological work in Jaisalmer, ONGC report (unpublished)

Narayanan V, Raju DSN (1996) Chronostratigraphy subdivision of Uttatur of Blanford (1862). In: Sahani A (ed) Cretaceous stratigraphy and palaeoenvironments: memoir number 37, Geological Society of India, pp 209–212

Naumann CF (1866) Lehrbuch der Geognosie, vol 3. Engelmann, Leipzig

Oldham RD (1885) Notes on the geology of the Andaman Islands. Rec Geol Surv India 18(3), 135–145

Pandey J (1972) Comment on the report 'Microfauna and age of the Baratang Formation of Andaman Islands', ONGC report, Unpublished, IPE/Pal/72/46

Pandey J (1986) Chronostratigraphic correlation of the Neogene sediments of Western Indian shelf Himalayas and Upper Assam. Paleontol Soc India, Spec Publ 1:95–129

Pandey J, Dave A (1996) Early Paleogene smaller benthic foraminifera of Indian basin. Contros. XV Indian Colloq Micropal Strat, Dehradun, pp 99–132

Pandey J, Dave A (1998a) Stratigraphy of Indian petroliferous basins. In: Pandey J, Agarwal RP, Dave A, Maithani A, Trivedi KB, Srivastsava AK, Somgj DN (1992) Geol Andaman, ONGC Bull 29(2):19–103

Pandey J, Dave A (1998) Stratigraphy of Indian petroliferous basins. XVI Indian Colloquium on Micropaleontology and Stratigraphy. National Institute of Oceanography, Goa, p 248

Pandey VJ, Srivastava S (1996) Some aspects of Mesozoic and Cenozoic paleogeography of Cambay Basin. Contros. XV Indian Colloq Micropal Strat, Dehradun, pp 45–54

Pandey J, Agarwal RP, Dave A, Maithani A, Trivedi KB, Srivastava AK, Singh DN (1993) Lithostratigraphy of petroliferous basins, document XI, Andaman Basin, KDMIPE. ONGC Publication, pp 1–103

Pascoe EH (1964) A manual of geology of India and Burma, vol 3, New Delhi

Pekar SF, Harwood D, Deconto R (2004) Resolving a late Oligocene conundrum: deep-sea warming versus Antarctic glaciation: 32nd international geological congress. Florence, Italy

Pettijohn EJ (1949) Sedimentary rocks. Harper & Row publisher, NewYork

Phillips J (1841) Figures and descripition of Paleozoic fossils of Cornwall, Devon and east Somerset. Longman, Browns, Green and Longmans, London

Pipper DIW (1972) Sediments of Middle Cambrian Burgress Shale, Canada, Lethaia, vol 5, no 2, pp 169–175

Poddar MC (1959) Geology and oil possibilities in Kutch, Western India. In: Proceedings of the Symposium of Dev Petr Res, ECAFE, no 10, pp146–148

Prasad B, Dey AK (1986) The occurrence of Eocene sediments in Arunachal Pradesh: a palynological evidence. ONGC Bull 23(2):67–74

Raju DSN (2009) Stratigraphy of Mumbai offshore Basin. ONGC Bull 44(2):344–346

Raju DSN, Misra R (2009) Proterozoic and Phanerozoic integrated stratigraphy of South-East Asia: India, Pakisthan, Bangladesh. Myanmar and Srilanka. ONGC Bull 44(2):1–534

Raju DSN, Dwivedi RN, Misra R (2009) Stratigraphy of Maganadi Basin. In: Raju DSN, Misra R (eds) Proterozoic and phanerozoic integrated stratifraphy. ONGC Bull, India 44(2):296–298

Raju DSN, Peter J, Shankar R, Kumar G (2005) An overview of litho-bio-chrono-sequence stratigraphy and sea-level changes of indian sedimentary basin. Assoc Petrol Geol, Spec Publ-1

Ramachandra MR, Chatterjee PK (1963) Progress of the work done in part of Middle Andaman Island, Field Season 1962–63. Unpublished report. Oil & Natural Gas Commission, Directorate of Geology, no. 154/86.4, pp 1–25

Rana RS, Singh H, Sahni A, Rose KD, Saraswathi PK (2005) Early Eocene chiropterans from a new mammalian assemblage(Vastan Lignite Mine, Gujarat, Western Penninsular Margin): oldest known bats from Asia. J Palaeontol Soc India 50:93–100

Ranga Rao A (1983) Geology and hydrocarbon potential of a part of Assam Arakan Basin and its adjoining region, petrol. Asia J 6(4):127–158

Ranga Rao A (1986) North—West Himalayan foothills: Its stratigraphical record and tectonic phases. Bull oil Nat gas comm 23(2):109–128

Ranga Rao A, Dhar CL, Obergfell FA (1977) Badhaura formation of Rajasthan—its stratigraphy and age: IV International Gondwana Symposium, Calcutta, pp 481–490

Rao KLN (1969) Lithostratigraphy of the paleogene succession of southern Cambay basin. ONGC Bull 6(1):24–37

Rao GN (1990) Subsurface stratigraphic nomenclature of Krishna-Godavari basin, ONGC, unpublished report

Rao RP, Ganapathi S, Raju DSN, Rangaraj BP (1971) Progress report on studies in Cauvery Basin. (Field Season 70–71). Unpublished ONGC report

Rao RP, Nair KM, Raju DS, Masthan S, Subramanyam P, Ramanathan RM (1972) Progress report on studies in the Cauvery and Godavari basins. (Field Season 1971–72), unpublished ONGC report

Renevier E (1873) Tableau de terrains sedimentaires formes pendant les époques de la phase organique du globe terrestre. Bulletin de la Societe Vaudoise des Sciences Naturelles, Laussane 12:218–252

Roychoudhary SC, Mathur RB, Mishra GS (1972) Subsurface stratigraphy of Tharad-Serau area, North-west Gujarat, ONGC Bull 9(2)

Sahni A, Saraswathi PK, Rana RS, Kumar K, Singh H, Alimohammadian H, Sahni N, Rose KD, Singh L, Smith T (2006) Temporal Constraints and depositional paleoenvironments of the Vastan Lignite Sequence, Gujarat: analogy for the Cambay Shale Hydrocarbon Source Rock. Indian J Petrol Geol 15:1–20

Samanta BK (1971) Early tertiary stratigraphy of the area around Garampani, Mikir-north Cachar hills, Assam. Geol Soc India 12(4):318–327

Samanta BK (1989) The study of foraminiferal biofacies and microenvironments of the Palaeogene sediments of Western Kutch, Gujarat, unpublished report submitted to KDMIPE, ONGC

Sengupta BK (1964) Teritary biostratigraphy of a part of Northwestern Kutch. J Geol Soc India 5:138–150

Shankar R, Singh G, Gopendra Kumar and Maithy PK (2005) Palaeogeographic evolution of India and Its Implications for Hydrocarbon resources: A new Approch. In: Raju et al. (ed) An overview of litho—Bio—Chrono—Sequence Stratifraphy and Sea level changes of Indian sedimentary basins. Association of Petroleum Geologists, Special Publication 1

Sharma DC, Bhandari A (1986) Late Eocene foraminifera from Sadewala structure, Jaisalmer Basin. Rajasthan. Xth Indian Colloquium of micropaleontology and strat, New Delhi

Sharma KK, Gupta KR (1983) Calc-alkaline island arc volcanism in Indus-Tsangpo suture zone. Geology of Indus suture zone of Ladakh, pp 71–78

Sharma VP, Ashraf Z, Singh RP, Khan EA, Singh OR, Singh K, Srivastava, VK (1990) Recent observations on the Liddar Group of Phanerozoic sequence in Kashmir Himalaya, India. Rec Geol Surv India 122(8):47–48

Shrivastava BP, Srinivasan S (1962) Geology of the area around Bap Nagaur, Western Rajasthan: progress report for the year 1961–62. ONGC report (unpublished)

Shrivastava BP, Srinivasan S (1963) Geology of Bikaner-Barmer area. Unpublished Report, ONGC, Dehradun

Shrivastava JP, Bansal SS, Gadkari SD (1974) Geology of part of Middle Andaman field season 1973–74. ONGC, Directorate of Geology

Shrivastava JP, Bhansal SS, Gadkari SD (1973) Geology of parts of Middle Andaman island, progress report of geological party no 16, Field season 1972–73, unpublished report, Oil & Natural Gas Commission, Directorate of Geology, Eastern Circle, Calcutta, report no 1490, pp 1–113 (Appendix Laboratory reports, field photographs)
Shrivastava JP, Gadkari SD, Bhasin AL (1972) Progress report of the work done in part of Middle Andaman and adjacent Island, field season 1971–72. Unpublished report ONGC, Instt. of Petroleum Exploration, Dehradun, Report No. 1376/86.10
Shukla SN, Katiyar GC, Mudiar B, Murlikrishna A, Mishra MM (1993) Lithostratigraphy of Indian Petroliferous Basins. ONGC Publication, Delhi, pp 1–98
Siddiqui HN, Bahi DP (1965) Geology of the bentonite deposits of Barmear distt. Rajasthan, GSI Mem. No. 96, 96p
Sigal J, Singh NP, Lys M (1971) The Paleocene-lower Eocene boundary in the Jaisalmer area India. J Foram Res 1:190–194
Singh NP (1976) Micropaleonotological controlling subsurface teritiary sequence of Jaisalmer Basin. IV, Indian Colloquium Micropaleonotological Stratigraphy, West Rajasthan, India, pp 259–278
Singh NP (1984) Addition to the tertiary biostratigraphy of Jaisalmer basin: Petroleum Asia J II(1):106–128
Singh NP, Baruah RM, Dave A (1986) Biostratigraphy of the Eocene sequence of upper Assam. ONGC Bull 23(2):45–66
Sowerby JDC (1840) Description of fossils from the Upper Secondary Formation of Cutch collected by CW Grant. Trans Geol Soc London 2(5)
Srikantia SV, Bhargava ON (1967) Kakara Series–a new Palaeocene formation in Simla Hills. Bull Geol Soc India 4(4):114–116
Sudhakar R, Basu DN (1973) A reappraisal of the Paleogene stratigraphy of southern Cambay Basin: ONGC Bull 10(1–2):55–76
Sundaram R, Rao PS (1986) Lithostratigraphy of Cretaceous and Paleocene rocks of Tiruchirapalli District, Tamilnadu, South India. Records Geol Surv India 115:9–23
Tewari BS (1957) Geology and stratigraphy of area between Waghopodar-Cheropadi, Kutch Western India. J Pal Soc India 2:136–147
Tipper GH (1911) Geology of the Andaman Islands with special reference to the Nicobar. Mem Geol Surv India 35(34) pt 3, 135–145
Tripathi C, Murti SKS (1981) Search for source rocks of alluvial diamonds in the Mahanadi valley. Geol Surv India Misc Pub No. 50, 205–212
Tripathi C, Satsangi PP (1982) Crustacean burrows from the Disang Group of Manipur. J Indian Min 36(3):24–26
Venkatachala BS, Rawat MS (1972) Palynology of the Tertiary sediments in the Cauvery basin. 1. Palaeocene-Eocene palynoflora from the subsurface. In: Proceedings of Seminar Paleopalynol Indian Stratigr (Calcutta), pp 229–335
Venkatarengan R, Prabhakar KN, Singh DN, Awasthi AK, Reddy PK, Mishra PK, Roy SK, Palakshi K (1993) Lithostratigraphy of petroliferous basins, documents VII, Cauvery basin, KDMIPE. ONGC publication, Delhi, pp 1–31
Wandrey CJ (2004a) Sylhet-Kopili/Barail-Tipam composite total petroleum system, Assam Geologic Province, India. Petroleum Systems and Related Geologic Studies in Region 8, South Asia, U.S. Geological Survey Bulletin 2208-D
Wandrey CJ (2004b) Bombay Geologic Province Eocene to Miocene composite total petroleum system, India. Petroleum Systems and Related Geologic Studies in Region 8, South Asia, U.S. Geological Survey Bulletin 2208-D
Wilson GF, Metre WB (1953) Assam and Arakan. Sci Pet 6(1):119–123
Wynne AB (1872) Memoirs on the geology of Kutch, to accompany the map compiled by a A.B. Wynne and F. Fedden, during the seasons 1867–68, 1868–69, Mem Geol Surv India 19(2):269
Zubkov IP, Naugolny IK, Zapivalov NP, Chandra PK (1966) Problem of correlation and distribution of hydrocarbons bearing horizons in the Eocene of Cambay basin. ONGC Bull 3(2), pp 9–13

Zutshi PL, Mittal SK, Shah L (1993a) Lithostratigraphy of Indian petroliferous basins. Document IV, Kutch saurashtra basin, KDMIPE. ONGC publication, Delhi, pp 1–50
Zutshi PL, Sood A, Mahapatra P, Ramani KKV, Divedi AK, Srivastava HC (1993) Lithostratigraphy of Indian petroliferous basins. Document Bombay Offshore basin, KDMIPE, ONGC publication, Delhi, pp 1–383

Paleogene Tectonic and Sedimentation History of the Andaman-Nicobar Accretionary Arc, Northeast Indian Ocean

P. C. Bandopadhyay and Andrew Carter

Abstract The Andaman-Nicobar archipelago in the northeastern Indian Ocean is a nonvolcanic outer arc island chain developed by tectonic accretion of sediments and ocean crust along the eastern margin of the subducting Indian lithosphere. The Paleogene stratigraphy of the island chain comprises olistostromes, olistoliths and coarse-grained volcaniclastic turbidite facies, and reefal limestones of late Paleocene-Eocene Mithakhari Melange and finer-grained siliciclastic turbidites (Andaman Flysch) of Oligocene age. The lower Paleogene sedimentary rocks were deposited in shallow-water basins formed on the upper trench slope and growing accretionary wedge and contain material sourced from a local volcanic arc and eroded ophiolite. By contrast, the Oligocene continent-derived siliciclastic sediments were originally deposited outside of the accretionary wedge as part of a large submarine fan system. Subsequent deformation and thrusting juxtaposed these different formations as trench rollback progressed and the accretionary wedge expanded westwards.

Keywords Paleogene · Turbidites · Arc volcanism · Andaman-Nicobar islands

Introduction

The Andaman-Nicobar archipelago consists of 319 islands spread over a length of 700 km and a width of 58 km between latitudes 6.45°–14°N and longitudes 92.15°–94°E. Formation of this island chain can be traced back to the late Cretaceous-Paleogene convergence and collisional events involving subduction, off scraping, underplating, accretion and uplift of ocean floor sediments and crust (ophiolites) along the eastern margin of the Indo-Australian plate. This created a north–south to east–west trending >5000 km long accretionary ridge that includes the Andaman-

P. C. Bandopadhyay (✉)
DST-Project, Department of Geology, University of Calcutta, Kolkata, India
e-mail: a.carter@ucl.ac.uk

A. Carter
Department of Earth and Planetary Sciences, Birkbeck, University of London, London, UK

S. Bajpai et al. (eds.), *The Indian Paleogene*, Society of Earth Scientists Series,
https://doi.org/10.1007/978-3-319-77443-5_2

Nicobar islands. The structure of Andaman-Nicobar islands comprises an accretionary prism formed by an imbricate stack of east dipping fault slices and folds that young to the west linked to a westward-migrating subduction zone (Roy and Das Sharma 1993). The geology of an accretionary wedge is complex, reflecting a dynamic environment that involved subduction, off scraping, accretion, folding and thrusting. Since the geological study by Oldham (1885), there have been a number of investigations centered on mapping and generalized descriptions of the lithostratigraphy, palaeontology and lithology of rocks of Andaman-Nicobar Islands (Tipper 1911; Gee 1927; Jacob 1954; Karunakaran et al. 1964a, b, 1968a, b; Chatterjee 1964, 1967; Haldar 1985). Studies noteworthy for providing improved constraints on stratigraphy, palaeontology, sedimentology, petrology, geodynamics, and isotopic ages of these rocks are those of Ray et al. (1988, 2013), Sengupta et al. (1990), Jafri et al. (1993), Bandopadhyay and Ghosh (1998), Chakraborty et al. (1999), Chakraborty and Pal (2001), Pal et al. (2003), Bandopadhyay (2005, 2012), Chakraborty and Khan (2009), Acharyya (1997, 2007), Pal and Bhattacharya (2010), Pal (2011), Allen et al. (2008), Pedersen et al. (2010), Sarma et al. (2010), Ray and Awasthi (2015) and many others; nevertheless the Paleogene tectonics and sedimentation history are poorly constrained. Here, we detail the field occurrences and attributes of the different Paleogene tectono-stratigraphic units and examine their structural/stratigraphic relationships, paleontology and geochemistry.

Tectonic Setting and Geological Framework

The present-day Andaman-Nicobar Archipelago can be viewed as part of a curvilinear belt of accretionary ridges that stretches from Sumba in Eastern Indonesia to Western Burma in the north (Moore and Karig 1980; Curray and Mungshinghe 1989) associated with the subduction of Indo-Australian oceanic lithosphere below an overriding Eurasian (Sunda) plate. The present-day regional tectonic map (Fig. 1) shows that the Andaman-Sumatra section is broadly aligned north-south and sub-parallel to the present-day motion of the Indo-Australian oceanic lithosphere resulting in an oblique convergence. The Andaman-Nicobar accretionary wedge lies between the Arakan-Yoma accretionary ridge of the Indo-Burma ranges to the north and outer arc islands (e.g. Nias, Mentawai) of the Sumatran arc-trench system to the south. This tectonic domain has been described as the western Sunda Arc (Curray and Allen 2008). The north-south running west Andaman Fault (WAF) to the east of Andaman Islands forms the most prominent morphotectonic feature of the Andaman Sea. It separates a shallow forearc to the west from a deeper backarc region to the east. East of the WAF are Quaternary inner arc volcanoes (Barren and Narcondam islands) and the late Miocene Andaman Sea spreading centre (ASSC) (Khan and Chakraborty 2005; Kamesh Raju et al. 2004). The 1200 km long Sunda trench defines the western limit of the Andaman-Nicobar island chain (Fig. 1) and in this sector the oblique

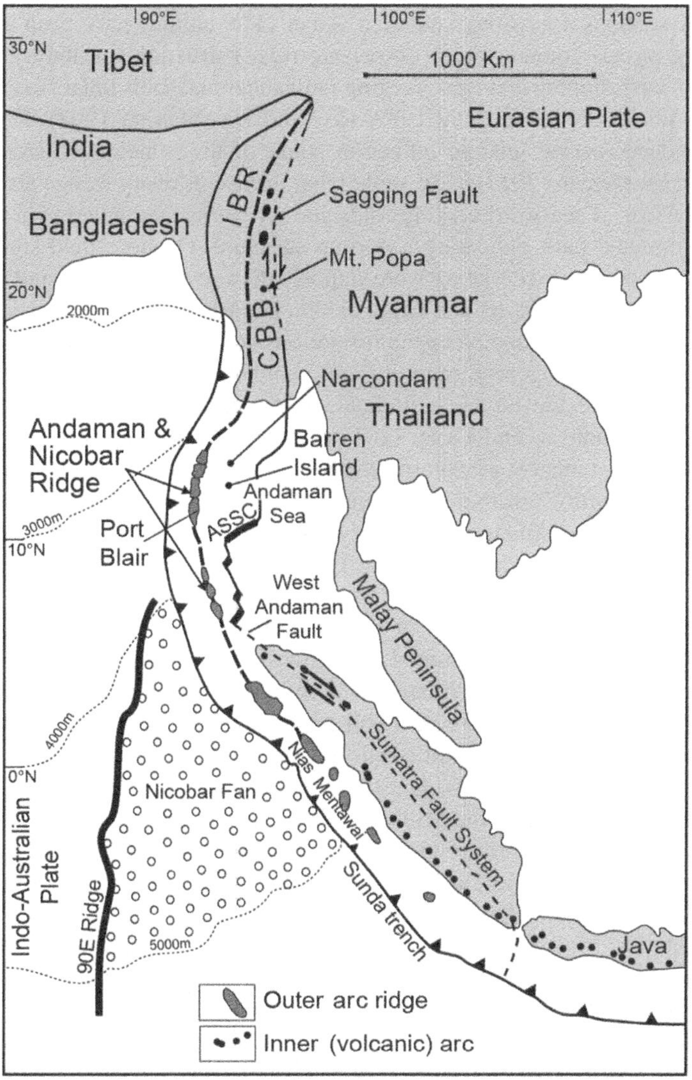

Fig. 1 Tectonic map showing Andaman-Nicobar accretionary ridge and the morphotectonic features surrounding the ridge. ASSC = Andaman Sea Spreading Centre, IBR = Indo-Burma Range, CBB = Central Burma Basin

convergence takes place at a rate of 43 mm/yr (Moeremans et al. 2014). During the Paleogene the subduction zone would have been located closer to the western edge of the Myanmar-Sumatra continental margin. Roy and Das Sharma (1993) has suggested that the site of the present-day Andaman forearc was the site of subduction during the late Cretaceous.

Seismic sections across the Andaman Sea at 11°N latitude have been interpreted as showing an accretionary prism (outer arc ridge) structure formed by imbricate thrusting of east-dipping and west-verging fault slices and folds linked to a westward migrating subduction zone (Roy 1983, Roy and Das Sharma 1993). Recent high resolution deep marine seismic reflection study of the Sunda subduction system (Singh and Moeremans 2017) has revealed that the accretionary wedge around 10°N latitude consists of sea-ward verging folds and thrusts accompanied by a landward-verging conjugate fault indicating a pop-up structure. Onshore field studies show that the sedimentary successions were deposited in several narrow and elongated trench-parallel basins that, by and large, exhibit a north-south strike and easterly dip with younger Paleogene and Neogene formations occurring in the west, structurally overlain by early Paleogene formations and a late Cretaceous ophiolite to the east. A more extensive review on the geological setting of the Andaman-Nicobar ridge is given by Bandopadhyay and Carter (2017a). The late Miocene-Pliocene pull-apart opening of the Andaman Sea is interpreted to have initiated in the south near Sumatra and propagated northwards (Kamesh Raju et al. 2004; Khan and Chakraborty 2005).

The geology of the Andaman-Nicobar accretionary wedge shows a complex network of metamorphic, sedimentary and igneous rocks (Fig. 2). Metasedimentary rocks thought to be older than the late Cretaceous ophiolites occur in the Andaman islands and on Tilanchang Island of the Nicobar Group but their genesis remain conjectural. The ophiolite, presumed to be the basement of the Tertiary sedimentary succession, is mainly represented by serpentinized mantle peridotites whereas the associated crustal rocks are rare and in the southern part of South Andaman Island where a near complete mantle-crustal transition has been recorded (Saha et al. 2010; Pal 2011; Ghosh et al. 2014). Stratigraphically, above the ophiolites are two important Palaeogene units, the Mithakhari Mélange and the Andaman Flysch. The Mithakhari Mélange comprises slumps, slides, olistostromes (debris flow conglomerates), olistoliths, turbidites, and reef limestones. The rocks are heavily fractured, folded, faulted and sheared, typical of mélange rocks of accretionary wedge of a subduction complex. Mud diapirism, is also a typical feature of the Mithakhari Mélange that is common amongst the accretionary settings of the Indonesian segment of the subduction zone (Barber 2013). Thick and extensive deposits of siliciclastic turbidites that were once part of an axially fed submarine fan represent the Oligocene Andaman Flysch (Fig. 2). Neogene (Mio-Pliocene and Pleistocene) rocks comprise shallow marine limestone, marls, calcareous sandstones and reworked felsic tuffs occur chiefly in both western and eastern offshore islands (Fig. 2). In comparison to the Paleogene, the Neogene sequences are less deformed. A wider description of the Neogene Archipelago Group can be found in Bandopadhyay and Carter (2017b). Salient features of the stratigraphy, sedimentology, lithology, palaeontology, isotopic ages, deformational structures, tectonic settings and sedimentary environments of the Cenozoic rocks of Andaman Islands are given in Table 1.

Table 1 Stratigraphy of the rocks of Andaman-Nicobar convergent margin accretionary ridge (Islands); compiled and modified after Chatterjee 1964, 1967, Karunakaran et al. (1968), Pandey et al. (1992), Rajshekhar and Reddy 2003 and Bandopadhyay (2012)

Age							
Biostratigrapjic	Isotopic	Group	Formation (Fo)		Structure/ Lithology	Sedimentary/ tectonic settings	Fossil record
Holocene- Pleistocene	14C-dating of beach rock (1350 to 4410 years B.P*	Archipelago	Neil with Neil Limestone Member and Chidiyatapu member (Rajshekhar and Reddy 2003)		Alluvium, mangrove, coral rags, (Holocene) beach rock and conglomeratic shell limestone (late Pleistocene)	Subaerial to intertidal	
Miocene to Pliocene		Archipelago (~400 m thick)	Karunakaran et al. (1968) 5. Jirkatang Limestone 4. Muralat Chalk 3. Melvile /Guiter Limestone 2. Round Chalk 1. Strait Conglomerate and shell-Sandstone	Srinivasan & Sharma, (1973) 5. Malacca Limestone 4.Sawai Bay 3. Long Siltstone 2. Nancowry Mudstone 1. Strait Sandstone	Cross-and parallel laminated bio-clastic limestone, marls, calcareous mudstone/siltstone, chalk, and felsic turbiditic tuffs	Shallow marine shelf	Foraminifers, *Lepidocyclina, Miogypsina, bryozoan and algae* (Srinivasan and Chatterjee 1981)
Late Eocene to Oligocene	Ar-Ar ages show deposition between 30 and 20 Ma**	1.Andaman Flysch (Karunakaran et al. 1968) 2. Port Blair Series (Chatterjee 1964) *(thickness estimation* varies from 300 to - > 3000 m)	3. Corbyn's Cove Fo 2.South Point Fo 1. Galathea Fo. (Ray 1982)	3. Greywacke Stage (unfossiliferous) 2. Grit Stage (Kirthar) 1. 1. Conglomerate Stage Nummulites of Kirthar *age are reworked* (Pandey et al. 1992; Bandopadhyay 2012)	Parallel-bedded quartz-rich greywackes-shale turbidites, Bouma sequences sole marks are abundant.	Deep sea fan	Barren of fossil (Chatterjee 1964; Allen et al. 2008)
Late Paleocene-Eocene	U-Pb and FT ages show deposition after 60 Ma and no later than 40 Ma**	Mithakhari Group*** (Karunakaran et al., 1968) Baratang Group (Mukherjee 1982) (thickness estimation varies from <700 to 1400 m)	(3) Namunagarh Grit including Wrightmyo Nummulitic Limestone member. 2. Hope Town-Conglomerate with Tugapur Limestone member 1. Lipa Black Shale		(3) Massive and locally graded and channelized beds of gritty and coarse grained volcanolithic sandstone, graded tuff, lithic-poor arkoses, and limestones. Trace fossils and shale flake conglomerates common	(3) shallow water basins perched on to the slopes of the accretionary wedge and fed by several small fans	*Atacicus Leymerie, Assilina papillata Nuttall, N. subatacicus Douville and Discocyclina* (middle Eocene); *Nummulites acutus, N. atacicus, Assilina papillata, Pelatispira, Biplanispira* (upper Eocene) (Chatterjee 1964)

(continued)

Table 1 (continued)

Age						
				(2) Matrix to clast-supported polymictic conglomerates and sandstone with shale and foraminiferal limestones	(2) Same as Namunagarh Grit	*Distichoplax biserialis* in limestones (late Paleocene) (Chatterjee 1964). *Assilina daveisi de Cizancourt* in conglomerate matrix (lower Eocene) (Chatterjee 1964).
				(1) Pyritiferous black shale with olistoliths in sheared argillite matrix	(1) Euxinic environment/ Trench setting	
Late Cretaceous to Paleocene	(95+- 1.3 Ma age of plagiogranite, Pedersen et al., 2010)	Ophiolite		Serpentinized harzburgite, peridotites, dunite, pyroxenite gabbro, diorite plagiogranite, basalt radiolarian chert, muddy carbonate, mudstone, tuffaceous siltstone	Uplifted and thrust slices of ocean floor turbidites and oceanic lithosphere forming accretionary ridge	*Spumellarion radiolarian, Globigerina eugubina, G. trilloculinsides, Globorotalia compressa* and *Globotruncana* (Roy et al., 1988)
Pre- Cretaceous rocks		Older metasedi-ments / Port Meadow Formation/ Prolog Group		Olistoliths of limestones, Mithakhari sandstones, red mudstone, breccias, pelagic red chert, quartz-chlorite-muscovite rock schist, garnet-pemontite-actinolite schist, crystalline limestones and banded quartzite	Represent elements of passive continental margin or metamorphic sole	

Rajshekhar and Reddy 2003*
Allen et al. 2008**
Mithakhari Group redefined as Mithakhari Mélange (Acharyya 1997; Bandopadhyay 2005, 2012) ***

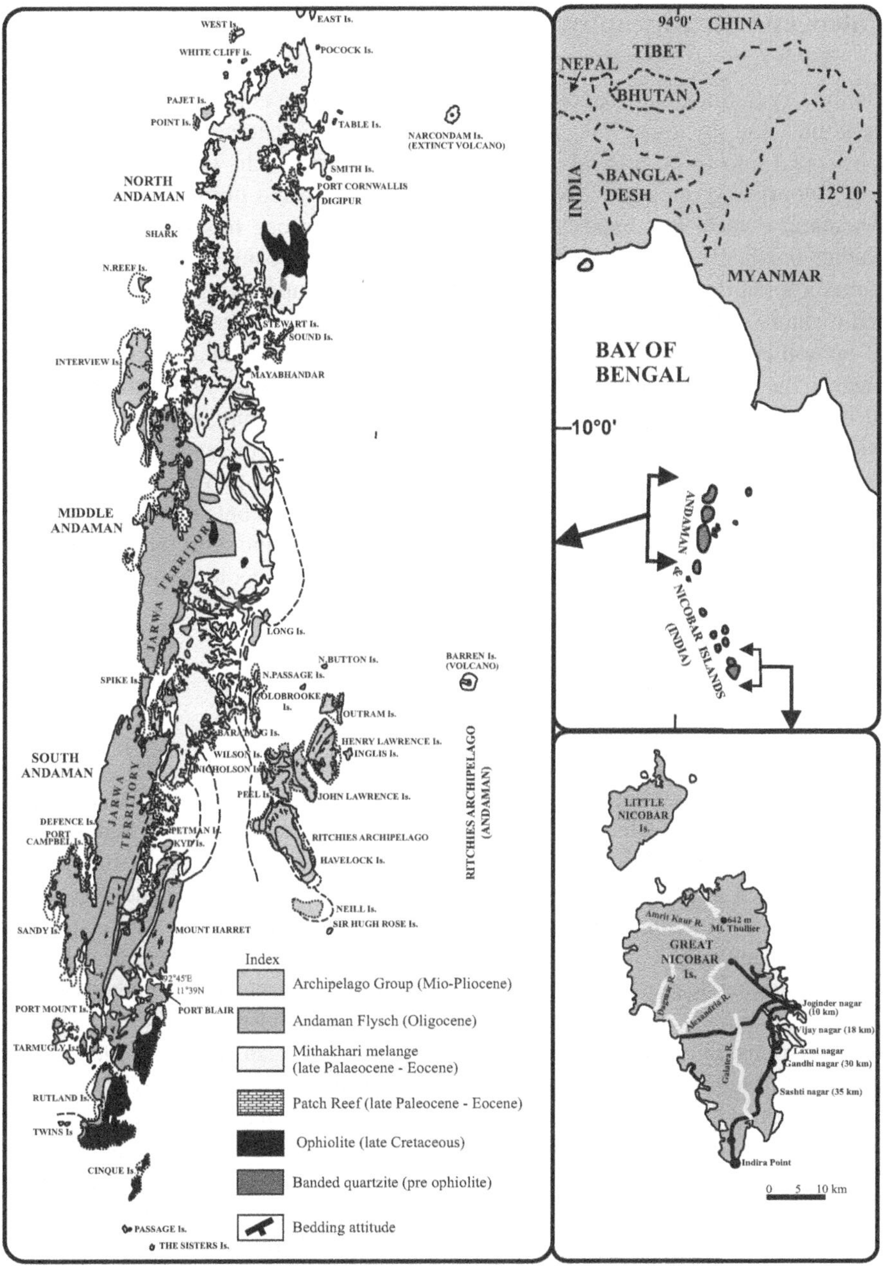

Fig. 2 Geological map of Andaman Island and Rutland showing distribution of major lithologies and structures. Inset at the top shows the geographic location of the Andaman-Nicobar ridge while the bottom one shows a geological map of Great and Little Nicobar islands which are mainly formed of Andaman Flysch

Paleogene Stratigraphy

On the larger Andaman islands (especially South Andaman Island), late Cretaceous ophiolites and the lower Paleogene formation are seen tectonically juxtaposed and telescoped into an N–S trending high-rising fold-thrust belt. The upper Paleogene unit comprises a thick succession of sandstone-dominated turbidites known as the Andaman Flysch. It is exposed almost continuously along the western part of the Andaman islands and is recognized as submarine fan deposits. The lower Paleogene formation (Mithakhari Melange) is a tectonic unit composed of variably deformed sedimentary sequences occurring in association with thrust slices (nappes) and blocks of ophiolites and pre- or sub-ophiolite rocks of meta-sedimentary composition. At outcrop the sedimentary sequences are discontinuous and show rapid lateral changes in lithology, thickness, and age and facies associations, implying deposition in a wide range of environments that best fit its description as a melange. Bandopadhyay and Carter (2017c) described and discussed the field and petrologic attributes of the Andaman Flysch and Mithakhari deposits in detail. We note that in some places deformation has not totally overprinted the original (remnant/ghost) stratigraphy and this offers the opportunity to learn about the pre-collision geology.

Mithakhari Melange

In this study, outcrops that contain a mix of boulders and blocks >2 m across with features indicative of emplacement by sediment gravity flows are classed as olistostromes. The Mithakhari Melange comprise boulders, blocks and bedded units of conglomerates with intercalations of gritty sandstones. The beds often tectonically steepen and in places become vertical (Fig. 3a, b). Other melange rocks are pervasively fractured and include brecciated massive sandstones, interstratified conglomerates-sandstones-shale (Fig. 3a), limestones and mudstones. In deformed units of interbedded sandstone-shale turbidites and sandstone beds, affected by layer parallel shearing and stretching, there are also necked sandstone blocks in scaly and foliated mudstone matrix that are similar to tectonic melanges (Lash 1985; Needham 1995; Maltman 1998; Ujiie 2002; Robertson and Ustaomer 2009) are common (Fig. 4). On the islands of Indonesia this type of deformation is also recorded in melange units formed by large-scale diapiric movement of plastic mud (Barber 2013). Rock units with preserved depositional sequences are classed here as coherent units of melanges, whilst rock units, such as interbedded sandstone-shales, overprinted by accretion related-deformation are classified as chaotic units (Cowan 1985).

Conglomerates marked on the geological maps of South and North Andaman islands are rarely seen as an individual outcrops instead they are more common as part of an interbedded conglomerates-sandstones-siltstone and shale mostly associated with ophiolites. The map of Middle Andaman (Fig. 5), instead, shows outcrops of conglomerates associated lithic grits and tuffs of the Mithakhari Melange and

(a)
(b)

Fig. 3 **a** Steep easterly dipping interstratified conglomerates-sandstone of the Mithakhari Melange, Chidiya Tapu, South Andaman. **b** Thick vertical beds of matrix to clast-supported conglomerates and intercalated gritty sandstones lenses, Middle Andaman

their occurrences are mapped as several linear and disconnected bodies bounded by the Oligocene Andaman Flysch to the west (Andaman Flysch-II on the map) and to the east by the sandstone-shale turbidites referred as Andaman Flysch-I. Parthosarathy (1984) who mapped the area described these conglomerates as fluvial

Fig. 4 Deformed interbedded sandstone-shale of the Namunagarh Grit shows layer parallel shearing and stretching in sandstones caused deformation of the beds that produce clasts and blocks of necked sandstones

occurring between two deep water facies. Bandopadhyay and Ghosh (1998) used a textural study of conglomerates on South Andaman to infer fluvial reworking of the conglomerate clasts before being emplaced into a deep water basin by subaqueous debris flows. The exposed sections of the Mithakhari Melange show an upward fining and thinning motif with evidence of slumping and soft sediment deformation. Bed contacts are generally sharp and planar and some evidence of fluvial channels can be found.

The conglomerates are polymictic, matrix-to locally clast-supported, ungraded to poorly graded and occur as boulders, blocks and bedded units. They often grade to pebbly sandstones. Although the framework clasts in the conglomerates are randomly oriented, planar clast fabric may also be observed in some rocks. The polymict conglomerates (Hope Town Conglomerate) and grits (Namunagarh Grit Formation) (Table 1) are interpreted as derived from the ophiolite and its pelagic cover sediments. Ultrabasic and basic rock fragments dominate the composition of the conglomerate clast population with subordinate to minor amounts of clasts derived from sedimentary sources. Basic rock fragments comprise clasts of, basalt, andesite, serpentinite and gabbro. The sedimentary clasts are mainly limestone, jasper and radiolarian chert with occasional platy clasts of mudstone. Rounded and well rounded granules and pebbles in conglomerate represent a sharp contrast to normally sub-rounded to angular coarse sand and granule sizes grains. It is also noted that the larger clasts are more rounded than the smaller ones.

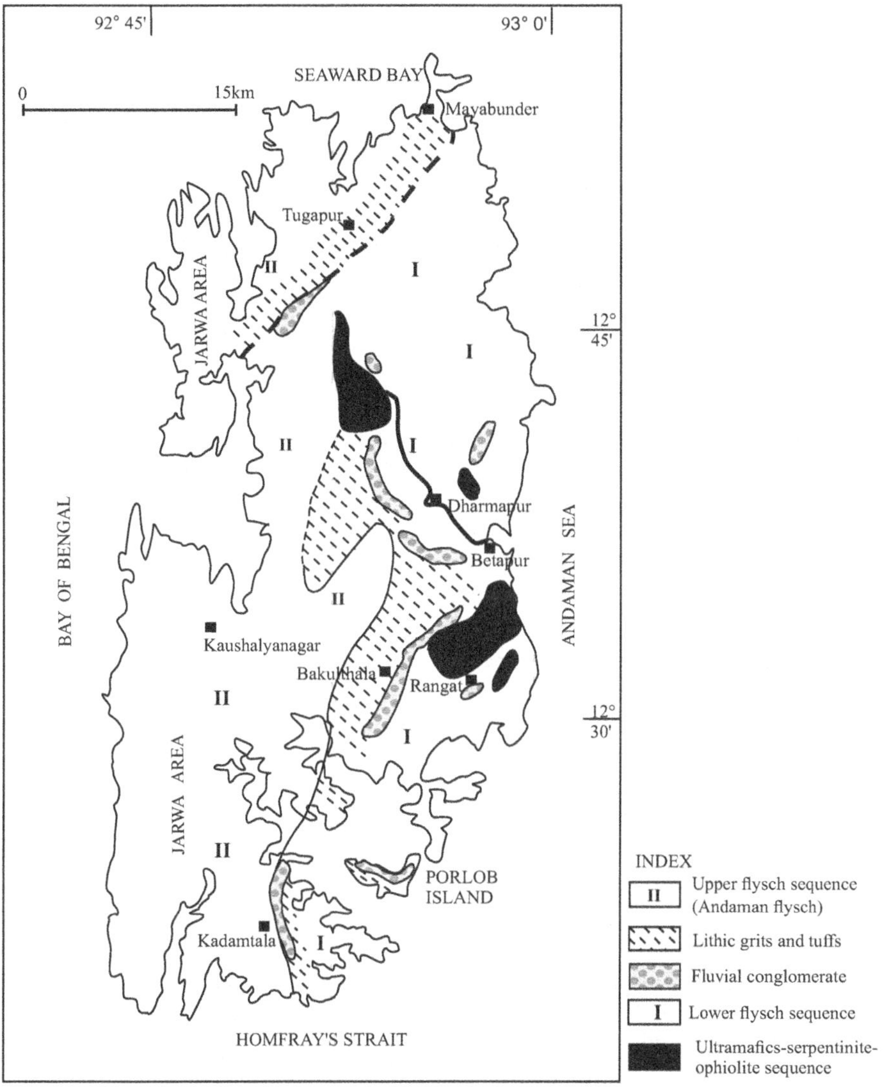

Fig. 5 Geological map of part of Middle Andaman (after Parthosarathy 1984) showing occurrences of conglomerate outcrops and their bounding lithofacies

Bandopadhyay and Ghosh (1998) described the clast compositions and textures of the conglomerates, and also recorded a decreasing trend of abundances and lithological changes in clast composition from north to south Andaman, implying the influence of local tectonics with diachronous uplift and variable supply of clasts from the conglomerate-source(s). These observations are consistent with thermo- and geo-chronological interpretations that suggested diachronous development of depositional basins associated with the Mithakhari Group (Allen et al. 2008). Chakraborty et al. (1999) studied the Mithakhari rocks from Middle Andaman and identified five facies associations (FA-1 to FA-5) inferred to represent: (1) subaerial alluvial fan, (2) shallow water, wave dominated shelf, (3) delta slope, (4) prodelta slope and, (5) submarine fan. The succession, which includes thin coals and gypsum, was deposited in a delta slope setting with facies associations ranging from subaerial alluvial plain to prodelta slope. The depositional age is not well defined due to the lack of distinct biostratigraphic evidence but shallow benthic foraminifera in the Hope Town Conglomerate including *Nummulites atacicus* constrain the age to between the late Ypresian and early Lutetian (Karunakaran et al. 1968) although many foraminifera are broken and abraded (i.e. reworked). The relationship of the Namunagargh Grits to the Hope Town Conglomerates is not clear although the Namunagarth Grit is presumed to be younger.

Gritty and massive sandstones of the Namunagarh Grit (Table 1) are characterized by coarse to fine-grained quartz-poor sandstones and siltstone. On South Andaman the type section and best exposures are found in quarry sections near Namunagarh village. The sandstones are 3–5 m thick, green, matrix-supported, unstratified to well stratified. Bandopadhyay (2005) identified coarse-grained to gritty unstratified and medium to fine-grained well stratified volcaniclastic sandstones facies. The coarse-grained beds, occurring at the base of the section, are $\geq$1 m thick, with sharp non-erosive contacts. The finer-grained beds consist of 4–8 cm thick beds of fine to medium grained volcaniclastic sandstones interbedded with thin mudstones. Bandopadhyay (2005) identified beds with abundant pyroclasts including vesiculated glass fragments, pumice clasts and shards, euhedral feldspars and angular lithic fragments from Namunagarh sandstones at Namunagarh quarry on South Andaman. Some sandstone beds invariably consists of >70% syn-eruptive volcanic constituents and are, therefore, tuffs derived from arc volcanism. Similar juvenile grains are recorded from Namunagarh volcaniclastic turbidite sandstones of the Mithakhari Melange that occur on Middle and North Andaman islands (Fig. 6a, b), implying a widespread volcanism. Bandopadhyay et al. (2009), Bandopadhyay (2012) and this study also confirm widespread occurrences of shallow water ichnotraces and rip-up clasts in sandstone and shale beds of the Namunagarh Grit of South, Middle and North Andaman islands. For more detailed observations of this unit see Chap. 8 of Bandopadhyay and Carter (2017c).

Andaman Flysch

The Andaman Flysch is considered to be Oligocene in age and is found exposed along the western margin of North and Middle Andaman as relatively thin turbidites that become thicker and alternate with rocks of the chaotic terrane on South Andaman (Bandopadhyay 2012). The turbidites cover virtually the whole of Great Nicobar Island. Only in the southern part of South Andaman are turbidites found on the eastern side. The rocks are typically quartzofeldspathic to feldspatho quartzose

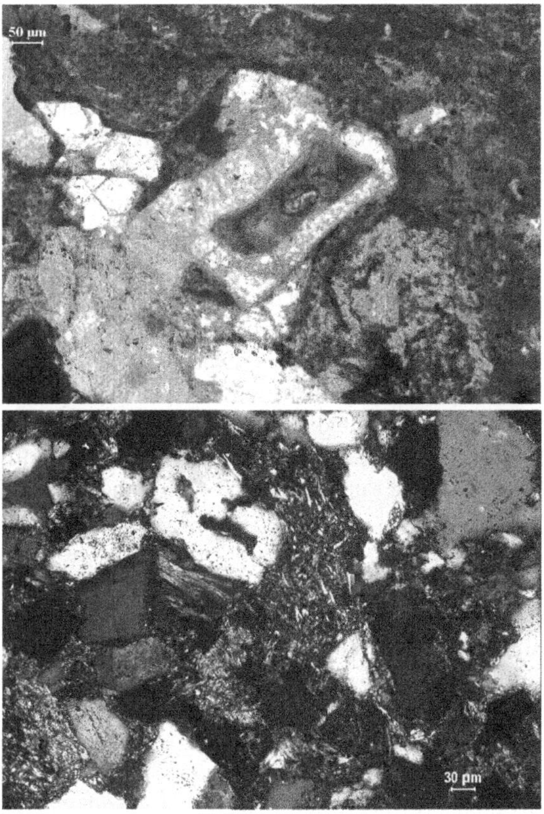

Fig. 6 Petrographic attributes of the volcaniclastic turbidite sandstones, Mithakhari Melange; **a** PPL microphoto of a juvenile crystals showing tabular shape, well developed crystal faces and inclusion of glass at the centre of the grain. **b** XPL microphoto of a volcaniclastic rock consisting of fresh juvenile grains of lithic and crystal fragments

metamorphiclastic sandstones in which the framework components are embedded in an argillaceous matrix. This represents a contrast to the chlorite rich matrix that is common in the Mithakhari sandstones. Current understanding of the Andaman Flysch highlights that the eastern and western exposures of the turbidites may not have the same petrogenesis and possibly different depositional ages (Bandopadhyay and Carter 2017c). These authors also mentioned that the term 'Andaman Flysch' is misleading since the Flysch in Andaman Island is not related to a deep marine facies in the foreland basin of a developing orogen.The term however, is deep-rooted in the literature and thus used here for convenience.

The best and most complete sequences of Andaman Flysch, documented in most studies (Karunakaran et al. 1964a; Bandopadhyay and Ghosh 1998; Chakraborty and Pal 2001; Allen et al. 2008), occur on the east coast of South Andaman around Corbyn's Cove. Coastal sections and inland outcrops in fresh road cuttings and stone quarries are also available near Brindavan, Danros Point, Collinpur and Wandoor on South Andaman. The Flysch turbidites across South Andaman Island together with their bedding attitudes has been interpreted as reflecting folding of the succession into an open anticline exposing the older Mithakhari Group at the core (Karunakaran et al. 1964a). Allen et al. (2008) mentioned that at Corbyn's Cove, steep dipping to vertical beds of Flysch are seen adjacent to the pillow basalt of the ophiolite sequence, although the nature of the contact is uncertain. Pandey et al. (1992) described the contact between the Mithakhari Melange and Andaman Flysch all along the mid-western side of the Middle Andaman Island as a major thrust called the Jarawa Thrust which trends N–S following the regional structural grain.

Though thick and extensive, the actual thickness of the Andaman Flysch is not known and estimates varies from 750 m (Roy 1983) to 3000 m (Pal et al. 2003). Classical Bouma sequences and fluidization structures are abundant in rhythmites exposed on the shore platforms at Corbyn's Cove. It is a sandstone-dominated turbidite consisting of meter-thick and unchannelized beds of sandstone that alternate with much thinner beds of shale. Popularly known as greywackes, the sandstone-shale rhythmites form 10–100 m thick and laterally continuous multi-storied units. The calcite cemented and very thick and massive beds contain spherical concretions of calcareous sandstone of diagenetic origin (Pettijohn et al. 1972). Good examples are seen in the sandstones exposed on the coast at Wandoor and Collinpur on South Andaman.

The sandstone beds show Bouma cycles that start with a massive or normally graded interval at the base grading upwards to a parallel laminated interval. These intervals correspond to Bouma Ta and Tb divisions. Soft-sediment deformation and slump folds are common in the more laminated intervals. A small-scale current ripple-laminated interval often overlies the parallel laminated or massive interval and represents the Tc division. However, the topmost two Bouma divisions (Td and Te) are difficult to separate in the field and the thick and laminated beds of dark grey shale are considered to represent theTd division. There is very little pelite division (Te) preserved. Chakraborty and Pal (2001) described top- and bottom-truncated Bouma cycles consisting of Ta-c, Tb-c, Tb-e and Td-e divisions. Very thick sandstone beds where loaded into the underlying shale beds produce ball and pillow structures and flame structures. Individual sandstone beds are marked by a sharp base with an assemblage of sole marks (flute and groove casts and flame structures) (Fig. 2 d, e, f of Bandopadhyay and Ghosh 2015) and a gradational top with the overlying shale. Directional structures in sandstone beds include flute casts, groove casts, pod marks and current bedding. Karunakaran et al. (1968) and Chakraborty and Pal (2001) recorded a southward palaeocurrent direction based on orientation of sole marks. They suggested sediment transport from the north and northeastern frontiers of Burma, consistent with sediment-provenance analysis based on detrital geochronology, sandstone-shale geochemistry and Sm-Nd isotope chemistry of shale (Bandopadhyay and Ghosh 2015; Allen et al. 2008). Allen et al. (2008) observed that the fine to very fine grained Flysch sandstones compare closely with modern sands of homologous grain size from the Irrawaddy delta in Myanmar.

Apart from a small area on the eastern side, both Great and Little Nicobar Island are entirely made up of Andaman Flysch (Fig. 2). Regional mapping revealed occurrences of long and parallel ridges of the Flysch formation formed by anticlinal folds (Karunakaran et al. 1975). Road cut sections near Campbell Bay, GreatNicobar Island expose cm to a few meters thick and parallel-sided beds of fine grained quartz-rich

Fig. 7 Field and petrographic attributes of Andaman Flysch from Great Nicobar Island. **a** Section of Andaman Flysch exposed at road cutting near Campbell Bay observed by Anindya Bhattacharya (senior geologist). Note the parallel-sided, unchannelized and internally structureless medium thick beds of sandstone alternate with dark grey shale. **b** XPL microphoto of sandstones reveals textural immaturity and poor sorting of the framework grains loosely packed in argillaceous matrix

greywacke sandstones interbedded with shale beds (Fig. 7a). The sandstone beds are mainly massive with thin intervals of current-ripple laminated Bouma divisions in the upper part. Orientation of current ripples reveals a southward paleocurrent direction which is similar to that recorded from South Andaman Island (Karunakaran et al. 1975). However, compared to South Andaman sandstones the Nicobar sandstones are finer grained and tend to be texturally less mature and relatively poorly sorted (Fig. 7b). Does this reflect a short distance transportation and a different source? The sandstone beds on Great Nicobar Island are mainly massive, parallel-sided, un-channelized, laterally continuous beds that are uniformly thick and contain thin (cm) intervals of current-ripple lamination (BoumaTc division) towards the top. Karunakaran et al. (1975) estimated ~5000 m thickness of the Nicobar turbidites and noted that they show an increased abundance of clay beds. The siliciclastic turbidites on the South Andaman and Great Nicobar Islands represent part of a sand-dominated submarine fan system, characteristic of an active margin setting (Mattern 2005). Sandstone-conglomerate associations representing facies A of Mutti and Ricchi Lucchi (1978) are absent. Trace fossils have not yet been identified in the Andaman Flysch turbidite sequences, including the Corbyn's Cove section.

Discussion

Several models exist for the Cretaceous-Cenozoic tectonic development of the Sumatra-Andaman-Myanmar region mainly based on study of the ophiolite sequences (Sengupta et al. 1990; Acharyya 2007; Pedersen et al. 2010; Morley and Searle 2017; Ghosh et al. 2017). Tectonic features of the modern Andaman-Sumatra subduction system, their various attributes and their control on geodynamics of this region have also been discussed in detail (Singh and Moeremans 2017). The early Cretaceous oceanic crust in the northeast Indian Ocean is overlain by a thick pile of sediments (Curray et al. 1979) deposited by the paleo Ganges and Brahmaputra drainages onto the shelf of the Bay of Bengal and later redeposited by gravity flows into the deeper basins including the Nicoar Fan. A similar situation lay to the east where a paleo Irrawaday deposited thick piles of sediment onto the extended margins of Sundaland.

The Andaman Sea is a complex back arc extensional basin formed by transtension soon after the collision of India with Eurasia during late Paleogene time (Peltzer and Tapponnier 1988) whilst Cenozoic paleogeographic reconstructions of Southeast Asia (Hall 2012) show that the Andaman Sea did not exist prior to the mid Miocene and that the location of Andaman and Nicobar Islands during the Paleogene was closer to continental margin, west of the Burma-Malaya-Sumatra peninsula. Interestingly, historical records described a similar setting. Sewell (1925) suggested that the Andaman–Nicobar Ridge had drifted toward the west away from the South–East Asian mainland, and had thus formed a pronounced curve with its apex in the region of 'Little Andaman' Island.

Following the breakup of Gondwana and the separation of India from Antarctica and Australia the early stages of India's northward drift took place at convergence rates that rose above 150 mm/yr (Copley et al. 2010). During the early stages of drift an intra-oceanic arc known as the Incertus-Woyla Arc underwent a polarity flip from south- to north-directed and the Ceno-Tethys began to subduct northwards. By the late Cretaceous the remnant arc had collided with the Burma-Sumatra margin (Hall 2012). It is in this setting that the Paleocene-Eocene-Oligocene geology, now exposed on the Andaman Islands, developed. Curray (2005) considered subduction along the nascent Andaman section to have initiated in the Cretaceous. U–Pb zircon dating of plagiogranite from the South Andaman ophiolite has yielded a 95 ± 2 Ma age, implying a late Cretaceous age for formation and emplacement of the Andaman ophiolite (Pedersen et al. 2010; Sarma et al. 2010). From this time onwards the Andaman accretionary wedge began to develop.

Paleocene-Eocene Tectonics and Sedimentation

Abundant basic and felsic (andesitic juvenile) sand grains in the Eocene Mithakhari sandstones indicate proximity to a volcanic arc, possibly on the western margin of the Malaya Peninsula forming part of the overriding Eurasia plate. The extensive and thick deposits of sedimentary rocks covering more than 90% of the onshore areas of Andaman-Nicobar accretionary ridge require proximity to continental regions. Studies have considered deposition of the Mithakhari sediments took place in small fault controlled trench-slope basins under submarine slope, shelf and alluvial plain environments (Ray 1982; Chakraborty et al. 1999). Bandopadhyay (2012) suggested that within the broad trench-slope tectonic setting Mithakhari rocks were deposited in several and discrete fault-controlled basins perched on the accretionary slope with reefal limestones deposited on the top of the slope. The discrete units of debris flow conglomerates are mainly lag deposits formed in submarine fluvial channels that cut into the underlying sandstones. Clast lithology in conglomerates sharply vary with one location to another: For example, at Chidiya Tapu, on the southern tip of South Andaman, the Mithakhari conglomerates contain abundant platy and rounded limestone clasts while around Port Blair rounded shale clasts are more common. In the Hope Town quarry section further north of Port Blair conglomerate clast lithology is polymictic yet significant amounts of basic rock fragments and rounded vein quartz are noticed. On Middle and North Andaman (near Bakultala, Betapur, Tugapur, Karmatang, Kalipur beach, Ramnagar beach) conglomerate clast lithologies are polymictic yet dominated by quartz with subordinate amounts of basic rocks with a few clasts of serpentinite. The most conspicuous feature, of the conglomerates as noted by Bandopadhyay and Ghosh (1998) on South Andaman, is the presence of rounded and well rounded grains of granules and pebbles that definitely indicate that these grains are reworked implying their derivation from recycled orogenic province. A late Paleocene-Eocene depositional age based on fossil contents is consistent with detrital geochronology data (Allen et al. 2008). Abundant shale pebble conglomerates and trace fossils indicate subaerial to shallow water deposition. The sheared and fractured blocks and phacoids of sandstones embedded in scaly, foliated mudstone recognised from the chaotic unit are typical of subduction zone accretionary complexes. Abundant neovolcanic andesite grains in sandstones are consistent with deposition in an active arc setting that witnessed felsic volcanism during the sedimentation in a forearc region that may include both forearc basin and/or accretionary basins. The significant strata disruption and discontinuity in stratified rocks of Mithakhari Melange favour deposition in accretionary basins.

Oligocene Tectonics and Sedimentation

Deposited between 30 and 20 Ma in an axially fed basinal scale submarine fan, the Andaman Flysch is devoid of ichnotraces and contains reworked body fossils. Shallow water sedimentary structures are also absent in the Flysch beds. All of these suggest a deep water depositional setting for the finer-grained siliciclastic turbidites composed of quartzo-feldspathic to feldspatho-lithic metamorphiclastic quartzwacke sandstones. Observed Bouma sequences are comparable to those found in submarine fan sequences (Pickering and Hiscott 2016) while the non-channelized parallel beds of dominantly fine-grained sandstones imply a mid-fan depositional setting and the predominance of sandstones in the Flysch turbidites compare well with turbidites formed in active margin settings (Mattern 2005). Directional structures in the sandstone beds and framework composition indicate that the detritus were derived from orogenic sources, which may be the western and eastern Burma blocks located to the north and northeast of Andaman-Nicobar Islands. We suggest that the sources of the Andaman Flysch were Himalyan derived including material from the Burmese arc-derived Paleogene Indo-Burman ranges (Allen et al. 2008; Limonta et al. 2017). The distribution of the Andaman Flysch and the Mithakhari Melange in the Andaman Islands (Fig. 2) and the Andaman Flysch covering virtually the whole of Great Nicobar Island suggest that the building out and up of the accretionary wedge peaked in the North Andaman sector. The accretionary process along the eastern side of the island chain spans the forearc high and likely rests on continental crust (Singh et al. 2013). By contrast the wedge that developed on the western side of the islands on ocean crust and represents an open deep sea fan that was thickest and widest in the Great Nicobar Island.

Conclusion

We conclude that the Mithakhari rocks were deposited when there was an active subduction, accretion and arc magmatism punctuated by intervals of tectonic quiescence that facilitated the growth of reefs and carbonate platforms in shallow water basins either starved of clastic material or received minimum supply. Turbidite petrology of the Mithakhari Melange indicates that an undissected magmatic arc was the main sediment-source with minor contribution classed as a recycled orogen. By contrast the Andaman Flysch was deposited during a time when there was a little or no arc magmatism, and a major part of this accumulation resulted from erosion of the regions affected by the India-Asia collision in the Eocene (Curray 1991). Associated tectonics would have produced an enhanced supply of terrigeneous clastic material to the continental shelf-slope where sediments were reworked by gravity flow sand redeposited in a deep sea fan. In addition, falling sea levels of the order of 140 m associated with early Oligocene global glaciations may have also contributed to the enhanced sediment supply through the re-mobilisation of shelf sediments (Curray and Allen 2008). Whilst the Andaman Flysch represents basinal scale deposition in a submarine fan, the Mithakhari Group was deposited in small and shallow water, fault-controlled slope basins formed on an accretionary complex, before being accreted to form an outer arc ridge. We suggest that the sources of the Andaman Flysch was dominantly Himalayan-derived but also included Burmese rocks affected by India-Asia collision.

Acknowledgement P. C. B. thankfully acknowledges the constant inspiration received from the convener of the national conference 'Paleogene of the Indian subcontinent' Dr. S. C. Tripathi, who kindly invited me to submit this paper.

References

Acharyya SK (1997) Stratigraphy and tectonic history reconstruction of the Indo-Burma-Andaman mobile belt. Indian J Geol 69:211–234

Acharyya SK (2007) Collisional emplacement history of the Naga-Andaman ophiolites and the position of the eastern Indian suture. J Asian Earth Sci 29:229–242

Allen R, Carter A., Najman Y, Bandopadhyay PC, Chapman HJ, Bickle MJ, Garzanti E, Vezzoli G, Ando S, Foster GL (2008) New constraints on the sedimentation and uplift history of the Andaman-Nicobar accretionary prism, South Andaman Island. GSA special paper in "Formation and Application of Sedimentary records in Arc -Collision Zone," No 436, pp 223–256

Bandopadhyay PC, Carter A (2017a) Introduction to the geological framework. In: Bandopadhyay PC, Carter A (eds) Andaman-Nicobar Accretionary ridge; geology, tectonics and hazards, Memoir 47 (Chapter 6). Geological Society, London, pp 75–93

Bandopadhyay PC, Carter A (2017b) The Archipelago Group; current understanding. In: Bandopadhyay PC, Carter A (eds) Andaman-Nicobar accretionary ridge; geology, tectonics and hazards, Memoir 47 (Chapter 11). Geological Society, London, pp 155–168

Bandopadhyay PC, Carter A (2017c) Mithakhari deposits. In: Bandopadhyay PC, Carter A (eds) Andaman-Nicobar accretionary ridge; geology, tectonics and hazards, Memoir 47 (Chapter 8). Geological Society, London, pp 111–132

Bandopadhyay PC, Ghosh B (2015) Provenance analysis of the Oligocene Andaman Flysch from South Andaman Island; a geochemical approach. J Earth Syst Sci 124(5):1019–1037

Bandopadhyay PC, Ghosh M (1998) Facies, petrology and depositional environments of the Tertiary sedimentary rocks around Port Blair, South Andaman. J Geol Soc India 52:53–66

Bandopadhyay PC (2005) Discovery of abundant pyroclasts in the Namunagarh Grit, South Andaman: evidence for arc volcanism and active subduction during the Paleogene in the Andaman area. J Asian Earth Sci 25:95–107

Bandopadhyay PC (2012) Reinterpretation of age and depositional environment of Paleogene turbidites in the Andaman and Nicobar islands, Western Sunda arc. J Asian Earth Sci 45:126–137

Bandopadhyay PC, Chakrabarti U, Roy A (2009) First report of trace fossils from Paleogene succession (Namunagarh Grit) of Andaman and Nicobar islands. J Geol Soc India 73:261–267

Barber AJ (2013). The origin of mélanges: cautionary tales from Indonesia. J Asian Earth Sci. https://doi.org/10.1016/j.jseaes.2012.12.021

Chakraborty PP, Khan PK (2009) Cenozoic geodynamic evolution of the Andaman-Sumatra subduction margin: current understanding. Island Arcs 18:184–200

Chakraborty PP, Pal T (2001) Anatomy of a forearc submarine fan: upper Eocene-Oligocene Andaman Flysch Group, Andaman Islands, India. Gond Res 4:477–486

Chakraborty PP, Pal T, Dutta Gupta T, Gupta KS (1999) Facies pattern and depositional motif in an immature trench-slope basin, Eocene Mithakhari Group, Middle Andaman, India. J Geol Soc India 53:271–284

Chatterjee PK (1967) Geology of the main islands of the Andaman. Proceedings, symposium upper mantle project. National Geophysical Research Laboratory, Hyderabad, India, pp 348–362

Chatterjee AK (1964) The tertiary fauna of Andamans. In: Sunduram RK (ed) International Geological Congress Report 22nd Session, New Delhi, pp 303–31

Copley A, Avouac J-P, Royer J-Y (2010) India-Asia collision and the Cenozoic slowdown of the Indian plate: implications for the forces driving plate motions. J Geophys Res 115:B03410

Curray JR (1991) Possible greenschist metamorphism at the base of a 22-km thick sedimentary section, Bay of Bengal. Geology 19:1097–1100

Curray JR, Allen R (2008) Evolution, paleogeography and sediment provenance, Bay of Bengal, Indian Ocean. Golden Jubilee Memoir. Geol Soc India (66):487–520

Curray JR, Moore DG, Lawver LA, Emmel FJ, Raitt RW, Henry M, Kieckhefer R (1979) Tectonics of Andaman Sea and Burma. In: Watkins JS, Montadert L, Dickerson PW (eds) Geological and geophysical investigations of continental margins. American Association Petroleum Geologists. Memoir 29, pp 189–198

Curray JR, Mungsinghe T (1989) Timing of intraplate deformation in northeastern Indian ocean. Earth & Planet Sci Lett 94:71–77

Curray JR (2005) Tectonics and history of the Andaman Sea region. J Asian Earth Sci 25:187–232

Gee ER (1927) Geology of Andaman and Nicobar Islands with special reference to Middle Andaman Islands. Record Geol Surv India 59(2):208–232

Ghosh B, Morishita T, Gupta BS, Tamura A, Arai S, Bandyopadhyay D (2014) Moho transition zone in the Cretaceous Andaman ophiolite, India: a passage from the mantle to the crust. Lithos 199:117–128

Ghosh B, Bandyopadhyay D, Morishita M (2017) Andaman-Nicobar ophiolites, India: origin, evolution and emplacement. In: Bandopadhyay PC, Carter A (eds) Andaman-Nicobar accretionary ridge; geology, tectonics and hazards, Memoir 47 (Chapter 7). Geological Society, London, pp 95–110

Haldar D (1985) Some aspects of Andaman ophiolite complex. Records Geol Surv India 115:1–11

Hall R (2012) Late Jurassic-Cenozoic reconstructions of the Indonesian region and the Indian Ocean. Tectonophysics 570–571:1–41

Jacob K (1954) The occurrence of radiolarian chert in association with ultramafic intrusive in the Andaman Islands and its significance in sedimentary tectonics. Rec Geol Surv India 53(2):397–422

Jafri SH, Balaram V, Govil PK (1993) Depositional environments of Cretaceous radiolarian chert from Andaman-Nicobar islands, north-eastern Indian Ocean. Mar Geol 112:291–301

Kamesh Raju KA, Ramprasad T, Rao PS, Ramalingeswara Rao B, Varghese J (2004) New insights into the tectonic evolution of the Andaman basin, northeast Indian Ocean. Earth Planet Sci Lett 221:145–162

Karunakaran C, Ray KK, Saha SS, Sen CR, Sarkar SK (1975) Geology of Great Nicobar Island. J Geol Soc India 16(2):135–145

Karunakaran C, Pawde MB, Raina VK, Ray KK (1964a) Geology of the South Andaman Island, India. Reports of the 22nd International Geological Congress, New Delhi, India, XI, pp 79–100

Karunakaran C, Ray KK, Saha SS (1964b) Sedimentary environment of the formation of Andaman Flysch, Andaman Islands, India. Reports of the 22nd International Geological Congress, New Delhi, India, XV, pp 226–232

Karunakaran C, Ray KK, Saha SS (1968a) Tertiary sedimentation in Andaman-Nicobar geosyncline. J Geol Soc India 9:32–39

Karunakaran C, Ray KK, Saha SS (1968b) A revision of the stratigraphy of the Andaman and Nicobar islands, India. Bull National Inst Sci, India 38:436–441

Khan PK, Chakraborty PP (2005) Two-phase opening of Andaman Sea: a new seismotectonic insight. Earth and Planet Sci Lett 229:259–71

Lash GG (1985) Accretion-related deformation of an ancient (early Paleozoic) Trench-fill deposit, central Appalachian orogen. Geo Soc Am Bull 96:1167–1178

Limonta M, Resentini A, Carter A, Bandopadhyay PC, Garzanti E (2017) Provenance of Oligocene Andaman sandstones (Andaman–Nicobar Islands): Ganga–Brahmaputra or Irrawaddy derived? In: Bandopadhyay PC, Carter A (eds) Andaman-Nicobar accretionary ridge; geology, tectonics and hazards, Memoir 47 (Chapter 10). Geological Society, London, pp 143–154

Maltman AJ (1998) Deformation structures from the toes of active accretionary prisms. J Geol Soc, London 155:639–650

Mattern F (2005) Ancient sand-rich submarine fans: depositional systems, models, identification, and analysis. Earth Sci Rev 70:167–202

Moeremans R, Singh SC, Mukti M, McArdle J, Johansen K (2014) Seismic images of structural variations along the deformation front of the Andaman-Sumatra subduction zone: implications for rupture propagation and tsunamigenesis, Earth and Planet. Sci Lett 386:75–85

Moore GF, Karig DE (1980) Structural geology of Nias Island, Indonesia: implications for subduction zone tectonics. Am J Sci 280:193–223

Morley AC, Searle M (2017) Regionaltectonics, structure, and evolution of Andaman-Nicobar Islands from ophiolite formation and obduction to collision and back-arc spreading. In: Ban-

dopadhyay PC, Carter A (eds) Andaman-Nicobar accretionary ridge; geology, tectonics and hazards, Memoir 47 (Chapter 5). Geological Society, London, pp 51–74
Mutti E, Ricchi Lucchi F (1978) Turbidites of the Northern Apennines. Int Geol Rev 20:125–166
Needham DT (1995) Mechanisms of mélange formation: examples from SW Japan and southern Scotland. J Struct Geol 17:971–985
Oldham RD (1885) Notes on the geology of Andaman Islands. Rec Geol Surv India 18(3):135–145
Pal T, Bhattacharya A (2010) Greenschist-facies sub-ophiolitic metamorphic rocks of Andaman Islands, Burma-Java subduction complex. J Asian Earth Sci 39(780):804–814
Pal T (2011) Petrology and geochemistry of the Andaman ophiolite: melt-rock interaction in a supra-subduction-zone setting. J Geol Soc London 168:1031–1045
Pal T, Chakraborty PP, Dutta Gupta T, Singh D (2003) Geodynamic evolution of outer arc-fore arc belt in the Andaman Islands, the central part of the Burma-Java subduction complex. Geol Mag 140:289–307
Pandey J, Agarwal RP, Dave A, Maithani A, Trivedi KB, Srivastava AK, Singh DN (1992) Geology of Andaman. Bull ONGC 29(2):19–103
Parthosarathy TN (1984) The conglomerates of the middle Andaman and their geologic significance. J Geol Soc India 25(2):94–101
Pedersen RB, Searle MP, Carter A, Bandopadhyay PC (2010) U-Pb zircon age of the Andaman ophiolite: implications for the beginning of subduction beneath the Andaman-Sumatra arc. J Geol Soc London 167:1105–1112
Peltzer G, Tapponnier P (1988) Formation and evolution of strike-slip faults, rifts and basins during the India-Asia collision: an experimental approach. J Geophys Res 93:15085–15097
Pickering KT, Hiscott RN (2016) Deep marine systems: processes, deposits, environments, tectonics and sedimentation. Wiley, American Geophysical Union. Earth Science Series, 672 pp
Pettijohn FJ, Potter PE, Siever R (1972) Sand and sandstone. Springer and Verlag, Berlin, p 241
Ray KK, (1982) A review of the geology of Andaman and Nicobar islands. Misc Publ Geol Surv India 41:110–125
Ray JS, Kumar A, Sudheer AK, Deshpande RD, Rao DK, Patil DJ, Awasthi N, Bhutani R, Bhushan R, Dayal AM (2013) Origin of gases and water in mud volcanoes of Andaman accretionary prism: implications for fluid migration in forearcs. Chem Geol 347:102–113
Ray JS, Awasthi N (2015) Geochemical provenance for Paleogene sediments in Andaman forearc: implication for Paleogene drainage in South Asia. In: National conference on Paleogene of the Indian subcontinent, Lucknow, 22–23 April. Abs vol, p 38
Ray KK, Sengpta S, Van Den Hul (1988) Chemical characters of volcanic rocks of Andaman ophiolite. J Geol Soc London 145:393–400
Robertson AHF, Ustaomer T (2009) Upper Palaeozoic subduction-accretion processes in the closure of Palaeotethys; evidence from the Chios melange (E. Greece), the Karaburan melange (W. Turkey) and the Teke Dere unit (SW Turkey). Sed Geol. https://doi.org/10.1016/j.sedgeo.2009.06.005
Roy SK, Das Sharma S (1993) Evolution of Andaman forearc basin and its hydrocarbon potential. In: Biswas SK et al. (eds) Proceeding of 2nd Seminar on Petroliferous Basins of India, vol 1. Indian Petroleum Publishers, Dehra Dun, pp 407–434
Roy TK (1983) Geology and hydrocarbon prospects of Andaman-Nicobar. In: Bhandari LL et al. (eds) Petroliferous basins of India. Petroleum Asia Journal, pp 3–50
Saha A, Dhang A, Ray J, Chakraborty S, Moecher D (2010) Complete preservation of ophiolite suite from south Andaman, India: a mineral-chemical perspective. J Earth Syst Sci 119(3):365–381
Sarma DS, Jafri SH, Fletcher IR, McNaughton NJ (2010) Constraints on the tectonic setting of the Andaman ophiolites, Bay of Bengal, India, from SHRIMP U-Pb zircon geochronology of plagiogranite. J Geol 118:691–697
Sengupta S, Ray KK, Acharyya SK (1990) Nature of Ophiolite occurrences along the eastern margin of the Indian plate and their tectonic significance. Geology 18:439–442
Sewell RBS (1925) The geography of the Andaman Sea basin. Memoir Asia Soc Bengal 9:1–26

Singh S, Moeremans R (2017) Anatomy of the Andaman-Nicobar subduction system. In: Bandopadhyay PC, Carter A (eds) Andaman-Nicobar accretionary ridge; geology, tectonics and hazards, Memoir 47 (Chapter 13). Geological Society, London, 195–206

Tipper GH (1911) Geology of Andaman Islands with reference to Nicobar. Memoir Geol Surv India 354:195–213

Ujiie K (2002) Evolution and kinematics of an ancient decollement zone, melange in the Shimanto accretionary complex of Okinawa Island, Ryukyu Arc. J Struct Geol 24:937–952

Geotectonic Evolution of the Paleogene Basins in and Around Peninsular India as Revealed in Seismic Sections and Deep Drilling

K. S. Misra

Abstract High resolution seismic data processed by Pre-Stacking and Depth Migration (PSDM) and Pre-Stacking and Time Migration (PSTM) processing techniques, have made on eloquent exposition of basin forming and modifying tectonics of Paleogene basins in and around peninsular India. These basins are characterized by prolonged extensional tectonics, development of rift and grabens, subsidence along nearly vertical faults, decompression melting and outpouring of enormous lava units during Cretaceous and deposition of uninterrupted Paleogene and Neogene succession. Cretaceous volcano-sedimentary sequence is invariably present in all the basins and is believed by the author to have formed due to down ward progression of faults associated with rift and grabens to critical limits to cause decompression melting. This sequence also forms technical basement for the deposition of Paleogene succession. Processing by PSDM and PSTM techniques have improved the resolution, due to which litho-units of Paleogene sequence could be identified. These litho-units have various proportions of limestone, shale and sandstone and are some of the finest source rocks of hydrocarbons in and around peninsular India. Extensional tectonics has deformed the entire Mesozoic and Tertiary succession; however, the Mesozoic rocks are more severely deformed mainly by nearly vertical faults. The effect of these faults progressively diminishes in over lying younger Paleogene and Neogene rocks. Most of these faults show upward or downward convergence, intersecting each other at the boundary between Paleogene and Neogene, thus forming an hour-glass structure. The prolific growth of coral reef complexes is also observed on volcanic platform, where gradual subsidence of certain blocks is ascertained.

Keywords Paleogene basins · Formation · Sedimentation
Hydrocarbons · Peninsular India

K. S. Misra (✉)
University of Petroleum and Energy Studies, Dehradun 248007, India
e-mail: drksmisra@gmail.com

S. Bajpai et al. (eds.), *The Indian Paleogene*, Society of Earth Scientists Series,
https://doi.org/10.1007/978-3-319-77443-5_3

Introduction

A sizable amount of non-renewable energy resources, in the form of Tertiary coal, lignite as well as oil and natural gas come from different Paleogene sedimentary basins of peninsular India. Earlier these basins were studied and described in localized areas. The present study incorporates all the basins in totality and enumerates the observed similarities and tectonic history. Many of the findings are quite significant for planning the exploration strategy. For regional study and mapping of large scale structures aerial photographs and satellite imagery, mostly on 1:50,000 are interpreted. This exercise was exceptionally useful in identification and mapping of remnants of lava channels and tubes, volcanic vents, cones, craters and calderas (Misra 2002, 2005). Mapping of dyke swarms has helped in identification of regions which experienced extensional tectonics, (Misra 2008a). Available aeromagnetic data of Kutch, Saurashtra and fringe areas, has led us to identify horst and grabens within contiguous basins. Vertical gradient derivative of aeromagnetic data has helped in demarcation of subsurface basin boundaries and basement blocks below the volcanic units, (Misra 2008b). The information gathered in land areas is integrated with the details obtained from the offshore and oceanic regions to get the complete picture. Seismic data and drill hole logs obtained by spending millions of dollars are utilized to map the extent of Paleogene basins, sedimentary succession in subsurface and offshore regions. Furthermore, seismic data processed by Pre-Stacking and Depth Migration (PSDM) and Pre-Stacking and Time Migration (PSTM) techniques has improved the resolution many fold. Several profiles are interpreted and were found extremely useful in understanding sub-basalt stratigraphy. The basin forming and modifying tectonics of each and every basin is discussed in light of new data generated and synthesized during the last few decades. The lithology and stratigraphy is correlated from seismic and drill log sections. Alignment and uniqueness of hydrocarbon pools is described and correlated with the subsurface features. New areas are added and prognosticated for planning exploration strategy in all the basins as well as in contiguous regions.

Regional Tectonic Setting and Paleogene Basins

Paleogene basins are located both in continental as well as oceanic regions around the peninsular India. On continents they occupy rift and grabens while oceanic regions are covered by uninterrupted Paleogene and Neogene succession. However, the most interesting ones are located in the off-shore regions where rift and grabens meet the oceans. These regions are located at intersections, where subsiding rift and grabens interact with coastal fault cluster cascading towards the oceans. The continental regions are transected by a number of rift and grabens which have developed due to extensional tectonics and were subsiding even during Permo-Carboniferous to accommodate coal bearing Gondwana sediments. These tectonic features are commonly described as grabens, rifts, tectonic and lineament zones are transecting Indian peninsula (Fig. 1). Preliminary description of few of these grabens has been given by Pascoe (1964). Narmada-Tapti Tectonic Zone trends in ENE–WSW, Koel-Damoder Graben in E–W, Mahanadi Graben in ESE–WNW and Pranhita-Godavari Graben in NW–SE direction. A gradual fanning out pattern is exhibited by these grabens, which host most of the Gondwana coal seams. On the western side, curvilinear Kutch Rift has E–W trend in west and attains NE–SW trend in the east. This rift is unique and has deformed marine Jurassic and Cretaceous volcanic succession. Narmada-Tapti Tectonic Zone divides the country into two tectonic entities and coincides with the trend of Kaladgi-Bhima Graben. This graben has been identified and found to extend in oceanic region (Misra and Misra 2010). In land it is marked by contrasting geology on either sides and is filled by Proterozoic sediments, while in oceanic extension it exhibits rapid rate of subsidence during Mesozoic with Cretaceous volcanic units at the top. Cambay Structure has NNW trend in north and attains N–S orientation in the south. These basement features have long geological history; have deformed Proterozoic successions in shoulder belts. Geological evidences suggest that they experienced dominant extensional tectonics from the beginning of the Phanerozoic times. Normal faults gave rise to progressively subsiding elongated basins, filled by Gondwana sedimentary rocks. The curvilinear East Coast Fault Zone runs mostly parallel to the coast. Successive blocks have faulted down in a cascading manner in SE direction. A few of these structures such as Kukdi-Ghod Lineament Zone and Manjeera Tectonic Zone did not develop as graben, but kept on experiencing extension tectonics (Misra 2007, 2008a). An excellent summary of stratigraphy of Paleogene basins is presented by Zutsi et al. (1993), Raju and Srinivasan (1983) Raju and Misra (2009). Tectonic deposition and evolution of these basins has been elaborated by Misra and Misra (2010, 2013). Later findings related to the basin forming and basin modifying tectonics is highlighted here. Tectonic details suggest profound effects on development of hydrocarbon pools. Firstly, in deposition of thicker sedimentary sequences, during the Paleogene times, and secondly effusion of volcanic units could have played significant role in generation and preferential accumulation of hydrocarbons.

The formation of Paleogene-Neogene basins is intricately associated with the tectonic history of rift and grabens cutting across the peninsular India (Fig. 2). Gradual subsidence along these rift and grabens has formed the basins while maximum modifying tectonics is witnessed in their intersectional regions. Paleogene basins have formed on top of Cretaceous volcano-sedimentary sequence. Paleogene and Neogene basins are also located on older Gondwana, Proterozoic and Achaean rocks. The significant Paleogene basins, along with associated tectonics are discussed here.

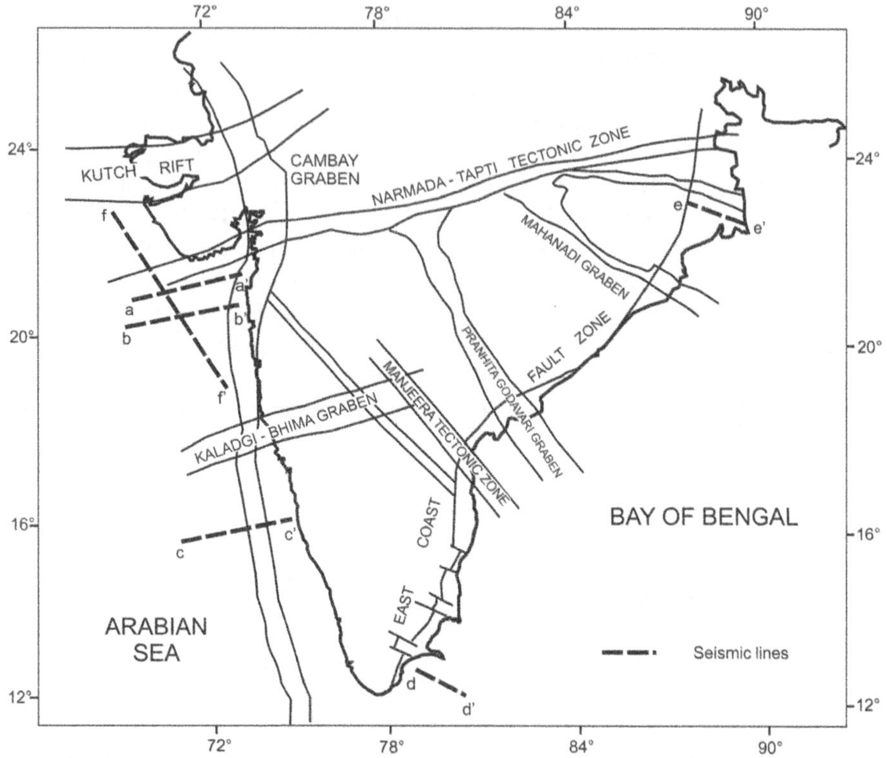

Fig. 1 Outline map showing disposition of major rift and grabens transacting Peninsular India. This map also shows the position of seismic lines interpreted in Figs. 3, 4, 6, 7, 9 and 10

Kutch Rift and Kutch Basin

Kutch rift is an ENE–WSW trending structure which attains almost NE–SW orientation in its eastern continuation. On the western side it continues in off-shore region. This basin is well known for marine Jurassic succession. Geological details of this basin have been given in detail by Biswas (1987). Cretaceous succession mainly comprises units of volcanic flows which are separated by fresh water inter-trappean beds in land areas, and by thick marine sediments in oceanic regions. This volcano-sedimentary succession, exhibits pronounced folding at the Cretaceous-Paleogene boundary. Furthermore, during this deformation the volcanic units are folded like any other sedimentary unit, thus suggesting fairly deep burial and later uplift. Paleogene succession lying over the volcano-sedimentary succession is largely confined to the fringe areas of Kutch region. However, very thick Paleogene marine sediments are logged in the contiguous oceanic regions of the Kutch basin. Misra (2008a, b) brought out transverse faults with strike-slip movement, which separate Kutch basin into several blocks. The rocks within the western block are horizontally disposed, in central

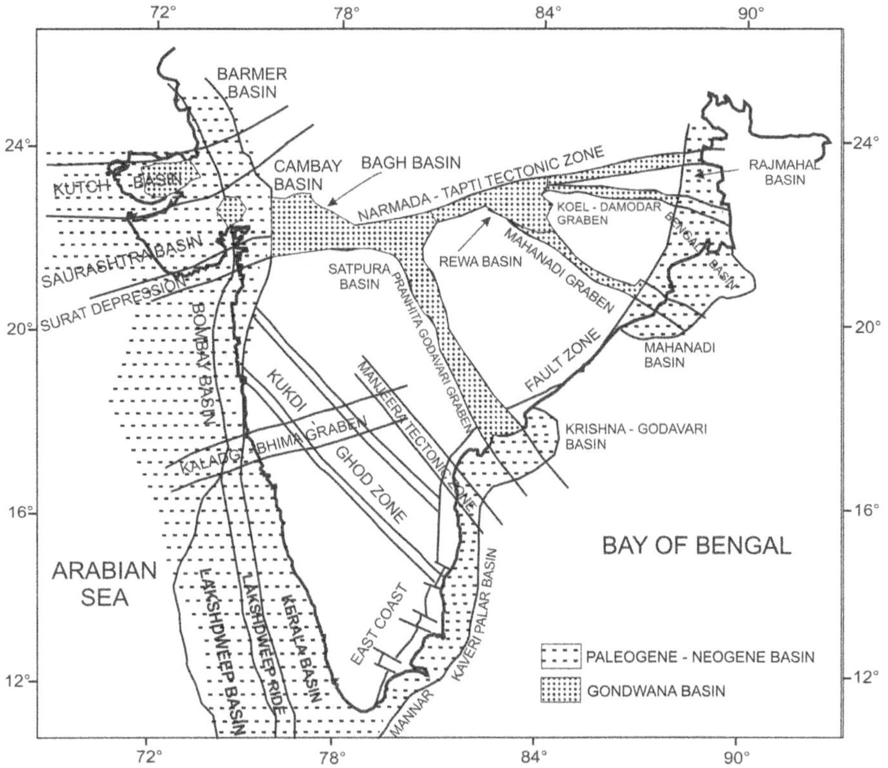

Fig. 2 Outline map of Peninsular India, showing Gondwana and Paleogene basins. The basins are located along the tectonic features and are more pronounced in the inter-sectional regions

block they are dipping gently in southerly direction, while in the eastern block the rocks are severely deformed and dip moderately in SE direction. The present study highlights the continued subsidence and sedimentation since Jurassic period, which progressed till the beginning of Cretaceous when lavas started outpouring. Like all other basins lava units are sandwiched between the two sedimentary sequences of pre-Cretaceous and Tertiary periods. Tectonic details of Kutch region are given in detail by Misra (1999, 2008a, b). Due to compounding effects of subsidence along Kutch rift and Cambay structure, thick Mesozoic and Tertiary sedimentary sequences are expected to be present in eastern part which includes Little Rann and Santalpur-Patan areas. Similarly Gulf of Kutch and western off-shore areas are also believed by the author, to have thick sedimentary sequences.

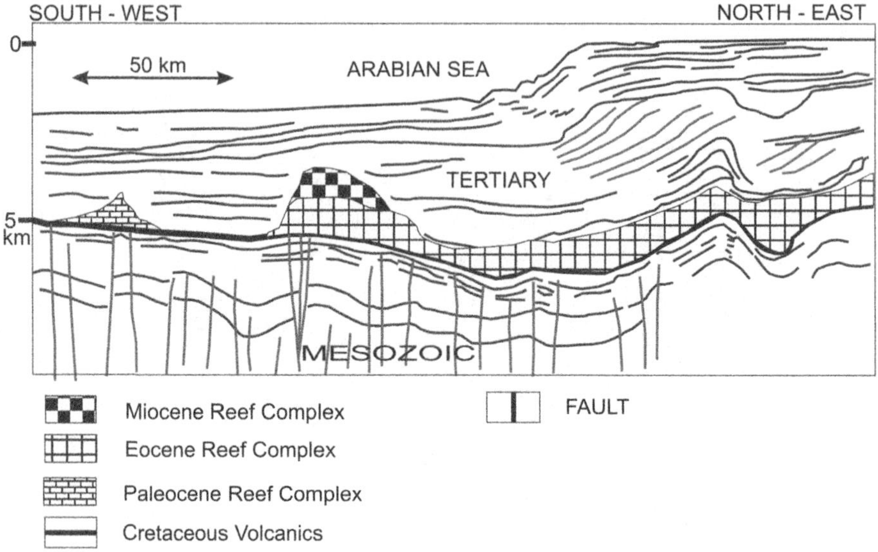

Fig. 3 Interpretation of PSDM profile across Saurashtra offshore region (a–a′). Cretaceous volcanic unit can be seen forming the base of Tertiary succession. Growth of coral reef complexes related to slow subsidence can be seen from the beginning of Paleogene and are related to different geological times. Termination of nearly vertical faults at the end of Cretaceous is also obvious

Saurashtra Basin

This basin occupies almost entire Saurashtra and contiguous offshore region. It presents nearly continuous succession of sedimentary rocks from Dharangadhra-Wadawan Mesozoic sequence exposed in the north eastern part. Neogene Gaj, Dwarka beds and Quaternary Miliolite limestone are mainly exposed in the coastal region. Paleogene sediments are completely absent from land areas. However, thick Paleogene sequences are interpreted in seismic sections in adjacent oceanic basins. Sandwiched between the pre-Cretaceous and Tertiary successions, thick volcanic units are mapped (Misra 1999). Like Kutch basin, this basin also has dipping Cretaceous volcano-sedimentary sequence. In southern coastal belt, it has moderate dips up to 30° in southerly direction. Interpretation of seismic profiles in offshore region, has revealed prolific growth of reef complexes (Fig. 3). Older Palaeocene complexes are located on the western side while younger Eocene complexes are located towards east. This suggests direction of shifting tectonics and accompanying subsidence.

Cambay Structure and Associated Basins

The Cambay structure is located on the western side of peninsular India and has NNW–SSE trend in the northern part and attains N–S trend in the southern continuation along the West Coast. This structure forms very important petroliferous basin of India. Earlier only the axial part with the Tertiary sequence was studied and described as Cambay Graben (Raju and Srinivasan 1993). Misra (2007, 2008a, b) and Misra and Misra (2010) suggested that it is much wider structure with Mesozoic Dharangandhra and Wadhwan on the western side and Himatnagar sequences in the east. These sequences seem to be deposited within progressively widening and subsiding structure. The lineament zones are observed further in the west passing through Cambay, Nal Sarovar and eastern margin of Little Rann of Kutch in NW direction. Similarly the margin of Mesozoic basin in Saurashtra towards the west and central Gujarat in the east are marked by the prominent lineament zones. Apart from these lineament zones, many other geomorphological features and hot springs are found associated with sub surface boundaries. Geological evidences also suggest that thicker Mesozoic rocks might be present within this structure underlain by the Deccan volcanic units. Due to compounding effects of subsidence along Kutch rift in the north and Narmada-Tapti Tectonic Zone in the south, much thicker Mesozoic rocks have been deposited. This sedimentation was followed by the Cretaceous volcanic activity. This structure also acted as effusive zone. The axial part of Cambay structure has faulted down to accommodate Paleogene succession.

Barmer Basin

Barmer Basin is northern extension of Cambay basin and is located between the NE-SW trending Jaisalmer Basin in the north and Kutch rift in the south. Felsic and mafic volcanic rocks occur in the shoulder area of this basin. They are also logged in drill holes forming the technical basement. It is filled by both Mesozoic and Paleogene sediments. Hydrocarbons are produced from Paleogene as well as fractured Cretaceous volcanic units or basement. Profound effect of tectonics and preferential accumulation of hydrocarbons is displayed. The oil and gas fields, such as Saraswati, Raageshwari, Kameshwari, Mangla, Aishvariya, Shakti, Bhagyam, Vijaya, Vandana and many other, are parallel to basin and associated faults. Of these Raageshwari field is producing hydrocarbons from underlying volcanic units.

Bombay Offshore Basin

Bombay offshore basin is located south of Surat depression and continues up to offshore extension of Kaladgi-Bhima graben. This basin is located on submerged Bombay platform, formed at the end of Cretaceous. The volcanic units have played significant role in levelling this platform and are sandwiched between Mesozoic and

Paleogene successions (Fig. 4). Vertical tectonics marked by vertical and upward bifurcating faults. These faults diminish at K-T boundary. Vertical tectonics culminated in effusion of Cretaceous units and controlled thickness and warping in Paleogene succession. Plotting of hydrocarbon pools, depending on the geological ages of reservoir rocks has been very rewarding. In offshore Bombay, the Paleocene, Eocene, Oligocene and Miocene pools have distinct relationship with geological setting (Fig. 5). Paleocene pools such as Panna and surrounding pools are elongated in N–S direction and appear to have developed due to subsidence along normal faults parallel to the extension of Cambay Structure. This succession started with the development of nearly N–S trending rift, on top of the Bombay platform. The presence of Paleocene rocks and hydrocarbon pools largely in southern part, suggests that the initiation of rifting started from south to north. Later, during the Eocene, this rifting became more prominent and continued in northerly direction right through the Cambay into Barmer region. This resulted in deposition of thick Eocene sequence in this basin. Ratna, Heera, Neelam, Bassein, Panna and surrounding smaller pools are associated with this sequence. Oligocene pools comprise cluster of north Tapti, mid Tapti and south Tapti fields, mainly localized in the southern part of Gulf of Cambay. These elongated pools correspond with the directionally similar submerged sand bars. These linear bars are affected by both longitudinal as well as transverse faults. Although they are submerged, even then their attitude can be ascertained from the satellite imagery. The unique crescent shaped Bombay high super giant hydrocarbon pool is also associated with lower Miocene carbonate reef complexes. Furthermore, a number of surrounding pools, as well as several new findings are also located in lower Miocene Gaj Formation.

Lakshadweep-Kerala Basin

Along the southern part of the west coast Lakshadweep and Kerala basins are located. Thnorthern limit of these basins is marked by the offshore extension of Kaladgi-Bhima graben, while on the southern continuation basins seem to turn around the southern tip of peninsular India. Longitudinally these basins are divided by a prominent rise, commonly referred as Lakshadweep ridge. Geomorphologic, N–S trending ridge is reflected as a chain of islands. Individual islands are however, trending in ENE–WSW. Interpretation of seismic data and drilling logs, have shown profound effects of vertical tectonics in Kerala basin. A series of echelon faults transect the gently folded Mesozoic succession. Some of them propagate upward into lower part of Tertiary, which show open warping (Fig. 6).

East Coast Fault Zone and Associated Basins

The present study suggests that East Coast Fault Zone forms 200–300 km wide belt with closely spaced nearly vertical faults. Prominent gravity anomaly is found asso-

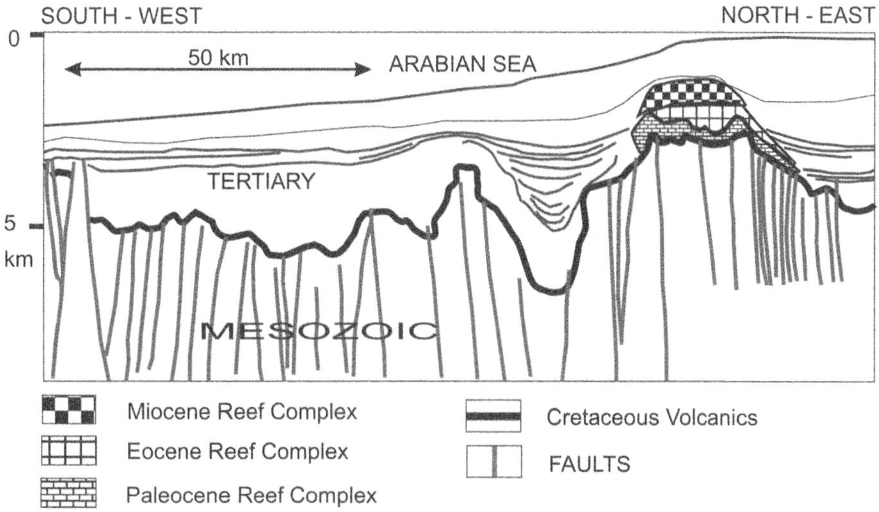

Fig. 4 Interpretation of PSTM seismic data (b–b′), showing Cretaceous volcanic unit sandwiched between older Mesozoic and Tertiary succession in Bombay offshore basin. Disposition of nearly vertical faults and coral reef complexes can also be seen. Upward bifurcation of nearly vertical faults and their role in forming the base of Paleogene is also illustrated

ciated with this zone. Geological evidences suggest that the land areas are rising constantly, while the oceanic region is subsiding along a hinge line, which corresponds to present day sea shore. Along east coast of India, sedimentary basins are located and are generally described as Mannar-Kaveri-Palar basins in the south, Krishna-Godavari and Mahanadi basins in the middle and Bengal basin in the north. The pronounced directional parallelism of these basins with the east coast of India and development of horst and grabens, strongly suggest that they are largely related to extensional forces operating across the NE–SW trending East Coast Fault Zone. Transverse block faulting is the most revealing fact emerged from the drill logs is that, the marker volcano-sedimentary sequence has successively faulted down to greater depths from approximately 2000 m in Mannar-Kaveri, 3200 m in K-G, 4000 m in Mahanadi and 5470 m in Bengal Basin (Misra 2005). The thickness of Tertiary succession also increases accordingly. The tectonic details of rift and grabens worked out in continental areas have facilitated understanding development of **oi** grabens worked out in continental areas have facilitated understanding development of oil reserves in off-shore region. Firstly deposition of thicker sedimentary sequences, both during the Mesozoic and Tertiary times, and secondly effusion of volcanic units, which have played significant role in formation and preferential accumulation of hydrocarbons.

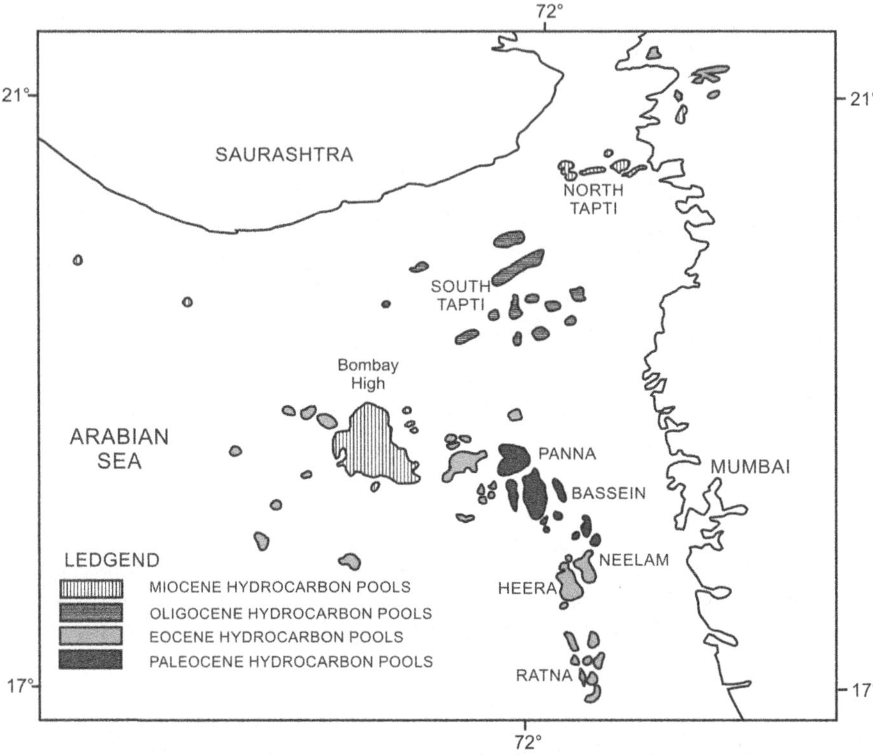

Fig. 5 Disposition of Hydrocarbon pools in Bombay offshore basin is shown as per the geological ages of reservoir rocks

Mannar-Kaveri-Palar Basins

A group of basins is located along the East Coast Fault Zone. The important ones are Mannar, Kaveri and Palar. They have formed by faulting of successive blocks in easterly direction. Mannar and Kaveri basins are separated by Palk Strait, which is largely submerged arcuate ridge connecting peninsular India with Sri Lanka. A series of emerging islands along this ridge suggests continuing tectonic movement. Among these basins, the Kaveri basin is largest and has very high hydrocarbon potential. In land areas a continuous Cretaceous succession is exposed and overlying volcanic units are absent. This succession has faulted down in offshore regions and is overlain by Cretaceous volcanic units. A number of horst and grabens have displaced basement blocks along with overlying sedimentary succession (Fig. 7). Cretaceous volcanic units are logged in both Mannar and Kaveri basins at an approximate depth of 2000 m. In Kaveri basin, volcanic units are located only in distal part of off-shore basin. Hydrocarbon pools are located in both land and off-shore regions, the significant ones are Adichapuram in land area and Ganesha in offshore region. This is one such basin where the source rock belong to Cretaceous and older succession.

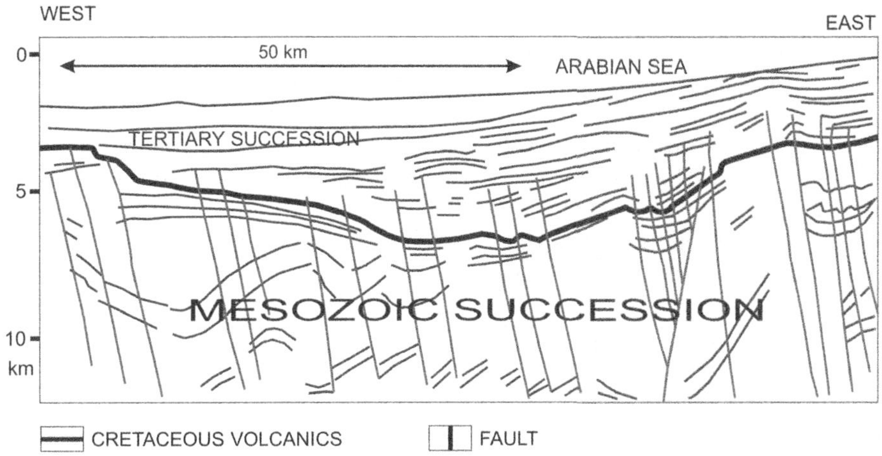

Fig. 6 Interpretation of PSDM seismic profile across Kerala basin (c–c′). Cretaceous volcanic units are sandwiched between uninterrupted Tertiary and deformed Mesozoic successions. Nearly vertical faults have formed the base of Paleogene basin and continue into the lower part of succession

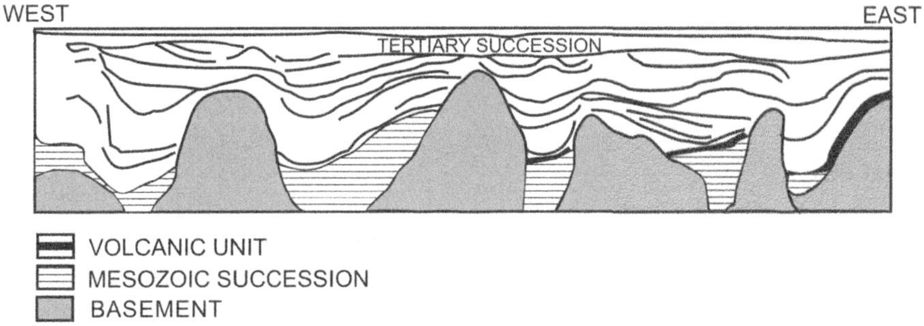

Fig. 7 Geological section across the Kaveri basin (d–d′). Cretaceous volcanic units can be seen confined towards east in far offshore region only. Older Mesozoic succession is confinedwithin the grabenportions formed by movement of basement blocks along nearly vertical faults related to East Coast Fault Zone

Krishna-Godavari Basin

Krishna-Godavari Basin is located in the intersectional region of East Coast Fault Zone and Pranhita-Godavari graben. Detailed stratigraphy of this basin is presented by Kumar (1983). Here the effect of tectonics associated with both the tectonic features are enumerated. Step faulting cascading in south–east direction along the East Coast Fault Zone, as well as echelon faulting towards axial portion along Pranhita-Godaveri graben are quiet apparent. Oldest rocks in this basin are of Permo-carboniferrous period. Cretaceous volcano-sedimentary sequence is well exposed in the land area has faulted down in offshore region. In land area the volcanic units are generally described as Rajahmundri volcanic, while in offshore they are known as

Rajol Formation. The hydrocarbon pools of K-G basin demonstrate an interesting relationship with the geological ages. The Cretaceous pools such as Mandapetta are elongated and largely parallel to curvilinear cost line and associated geomorphologic features, such as beach ridges and coastal dunes. Conversely the Paleogene and Neogene hydrocarbon pools are associated with the paleo-channel systems, having rather NW–SE trend. This anastomosing system has developed by Krishna and Godavari rivers and their distributary channels. The hydrocarbons are concentrated in pebbly and sandy channel courses, which are encased in impervious clay horizons. The prolific Dhirubhai and Deendayal pools are associated with such channel complexes.

Mahanadi Basin

Mahanadi basin is located in the intersectional area in the East Coast Fault Zone and Mahanadi graben. This basin has a small land component known as Cuttack depression and a large offshore extent. On southeastern side this basin gets terminated where East Coast Fault Zone brings Archean rocks close to the coast, however in the north eastern side it merges with the offshore component of the Bengal basin. Cretaceous volcanic units are logged at approximate depth of 4000 m (Jagannathan 1983). In Cuttack depression, no oil is found, but in off-shore region paleochannel complexes as well as Paleogene sediments host a number of small pools.

Bengal Basin

Bengal basin has land, subsurface and offshore components. On land the total succession is described as Rajmahal basin. The down faulted blocks are covered by thickest Paleogene, Neogene sediments and alluvium forming the main Bengal basin. This basin is very complex, as it is located in the region where East Coast Fault Zone divides it into two sub basins. Locally this tectonic feature is known as Hinge zone. A number of major structures are also identified. Koel-Damodar valley graben and Narmada-Tapti Tectonic Zone also continues into the Bengal basin (Fig. 8). The rate of subsidence has also been highest among all the Phanerozoic basins of peninsular India. If we take Cretaceous volcano-sedimentary sequence as marker horizon, the thickness of Tertiary rocks is >4000 m and increases drastically in southeast direction (Fig. 9). The Cretaceous volcanic activity has been described by (Roybarman 1983), as colossal and spasmodic. The hydrocarbon potential of Bengal basin has been rated as high because of high content of good quality organic matter and favorable thermal history. However, favorable trapping mechanism as well as seals is yet to be identified.

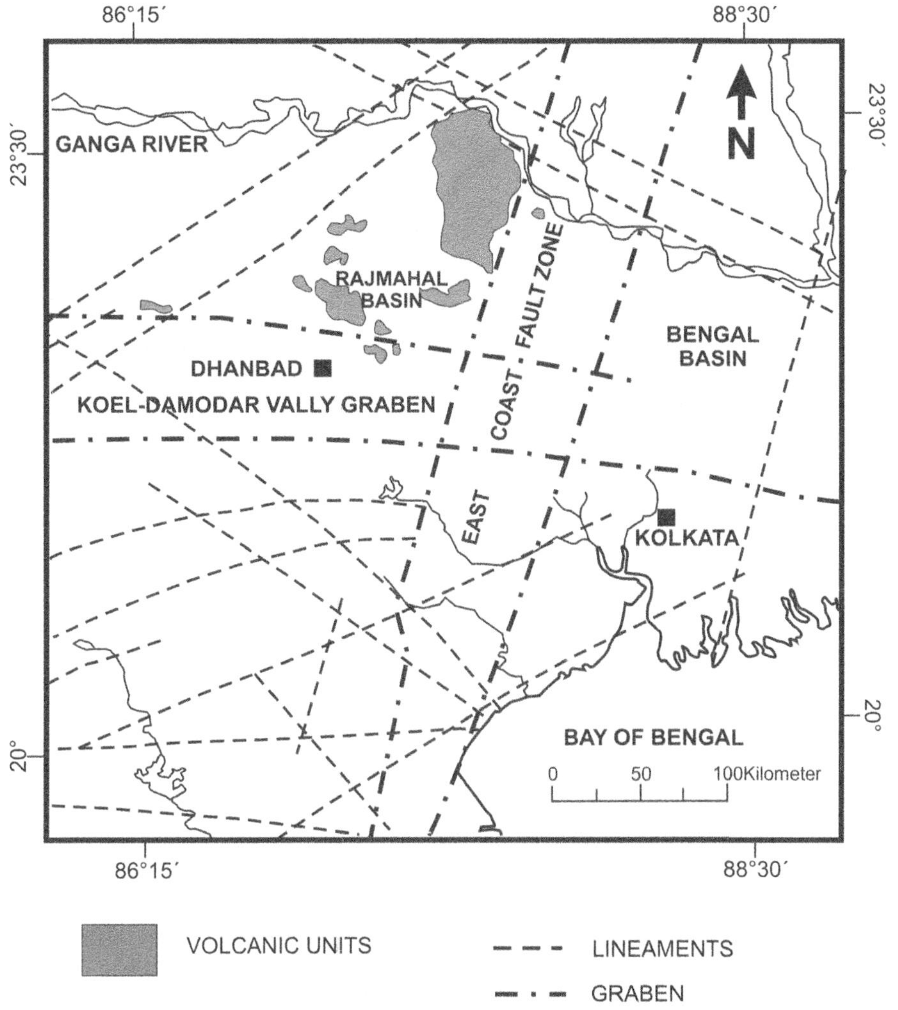

Fig. 8 Map showing interpretation of satellite imagery in the vicinity of Bengal basin. Disposition of East Coast Fault Zone, Koel-Damodar Valley Graben and other major lineaments can be seen. Distribution of Cretaceous volcanic units in the vicinity of Rajmahal basin are also shown

Narmada-Tapti Tectonic Zone and Associated Basins

The Narmada-Tapti Tectonic Zone trends in ENE–WSW direction, and is of continental significance. It is oldest rift which divides peninsular India into northern and southern blocks. West (1962) considered it as Precambrian deep crustal fracture zone. Radhakrishna (1989) described Proterozoic mobile belt activity along this zone in central India. Misra (2008b) found geological evidences to suggest that along this zone, extensional tectonics continued, a dyke swarm was emplaced and acted as

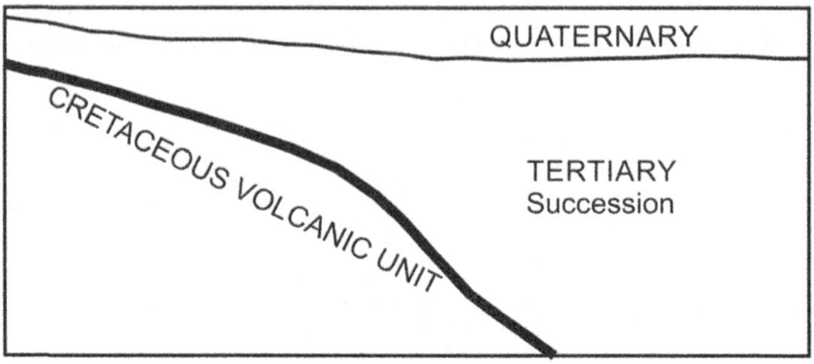

Fig. 9 Cretaceous volcanic unit can be seen forming the platform over which Tertiary succession is deposited in Bengal basin (e–e′). The exponential increase in the thickness of Tertiary succession towards east can be observed

major effusive zone during the eruption of Cretaceous volcanism. From west to east several basins have developed and are Surat depression, Bagh, Satpura, Rewa and Bengal basins.

Surat Depression

This unique submerged geomorphological feature is trending in ENE–WSW direction and is located south of Saurashtra. The existence of several kilometer deep valley in sea bottom, suggests that the rate of subsidence is higher than the sedimentation. A number of nearly vertical faults are largely responsible for down throw of successive blocks. The volcano-sedimentary succession of Cretaceous period has faulted down, over which thicker Tertiary succession is deposited towards the southeast (Fig. 10). This disposition suggests that the Surat depression existed from the beginning of Tertiary as a sheltered basin and would have played significant role in formation of hydrocarbon.

Cretaceous Volcanic Units

Cretaceous volcanic units are occurring on land, sub-surface, off-shore and oceanic regions. Data collected from different sources is collated with new data generated from seismic profiles and drilling logs, during the present study is presented here (Fig. 11). Most significant finding is that these volcanic units are invariably present in all the petroliferous Paleogene basins. These units also form technical basement over which uninterrupted Paleocene sediments are deposited. Several units are inter-layered with sediments and together form a well-defined volcano-sedimentary sequence ranging from the beginning to the end of Cretaceous. Earlier the study

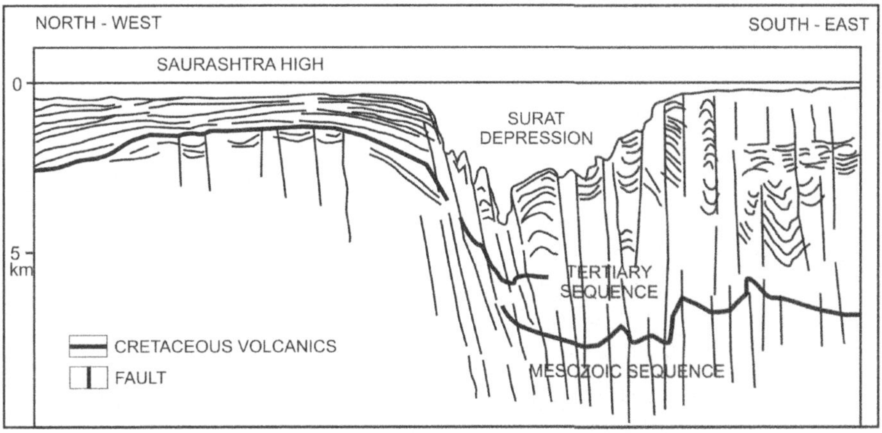

Fig. 10 Interpretation of seismic profile across Narmada-Tapti Tectonic zone in offshore region (f–f′). Subsidence of blocks along vertical faults has increased the thickness of Tertiary sequence. The very presence of elongated depression in the ocean floor

of these volcanic units was confined to the Deccan region of western Maharashtra where they were described as Deccan Traps.

In this region, only certain units are exposed and neither lower nor upper stratigraphic controls are present. It has also been difficult to estimate the thickness of volcanic units which would have been eroded off. Models were made based on study in limited areas and sections. Assumption was made that the entire volcanism has taken place from single central type of activity. Later studies by Misra (1999, 2002, 2005, 2008b) demonstrated multi central volcanic eruption where volcanic fields with enumerable plugs, cones, craters and calderas are located Satellite imagery was extremely useful in recognizing these features (Fig. 12a). Similar cone and craters were reported from Mount Pavagadh (Sheth et al. 2004). Field mapping, assisted by satellite imagery interpretation has brought out innumerable centers of eruption from all over the volcanic country (Fig. 12b). Here a huge volcanic caldera can be seen from where lava flow has flowed in easterly direction. The variation in weathering characteristics, has accentuated them to be picked up in satellite imagery. Furthermore, an integrated system of lava channels and tubes has been brought out by Misra (2002). Mapping of their remnants has shown that they are emanating from effusive centers (Fig. 12c, d). The most impressive center is located at Asoliadongar near Kalamdari village, close to Chandwad on Bombay–Agra road.

The initial volcanism was of felsic nature and rhyolites, obsidian, trachyte have erupted. Felsic sequences dip moderately towards southeast in southern Saurashtra (Fig. 12e). Dips in sequences have significancant effect on thickness pattern and disposition of overlying Paleogene sediments. It has been realized that even very low gradients play important role in movement and preferential accumulation of hydrocarbons. Furthermore, this felsic sequence was accompanied by explosives activity where volcanic bombs, agglomerates and tufts have ejected (Fig. 12f). Such

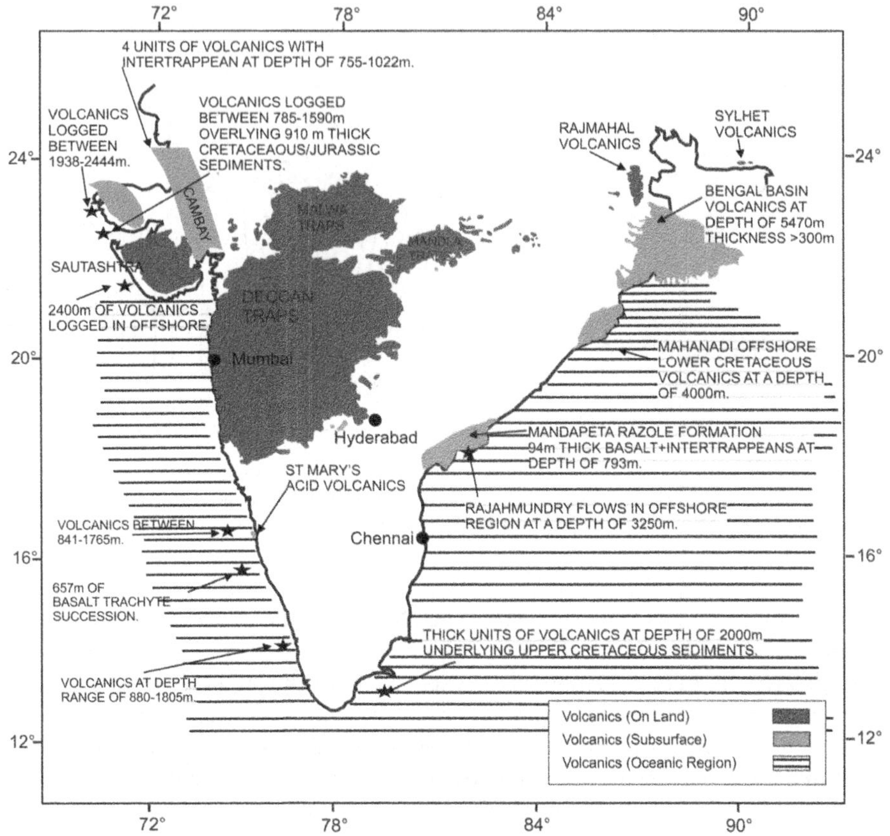

Fig. 11 Map of peninsular India and surrounding oceanic region, showing distribution of Cretaceous volcano-sedimentary succession on land, subsurface, off-shore and oceanic regions

fields have cropped up at different places, not only within the Deccan region but also in far off areas in the entire peninsula and surrounding oceanic regions. The most interesting volcanic units outside the main Deccan region are Rajahmundri in Andhra Pradesh. They are disposed in northeast southwest direction, which is rather parallel to the coast. However, the effusive centers identified are oriented in NW–SE direction, which corresponds with the trend of Pranhita-Godavari graben (Fig. 12g). In K-G basin it has not been possible to resolve different volcanic units in drilling logs. Field mapping has, however, shown that they have units separated by prominent inter-trappean beds (Fig. 12h).

The study enumerates evidences to support the Cretaceous volcanism is rift related and attributes to decompression melting. Thickness patterns of these volcanic units and sedimentary successions reveal tectonic history. They arerelatively thick along the rift and grabens and reach to enormous thickness of several kilometers in intersectional areas. This suggests that lava coming out along the rift and grabens was filling

Fig. 12 **a** Satellite imagery of Central Kutch near Nakhatrana showing several volcanic plugs, cones and craters. **b** Satellite imagery in the vicinity of Aurangabad Maharashtra, A huge caldera, from which lava has followed in easterly direction can be seen. **c** Gunjale-Vambori area near Ahmednagar, depicting remnants of levee portions of lava channel. **d** Asolia dongar near Chandwad in Maharashtra, emanating lava tube can be seen on the eastern side. **e** Interlayered succession of felsic volcanic rocks. The light colored bands are of rhyolite and darker ones are of obsidian. **f** Spherical volcanic bombs of rhyolite embedded in obsidian, near Loingdi village in southern Saurashtra. **g** Quarry section in Rajahmundri volcanic units, depicting a number of hard compact volcanic vents oriented in NW–SE direction. **h** Inter-trappean limestone beds with volcanic units

the subsiding basins. Maximum thickness of more than three kilometers is logged in Ankaleshwar Deep well, in the intersectional area of the Narmada-Tapti tectonic zone and Cambay graben. Furthermore, the thickness patterns of the sedimentary sequences both above and below the volcanic units also confirm this conclusion.

Cretaceous-Paleogene Boundary

Detailed records of significant geological events, during the Cretaceous and Paleogene period are best preserved within the sedimentary basins in and around peninsular India. Extensional tectonics, which dominated during the Mesozoic and resulted in rampant and cyclic volcanism ceased at the end of Cretaceous (Fig. 13). However, extensional forces continued throughout Tertiary with reduced magnitude. Accompanying subsidence along the rift and basins as well as oceanic regions is attributed to the formation of Paleogene basins. Furthermore, the Cretaceous and older sequences are more severely deformed and faulted along nearly vertical faults. The intensity of deformation and displacement along faults gradually diminishes upward in younger formations and only gentle warping is recorded. As can be seen in interpreted seismic profiles the faults exhibit bifurcation during upward progression. Although most of the faults get terminated at the Cretaceous-Paleogene boundary a few prominent ones are seen continuing upwards.

Conclusions

The study has led to following conclusions and recommendations:

Geological details emerged from field mapping, interpretation of high resolution satellite imagery, seismic data and drilling data have significantly improved understanding of Paleogene basins.

Paleogene basins in and around peninsular India have developed in response to extensional tectonics. Prolonged extension has given rise to the rift and grabens.

Basin forming tectonics involved subsidence of basement blocks along nearly vertical faults. In most of the cases successive blocks have faulted down towards axial portion in an en-echelon manner.

Faults related to rift and grabens progressed in downward direction and reached to critical limits at the beginning of Cretaceous period to cause decompression melting, magma generation and widespread effusion of volcanic units. The volcanic units are interlayered with sediments and forms technical basement for deposition of uninterrupted Paleogene and Neogene succession on land, offshore and oceanic regions. The study of Cretaceous volcanism has been very rewarding in innumerable ways. Since volcanic units are sandwiched between the two sedimentary sequences on either side of Cretaceous-Tertiary (KT) boundary, they have been utilized as prominent marker horizons. Their disposition has suggested that successive blocks have faulted down

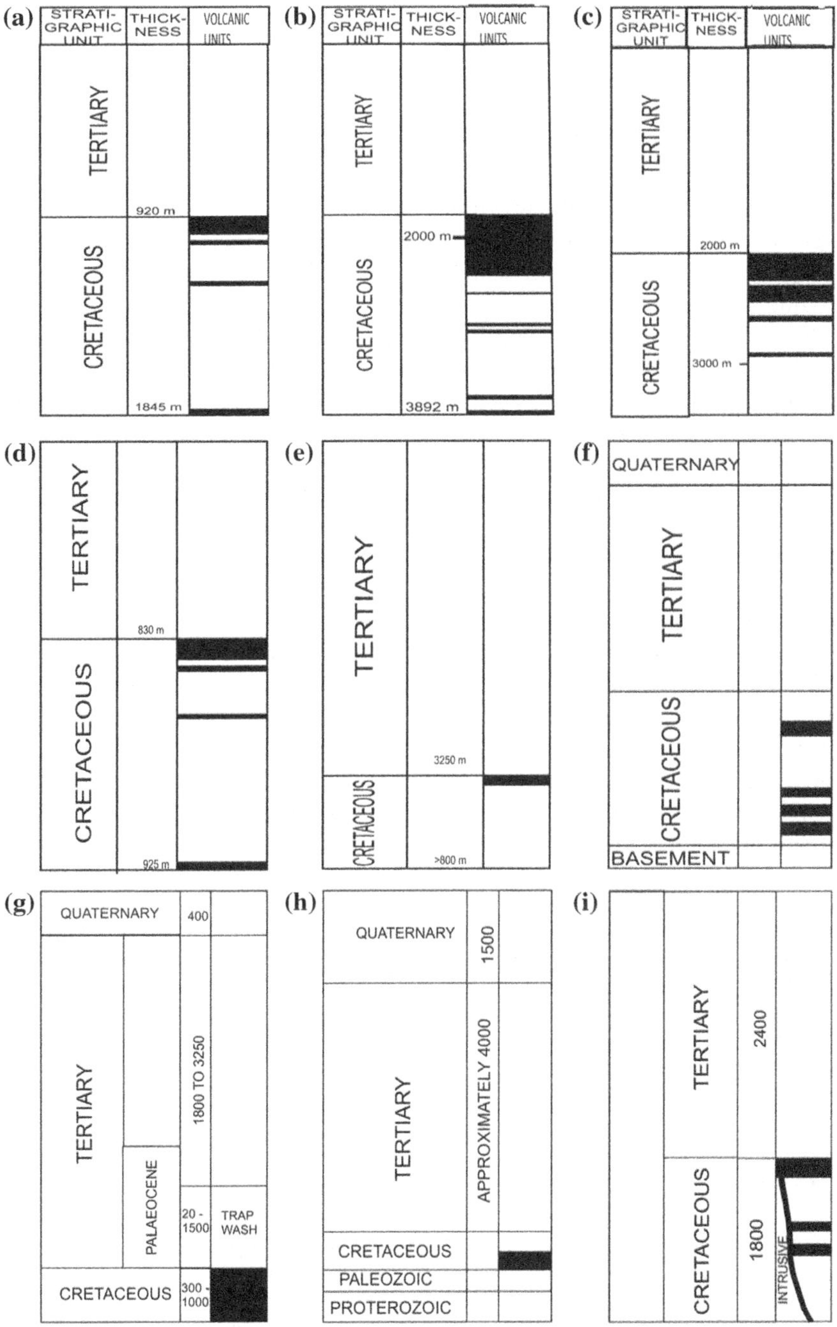

Fig. 13 Generalized well logs of different basins added and modified after Berger et al. (1983), **a** Off-shore Kutch; **b** On-land Kutch; **c** Off-shore Saurashtra; **d** Lakshadweep basin; **e** Krishna-Godavari basin; **f** Off-shore Mahanadi basin; **g** Cambay basin; **h** Bengal basin; **i** Andaman basin

from a depth of 2000 m in Mannar-Kavery, 3250 m in K-G, 4000 m in Mahanadi and 5470 m in Bengal basin.

Study incorporates tectonics as well as interlayered volcanic units in totality. Data sets acquired during hydrocarbon exploration at exponential cost are amalgamated. The volcanic units have posed insurmountable problems in gathering information about underlying Mesozoic and older rocks; because seismic waves normally do not penetrate through the thick basaltic units. Drilling through the basalt is also difficult and costly particularly in offshore and oceanic regions. In this case study we suggest an integrated methodology along with PSDM and PSTM processing technique to get the required information.

The extent and distribution; stratigraphic position, physical volcanology and tectonic disposition of Cretaceous volcanic units is very significant to understand tectonic history, magma generation and effusion. This in turn has explained evolution of older and younger basins, thermal history, and development of seals and preferential accumulation of hydrocarbons within Cretaceous and Tertiary succession.

The study recommends intersectional regions of Barmer-Jaisalmer, Barmer-Cambay and Cambay with Narmada-Tapti Tectonic Zone for detailed study and exploration. Similarly extensions of Narmada-Tapti Tectonic Zone in the form of Surat depression, Kaladgi-Bhima graben in offshore and Narmada-Tapti Tectonic Zone as well as Koel-Damodar Valley graben in Bengal basin, deserve special attention. Shoulder areas of Cambay graben, offshore regions of Kutch and Saurashtra, are also recommended for detailed exploration. Further drilling of wells terminated on reaching volcanic units, is suitably suggested for production of natural gas from volcanic units.

Pools associated with the rifts and grabens display very strong tectonic control and are elongated parallel to their axes. In intersectional areas of these structures, the pools are parallel to super imposed tectonics.

Acknowledgements I will like to profusely thank Dr. Satish Tripathi, Director, Geological Survey of India for invitation to attend the seminar, participate in valuable discussions and finally submit this manuscript for publication in prestigious Springer special volume. Seismic profiles processed by PSDM and PSTM techniques were provided by Ms. Sujata Venketraman, Program Director G X Technologies, Houston, Texas, were of great value in understanding Paleogene basins. Innumerable anonymous well site geologists who have meticulously logged hundreds of drill holes from different basins are thankfully acknowledged.

References

Berger P, Cerny P, Hook JC, Jaffex Kabir CS, Kumar R, Minne JC, Pinningtou DJ, Serra O, Shukia S, Sibbit, Visvanath SN (1983) Petroleum geology of India. Schlumberger well evaluation conference, New Delhi, pp 1–26

Biswas SK (1987) Regional tectonic framework, structure and evolution of the western marginal basins of India. Tectonophysics 135(4):307–327

Jagannathan CR, Ratnam C, Baishya, Das Gupta U (1983) Geology of the shore Mahanadi basin. Pet Asia J (Spl Issue Petroliferous Basins of India):101-104.

Kumar SP (1983) Geology and hydrocarbon prospect of Krishna-Godavari and Cauvery Basin. Pet Asia J 4:57–65
Misra KS (1999) Deccan volcanics in Saurashtra and Kutch, Gujarat, India. Mem Geol Soc India 43:325–333
Misra KS (2002) Arterial system of lava tubes and channels within Deccan volcanics of Western India. J Geol Soc India 59:115–124
Misra KS (2005) Distribution pattern, age and mode of eruption of Deccan and associated volcanics. Gondwana Geol Mag 8:53–60
Misra KS (2007) Tectonic history of major geological structures of peninsular India, development of petroliferous basins and eruption of Deccan and associated volcanics. J Geophys 27(3):3–14
Misra KS (2008a) Dyke swarms and dykes within the Deccan Volcanic Province, India. Indian Dykes, Narosa Publishing House Pvt. Ltd. New Delhi, India, pp 57–72
Misra KS (2008b) Cretaceous volcanic sequences and development of hydrocarbon pools in and around Peninsular India. Petroview 2(3):208–320
Misra KS, Misra A (2010) Tectonic evolution of sedimentary basins and development of hydrocarbon pools along the offshore and oceanic regions of peninsular India. Gondwana Geol Mag 12:165–176
Misra KS, Misra A (2013) Hydrocarbon exploration in sub-basalt basins around peninsular India. Search and Discovery, AAPG Article# 50804
Pascoe EH (1964) The geology of India and Burma (V. III). Government of India Publication
Radhakrishna BP (1989) Suspect tectono-stratigraphic terrane elements in the Indian sub-continent. J Geol Soc India 34:1–24
Raju DSN, Misra R (2009) Proterozoic and Phanerozoic integrated stratigraphy (South-East Asia). ONGC Bull 44
Raju ATR, Srinivasan S (1983) More hydrocarbon form well explored Cambay basin. Pet Asia J 4(4):25–35
Raju ATR, Srinivasan S (1993) Cambay basin-petroleum habitat. In: Proceedings of the second seminar on the petroliferous basins, Dehradun, pp 33–78
Roybarman A (1983) Geology and hydrocarbon prospects of West Bengal. Pet Asia J (Spl Issue) Petroliferous Basin of India, 51–56
Sheth HC, Mathew G, Pande K, Mallick S, Jena B (2004) Cones and craters on Mount Pavagadh, Deccan traps: rootless cones? Proc Indian Acad Sci (Earth Planet Sci) 113(4):831–838
Zutsi PL, Sood A, Mahapatra R, Raman KKV, Divedi AK, Srivastava HC (1993) Lithostratigraphy of petroliferous basins. Document V, Bombay offshore Basin, ONGC Publication, pp 1–358

Foraminiferal Effects of Regional Fire and Attendant Paleoenvironment During K/Pg Transition: Organo-Chemical Evidence from the Um Sohryngkew River Section, Meghalaya, India

Sucharita Pal, J. P. Shrivastava and Sanjay K. Mukhopadhyay

Abstract Palaeoenvironmental studies based on organic matter associated with the biostratigraphically continuous Cretaceous-Paleogene transition section of the Langpar Formation, Meghalaya promise to provide crucial evidence of foraminiferal effects close to K/Pg boundary. Present work primarily focuses on the upper Maastrichtian *Pseudoguembelina hariaensis* Zone (= Zone CF3) that records incidence of 'regional fire' to affect paleoenvironment, which facilitated planktonic foraminiferal disappearance. Incidence of coaly matter towards the upper part of the *Racemiguembelina fructicosa* Zone (= CF4 Zone) endorses that the incidence of 'fire' prevailed for a longer period, encompassing the upper Maastrichtian biozones CF4-CF3. The yellowish brown clay layer in biozone CF3 is marked with well established excursions in HMW PAH compounds which coincide with the Ce anomaly layer occurring below the PGE anomaly layer around the contact between *Pseudoguembelina palpebra* Zone (= CF2 Zone) and *Plummerita hantkeninoides* Zone (= CF1 Zone) and below the planktonic foraminiferal change at the K/Pg. Significantly, strong (~4 fold increase) peaks were noticed in biozone CF3 (sample JP12) as marked by sudden increase in the total PAHs—4, 5, 6 ring PAH compounds and few 3 (anthracene, fluorine) ring PAH compounds; biozone CF2 (sample JP-13) also shows excursions in 3 ring PAHs—phenanthrene, 3-methylphenanthrene, 2-methylphenanthrene, 9-methylphenanthrene, 1-methylphenanthrene, but the former shows *n*-alkane and fatty acid (including their CPI values) excursions. Besides, remarkably high amount of combustion marker PAHs [fluoranthene, (10.46 μg/gc), pyrene (7.20 μg/gc), chrysene (8.28 μg/gc), benzo(a)anthracene (9.92 μg/gc)] in the biozone CF3, similar to those of well studied K/Pg boundary sections of Stevns Klint (Denmark), Gubbio (Italy), Woodside Creek (New Zealand), and Arroyo el Mimbral, Tamaulipas, (Mexico) suggesting correspondence in the incidence of 'fire' in India where it was triggered presumably by the heat radiating from the epigenic plumes of Abor vol-

S. Pal · J. P. Shrivastava (✉)
Department of Geology, University of Delhi, Delhi 110007, India
e-mail: jpshrivastava.du@gmail.com

S. K. Mukhopadhyay
Plot 2, Nabaroon Co-Operative Housing Society, Santoshree
Palli, Thakurpukur 700063, Kolkata, India

S. Bajpai et al. (eds.), *The Indian Paleogene*, Society of Earth Scientists Series,
https://doi.org/10.1007/978-3-319-77443-5_4

canic and Ninetyeast Ridge, and greenhouse effects of Deccan volcanism. The fire seemingly facilitated step-wise disappearances of planktonic foraminifera during biozones CF4-CF3 and instigated migration of some forms from the warm water environment. A very strong proton peak at δ1.51 ppm in yellowish brown clay layer (JP-12) indicating the presence of CH_2 group of alkane compound in the sample. FTIR spectra shows weak bands in the region of 2304–2370 cm^{-1} and attributed to the presence of traces of CO_2 and CO, which are related to the combustion incidence. Such environment was possibly created in the interspaces of the sediments by burning of organic matter in the presence of oxygen whereby CO_2 formed and water released in the environment. High abundance of combustion derived PAH in the yellowish brown layer of biozone CF3 of the succession endorses this observation. The study considers 'regional fire' as one of the factors for step-wise disappearance of planktonic foraminifera prior to the advent of the K/Pg boundary in the shallow shelf of Meghalaya.

Keywords Organic compounds · Biozone CF3
K/Pg transition · Palaeoenvironment · Langpar Formation · Meghalaya

Introduction

Fine resolution biostratigraphy based on planktonic foraminifera revealed that the Um-Sohryngkew river section (Fig. 1) is continuous across the Cretaceous-Paleogene (K/Pg) boundary (Mukhopadhyay 2008); thus, the section is significant to study paleoenvironment that existed during upper Maastrichtian. Shrivastava et al. (2013) based on TOC contents, together with δCe and δEu anomalies, assigned suboxic/anoxic—suboxic—suboxic/anoxic—oxic conditions for bottom seawater sediments spanning the biozones CF4 (= *Racemiguembelina fructicosa* Zone) to Pla (*Parasubbotina pseudobulloides* Subzone) across the K/Pg boundary. Pal et al. (2015c) reported significant increase in the kaolinite contents as well as layer and inter-layer charges in the illites occurring in the yellowish brown clay layer which is comparable with the clays associated with the well established K/Pg boundary layers of Agost, Caravaca, Petriccio and El-Kef sections; they (Pal et al. 2015c) recorded sudden rise in temperature (>140 °C) in the upper part of the biozone CF3 (sample JP-12) which is analogous to the attributes of the clays associated with the K/Pg boundary layer of the Caravaca section, Spain. This implies high temperature prevailed during the formation of the yellowish brown clay layer. They also observed wide variation in climate from humid tropical to arid-semiarid, possibly due to heat radiation from the shallow plumes of Abor and Ninetyeast Ridge magma and greenhouse effects of Deccan volcanism. Some of these environmental changes are coeval with different eruptive phases of Deccan volcanism and agree with the planktonic foraminiferal extinction events (cf. Mukhopadhyay 2008), Ir anomaly incidence of Bhandari et al. (1987) and Au, Pt and Pd anomalies occurring near the K/Pg boundary (Mukhopadhyay 2009). It is not well known whether the Deccan volcanism, a global thermal event directly interfered in the late Maastrichtian-early Paleocene sedimentation of Meghalaya. Organic geochemical studies on the K/Pg transition sediments recorded evidence of 'wildfires' (Wolbach et al. 1985, 1988; Venkatesan and Dahl 1989; Heymann et al. 1994a, b, 1998; Mita and Shimoyama 1999; Arinobu et al. 1999) whereas Wolbach et al. (1985, 1988), Venkatesan and Dahl (1989), Heymann et al. (1994a, b, 1996) linked the 'wild fire' with bolides (extraterrestrial) impact, a hypothesis which has a competitor in Deccan volcanism to explain the catastrophic events during the K/Pg transition. Although a direct influence of extraterrestrial bolide impact on the K/Pg boundary sediments of Therriaghat is yet to find, evidence of increased temperature in the succession needs proper explanation.

Yamamoto et al. (1996) discussed that the biomarker distribution study helps to identify the organism contributing to sedimentary organic matter and regarded it as another complementary palaeontological approach. Previous organic geochemical studies on K/Pg boundary sediments have found evidence of wildfires (Wolbach et al. 1985, 1988; Venkatesan and Dahl 1989; Heymann et al. 1994a, b, 1998; Mita and Shimoyama 1999; Arinobu et al. 1999). Other organic geochemical studies (e.g. Meyers and Simoneit 1989) examined microbial hydrocarbons, fatty acids and terrestrial resin acids in the sediments above and below the claystone at Stevns Klint, Denmark. Formation of organic macromolecular structure is complex; therefore, recognition as well as variation in the distribution of organic molecules if recorded systematically across the succession would unveil valuable information about depth, temperature of burial, oxidation state and other palaeoenvironmental variables accountable for their formation. Such studies being significant in this context need to be carried out in this area. Since the yellowish brown clay layer (sample JP-12) of the biozone CF3 containing abnormally high amount of combustion-derived PAH compounds (Pal et al. 2015a, b) and Ce anomaly layer (Shrivastava et al. 2013) occurring below the bed having PGE anomaly layer in biozone CF2 (Mukhopadhyay 2009) as well as the bed recording planktonic foraminiferal change of the K/Pg boundary (Mukhopadhyay 2008), gains impetus to be focused along with its overlying beds (samples JP-13, JP-14). This study while recording systematic variation in the distribution of organic molecules associated with kaolinite-illite complexes in the late Cretaceous-early Paleocene marine succession of the Um-Sohryngkew river, Meghalaya and 'incidence of fire' in the upper Maastrichtian part, aims at comprehending the palaeoenvironmental reasons for the derivation of the molecules from the original organic source matter, and finding out the cause of the fire and its effect on upper Maastrichtian planktonic foraminifera.

Materials and Methods

Systematic samples (JP1-JP16) were collected from about 10 m. thick measured litholog (Fig. 1) spanning late Cretaceous-early Paleocene across the K/Pg boundary and occurring towards the middle part of the Langpar Formation. The samples covered all the biozones from Zone CF4 to Subzone Pla in ascending order. The

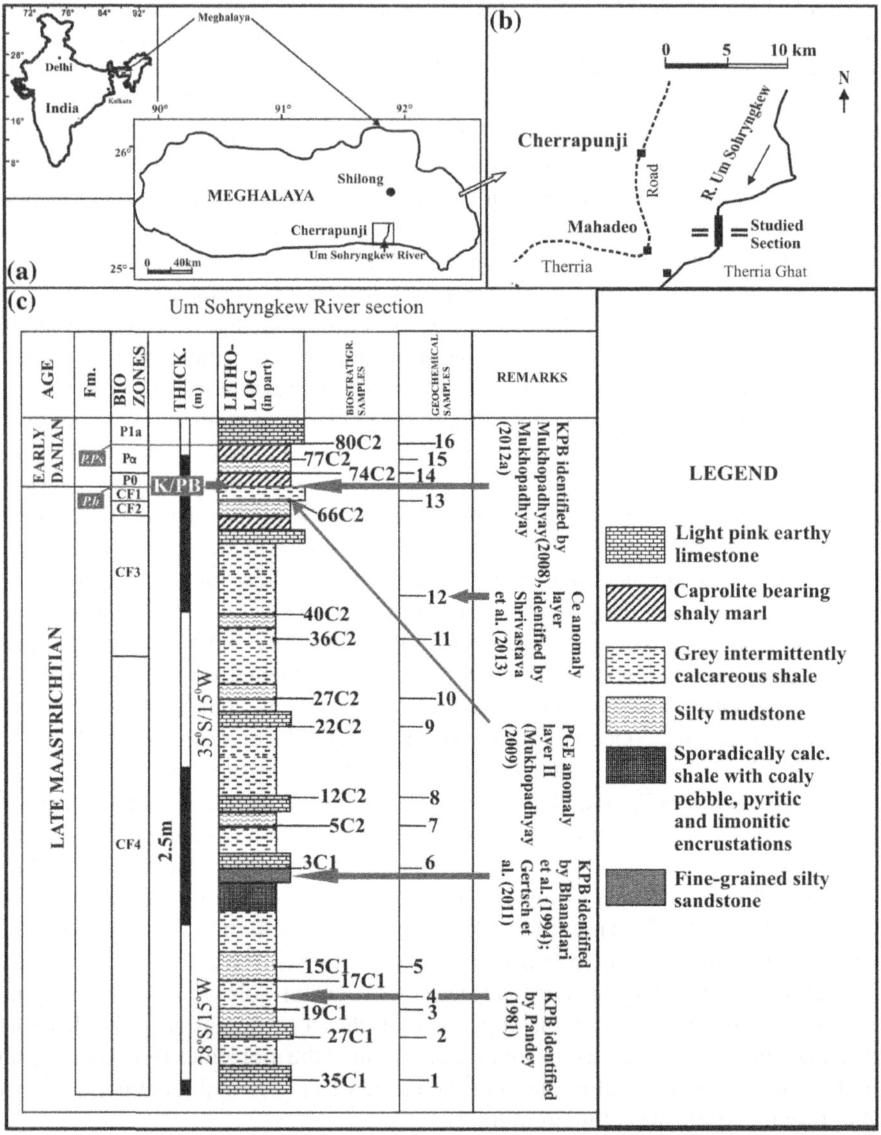

Fig. 1 Lithostratigraphy (modified after Mukhopadhyay 2008) and sample locations on the Um-Sohryngkew river section. *Insets* **a** location of Meghalaya with respect to India, **b** location of the Therriaghat section around the river Um-Sohryngkew and **c** lithological log at the K/Pg interval in the Um-Sohryngkew river section. *Note* Samples 1–16 collected from the section is referred to JP1-JP16 in the text

samples were collected from 25 cm groove; for friable sediments hollow steel pipes (3.5 cm diameter with one end pointed) were used as an auger to avoid contaminations. Fraction of a sample weighing 50–70 gm was processed in the laboratory initially by breaking into small pieces (2–4 mm) with a steel hammer. Interestingly, it has been noticed that the primary colour of sediments changes from grey to light pink as ascending in the succession.

Organic matters were isolated by following combination of mechanical and chemical procedures (cited Fig. 2; Pal et al. 2015a). Approximately, 20 gm sample chips were cleaned twice by ultrasonic agitation with solvent mixture of benzene and methanol (6:4) for 5 min and allowed to freeze dry to dryness, finally powdered in the agate mortar pestle to pass through 200 mesh. Samples (15 gm) were extracted three times with a solvent mixture of methanol and dichloromethane (1:1) by a solvent extractor with ~80% recovery. After removing elemental S (by adding Cu foil and sonicating for 1 h), 15 ml of (10%) KOH-methanol solution and 2 ml water were added to the 10 ml extract. The extracts were fractionated into neutral (*NF*) and acidic (*AF*) compounds (by liquid-liquid separation) by dissolving in *n*-hexane + diethyl ether (9:1) and *n*-hexane + dichloromethane (9:1) mixtures, respectively. Neutral compounds were fractionated into aliphatic hydrocarbons, PAHs, and other group of compounds using silica-gel (heated at 200 °C, 1% deactivated) column [borosilicate glass column size—100 cm (length) and 0.8 mm (internal diameter)] chromatography. After the methylation [15 ml of MeOH: HCl (95:5), mixture was added to *AF*. After heating gently in the water bath at 70 °C for 12 h, acid will form methyl ester or diazonium salt (according to derivatizing agent). Boron triflouride forms a complex with methanol and it reacts with an acid and active hydrogen from acid (–OH) is replaced by methyl group, forming a methyl ester. For extraction of fatty acid, 2 mg methyl ester standard in 100 cc dichloromethane, 0.5 ml was poured by syringe in *AF*.], FAMES (fatty acid methyl ester) were extracted from acidic compounds with a *n*-hexane and diethyl ether solvent mixture (9:1). Acidic compounds were fractionated into fatty acids and remaining parts of the aliphatic hydrocarbons and PAHs by silica-gel column chromatography. The characterization of these compounds was made using Fourier Transmission—infrared (FTIR) and nuclear magnetic resonance (NMR) spectroscopic techniques. FTIR spectra were recorded on Nicolet MAGNA—FTIR 750 Spectrometer Series II. ^{1}H and ^{13}C NMR spectra were re-corded on by using a JEOL JNM ECX-400 P 400 MHz instrument, and the data were processed by using the JEOL DELTA program version 4.3.6.

Results

TOC data show wide variation (0.2777–0.9786 wt%) from Biozones CF4 to Pla, but a spike is observed in the lower level of the biozone CF3. The carbon number distribution of *n*-alkanes from biozone CF4 to Pla mainly ranges from C_{11}–C_{36}. The organic compounds such as pristane and phytane are present in significant amount. Concentration (Appendix 1; Pal et al. 2015b) of total *n*-alkane varies between 1.027

and 14.15 μg/gc. Amongst these, long (C_{27}–C_{33}), mid (C_{20}–C_{25}) and short (C_{14}–C_{22}) chain *n*-alkane concentrations ranging from 0.111 to 0.869, 0.163 to 1.722 and 1.051 to 8.622 μg/gc, respectively. The carbon number for *n*-fatty acids is ranging from C_6–C_{32} and their concentration varies between 1.22 and 7.33 μg/gc in which odd carbon numbered *n*-fatty acids occur predominantly (1.25–5.90) over even carbon numbered fatty acids. Out of these, long (C_{20}–C_{30}) and short (C_{14}–C_{24}) chain *n*-fatty acid concentrations ranging from 0.00 to 0.20 and 0.97 to 5.76 μg/gc with CPI value of 0.86–24.75 and 15.54–37.24, respectively (Appendix 2; Pal et al. 2015b). The GC-MS spectra (cited Fig. 8; Pal et al. 2015b) show presence of saturated (from C_6 to C_{32}) and unsaturated (C_{14}–C_{18},) *n*-fatty acids. The short chain *n*-fatty acids are abundantly present in all the samples. The prominent peak of C_{16} that corresponds to the palmitic acid ($C_{16:0}$) is present in all the samples. The even carbon numbered compounds (from C_{12}–C_{20} with CPI values 15.49–37.09) are predominant throughout the succession. The long chain fatty acids (beyond C_{20}) are either absent or occur in minor amount (0.01–0.27) in few samples (cited Fig. 8; Pal et al. 2015b). A large number of non-volatile PAH compounds have been found in this section (Appendix 3; Pal et al. 2015b). The total PAH concentration varies between 1.296 and 109.53 μg/gc. The HMW PAH compounds predominate in the sample JP-12 of biozone CF3. The LMW PAH compounds show dominance in the biozone CF1 (sample JP-13) whilst, the HMW PAH compounds predominate in biozone CF3 (sample JP-12). The high PAH values are indicative of soot particles associated with the sediments coming from atmospheric transportations. The phenanthrene and chrysene show their maximum concentrations of 8.28 μg/g TOC and 5.72 μg/g TOC in the biozone CF3 (JP-12) and in the biozone CF1 (JP-13), formed during diagenesis as biological precursors (Killops and Massoud 1992). Most of the samples are of pyrolytic origin.

FTIR spectra (Fig. 2) from the yellowish brown clay layer (JP-12), greenish yellow layer (JP-13) and shaly marlite layer (JP-14) from biozones CF3, between CF2 and CF1, and P0 (JP-12, 13 and 14) exhibit almost similar absorption bands, but, show little difference in their relative intensities. The major difference between these three spectra deals with the relative intensity of the bands corresponding to 2920, attributed to asymmetrical CH_2 stretching which is relatively intense in the biozone CF3 (JP-12). Both the methyl (CH_3–) and methylene (–CH_2–) legends have their specific deformation vibrations which appear in the region of 1500–1300 cm^{-1}. A vibration band found in the FTIR spectra of second derivative in the JP-13 is assigned to the band around 1450 cm^1 to the symmetric deformation (scissor) of CH_2 groups. The absorption bands of organic compounds recorded show- stretching of aliphatic C–H groups in the region from 3000 to 2800 cm^{-1}, aliphatic C–H stretching region (3000–2800 cm^{-1}) resolved into three spectral bands at 2950 (asymmetrical CH_3 stretching), 2920 (asymmetrical CH_2 stretching) and 2850 cm^{-1} (symmetrical CH_2 stretching). The FTIR spectra with increasing maturity exhibit a decrease of absorption in the aliphatic regions at 1375, 1450, and 2800–3000 cm^{-1} (Lis et al. 2005). The broad band between 3700 and 3000 cm^{-1} indicates a wide range of hydrogen bond lengths and orientations, mostly corresponding to the OH and NH groups. The presence of CH_2 and CH_3 alkyl groups is identified by a sharp stretching absorption around 2922 cm^{-1} and a weaker one around 2852 cm^{-1}. The aliphatic C–H

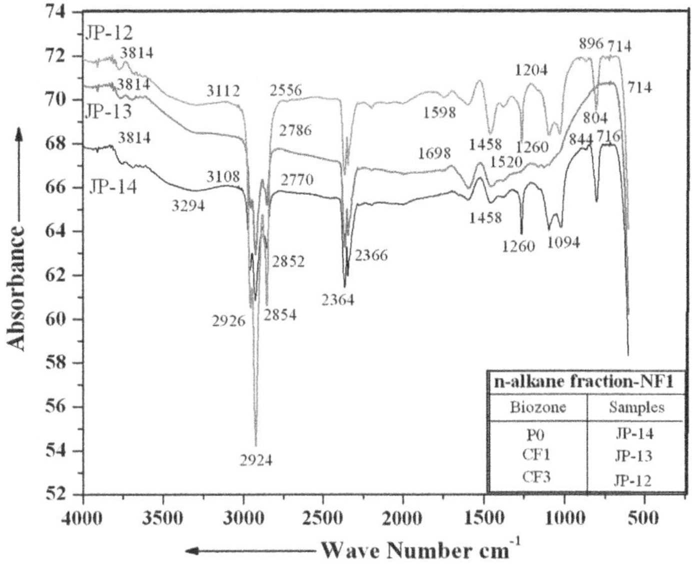

Fig. 2 FT-IR spectra of *n*-alkane fractions (NF1) as obtained from JP-12, JP-13 and JP-14 samples, representing biozones CF3, CF1 and P0, respectively

deformation vibrations of the methyl and methylene groups are possibly contributing to the bands around 1430 cm^{-1}. Significant differences observed between these three spectra of the Um-Sohryngkew river section are mostly linked to the variations in the relative contributions from the three major regions of the spectrum, namely, (i) decrease in the O–H/N–H stretching band intensity between 3000 and 3700 cm^{-1} and (ii) increase in the C–H stretching band intensity between 2800 and 2950 cm^{-1} (Table 1). The mid-infrared spectra, which is dominated by emission features at 3030, 1610, 1310, 1160, 890 cm^{-1}, commonly known as the aromatic infrared bands. The Polycyclic Aromatic Hydrocarbon compounds (PAHs) are widely accepted to be the carriers of these mid-IR emission bands (Allamandola et al. 1999). The stretching of aromatic C–H groups in the region from 3100 to 3000 cm^{-1}, bending of oxygenated and aromatic groups in the region from 1550 to 1750 cm^{-1}, bending modes of CH_2 and CH_3 groups at 1450 cm^{-1}, and aromatic out-of—plane C–H deformation region from 700 to 900 cm^{-1} and increasing absorption in the aromatic regions (3000–3100 cm^{-1}, 1600 cm^{-1} and 700–900 cm^{-1}) observed in samples. In the present samples, the low thermal evolution of organic matter is confirmed by the absence of peaks in the three aromatic regions (Fig. 3). This explanation agrees with the GC-MS data that put in the additional evidence for the general immaturity of the organic remains (Pal et al. 2015a). There are several peaks with medium to weak intensities present in all the spectra (Fig. 3) and corresponding values are presented in Table 2.

PAH compounds separated from sample JP-12, JP-13 and JP-14 was also analyzed by ^{13}C neutron magnetic resonance and ^{1}H neutron magnetic resonance techniques.

Table 1 Frequency ranges and absorption band assignments for *n*-alkane (*NF1*) fractions obtained from JP-12, JP-13 and JP-14 samples representing biozones CF3, CF1 and P0

Frequency ranges (cm^{-1})	Band assignments	Observed frequencies in samples			Comments
		JP-12	JP-13	JP-14	
2935–2915	Methylene C–H asymetrical streching	2924	2924	2926	Asymmetrical CH_2 streching
2865–2845	Methylene C–H band	2854	2852	2852	Symmetrical CH_2 stretching
1450 and 1375	CH_3 band		1450		
1485–1445	Methylene (CH_2) band	1458		1458	
1300–1000	In plane C–H bending vibration	1030, 1204, 1260		1094, 1210, 1260	
900–690	Out-of-plane C–H bending vibration	804, 896		844	

Table 2 Frequency ranges and absorption band assignments for PAH (*NF2*) fractions obtained from JP-12, JP-13 and JP-14 samples representing biozones CF3, CF1 and P0

Frequency (cm^{-1})	Band assignments	Observed frequencies in samples			Comments
		JP-12	JP-13	JP-14	
3100–3000 (m)	C–H stretch		3008	3008	C–H group
~3030 (v)	Aromatic C–H strech		3030	3030	
3050–3010	= C–H stretch				
860–680 (s)	Aromatic C–H bending				
1700–1500 (m, m)	Aromatic C=C bending	1594	1538, 1600	1548, 1600	Aromatic group
1600, 1475	C=C stretching		1470	1474	CH_2, CH_3 group
900–690	= C–H bending	914	728, 916	728, 916	C–H deformation

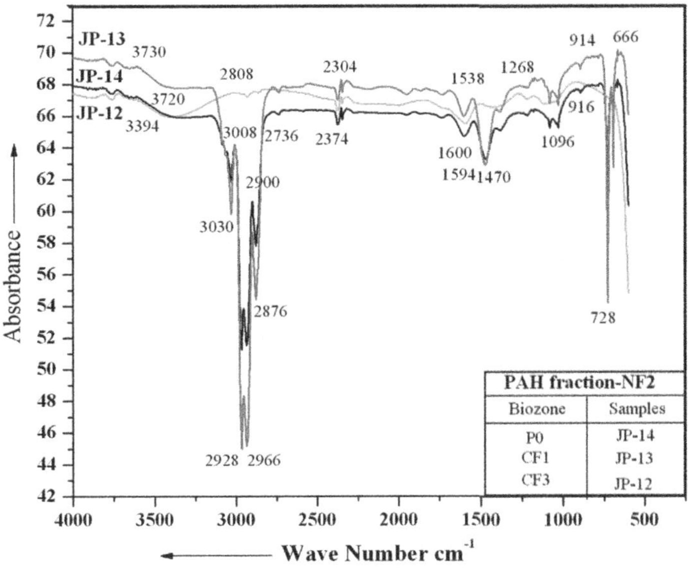

Fig. 3 FT-IR spectra of PAH fractions (NF2) as obtained from JP-12, JP-13 and JP-14 samples representing biozones CF3, CF1 and P0, respectively

In yellowish brown clay layer (JP-12), one very strong proton peak at δ1.51 ppm (Fig. 4) was observed; indicate presence of CH_2 group of alkane compound in the sample (Silverstein et al. 2005). In sample JP-13, two isomers of methylphenanthrene found at δ2 and 1.8 ppm. The proton peak at δ6.41, 7–7.24 ppm showed presence of aromatic group. Proton peak noticed at δ8 ppm shows presence of anthracene. In sample JP-14, one proton peak at δ1.51 ppm indicates presence of aliphatic compound in the sample. Sample JP-12 shows peak at δ40 ppm, indicates presence of CH_2 group, supportive of the proton NMR data.

Discussion

Study of ^{1}H and ^{13}C NMR spectra shows presence of $–CH_2–$ alkyl group attached to *n*-alkanes in biozones CF3 and P0 (samples JP-12 and JP-14), but do not show presence of aromatic compounds, however, significant presence of low molecular weight aromatic 3 ringed methylphenanthrene compound is noticed in sample JP-13 corresponding to biozones CF2-CF1. These compounds were derived from the complex organic molecules, formed as a result of sudden increase in the temperature, but at low-pressure. This anomaly with respect to biozone CF2 (sample JP-13) corresponds to 65.45–65.3 Ma which spans the most vigorous second phase of Deccan volcanic eruption and coeval with the development of the brownish yellow layer

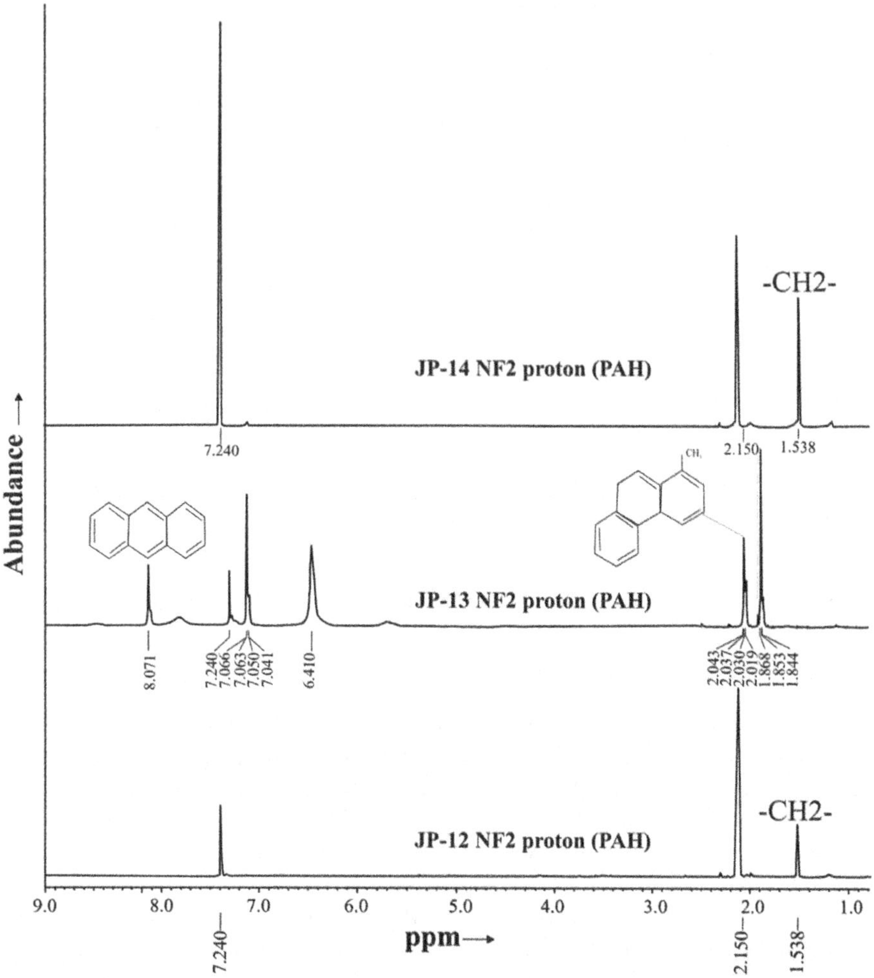

Fig. 4 ^{1}H NMR spectra of PAH fractions (NF_2) as obtained from JP-12, JP-13 and JP-14 samples representing biozones CF3, CF1 and P0, respectively

of the biozone CF2 ranging in age from 65.4 to 65.2 Ma (Barrera 1994; Courtillot et al. 1996; Hoffman et al. 2000; Keller 2001) and marked by the re-appearance of *Pseudotextularia elegans.* The two peaks at δ20 and 40 ppm in the 13C NMR spectrum (Fig. 5) show presence of CH_3 and CH_2 groups (Von Wehrli and Wirthlin 1976). Presence of aromatic group in sample JP-13 is supported by its peaks at δ120 and 130 ppm (Fig. 5). One peak observed at δ170 ppm in sample JP-13 shows presence of carbonyl (ester, acid) groups (Ludemann and Nimz 1974; Wilson 1987; Knicker 1993). It is possible that the aliphatic organic compounds present in clay inter-layers had most likely been transported there by pulveric acid or other humic materials

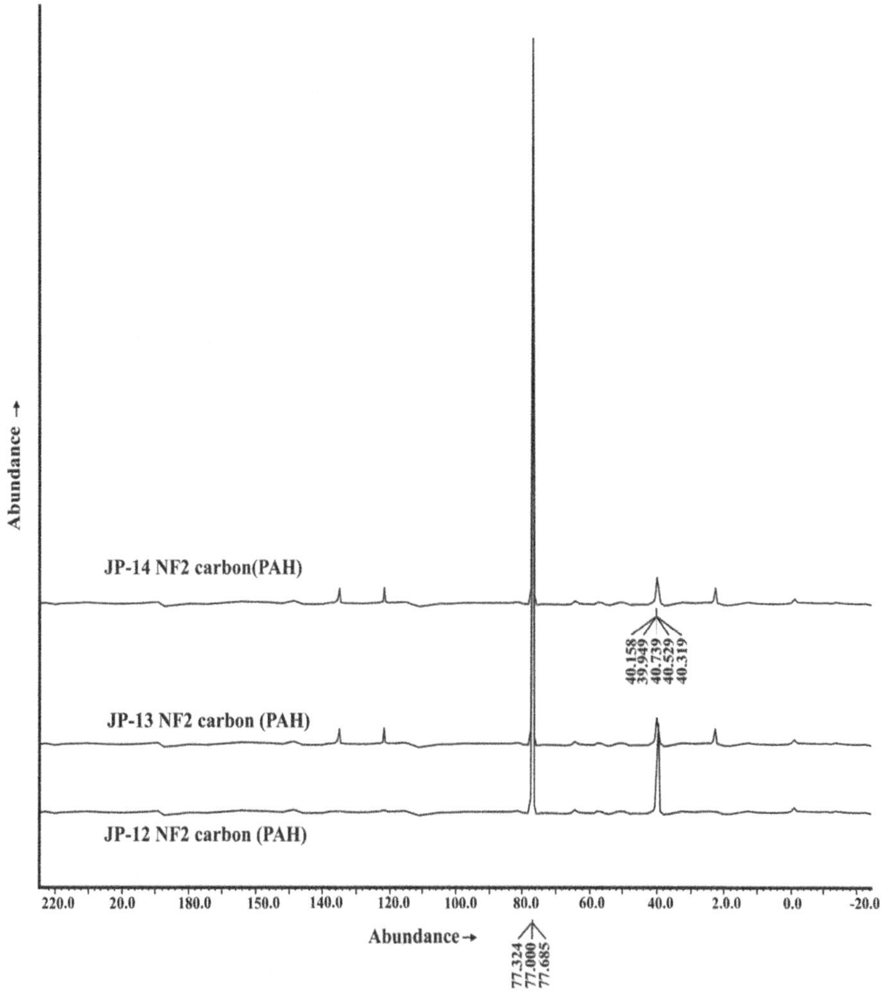

Fig. 5 ^{13}C NMR spectra of PAH fractions (NF_2) as obtained from JP-12, JP-13 and JP-14 samples, representing biozones CF3, CF1 and P0, respectively

within which they were captured and acted as vehicles for the transport of hydrophobic compounds within the sediments.

Weak bands in the region of 2304–2370 cm^{-1} are ascribed to the presence of traces of CO_2 and CO which is related to combustion environment. Such environment was possibly created in the interspaces of the sediments by burning of organic matter in the presence of oxygen, thus, formed the CO_2 and water in the environment. This observation is also supported by high abundance of combustion derived PAH in yellowish brown layer of the Biozone CF3 of the succession, which is comparable with the well studied K/Pg boundary sections of Stevns Klint (Denmark), Gubbio

(Italy), Woodside Creek (New Zealand), and Arroyo el Mimbral, Tamaulipas, (Mexico) (cited in Fig. 6, Pal et al. 2015a). Their pyrolitic signature suggests prevalence of 'global fire' that might have been triggered by the radiating heat from the shallow plumes of the Abor volcanic and Ninetyeast Ridge, and the greenhouse effects of Deccan volcanism (Table 3). This regional incidence might have caused distress to biota during CF3. The association of Ir anomalies with the basal coal layer at the K/Pg boundary site of Raton in Colorado (Tschudy et al. 1984), soot layers at widely separated geographical boundary sites (Wolbach et al. 1985, 1988, 1990a, b; Heymann et al. 1998) and spike in the polycyclic aromatic hydrocarbons at the K/Pg boundary, in Caravaca, Spain (Arinobu et al. 1999), provided evidence of 'global fires' during the K/Pg transition. In Um-Sohryngkew river section, the occurrence of coaly matter towards the upper part of the biozone CF4 endorses that the incidences of regional fire, prevailed for a longer period, encompassing the upper Maastrichtian biozones CF4-CF3. The 'fire' facilitated step-wise disappearances of planktonic foraminifera during biozones CF4-CF3 and instigated some forms to migrate out of the warm water environment (Table 3).

Conclusion

Our study reveals that the Meghalaya area experienced very warm climate during most of the upper Maastrichtian. Organo-geochemical results indicate prominent (positive as well as negative) spikes in the *n*-alkanes (C14−22), *n*-fatty acids (C14−18) and high molecular PAH compound concentrations across the Um-Sohryngkew River succession. Presence of anomalously high concentration of HMW PAH compounds in the late Maastrichtian biozone CF3 and based on pyrolitic signatures the cause is considered as regional terrestrial fire. Their complex molecular structures show presence of traces of CO_2 and CO which is related to combustion environment, possibly created in the interspaces of the sediments by burning of organic matter in the presence of oxygen, thus, formed the CO_2 and water in the environment. Among many factors that caused to warming of the climate, regional fire is presumed to be one of the important factors. Its evidence cause and effects are summarized.

a. **Evidence of fire**: Patchy coal in the Mahadeo Formation (Mukhopadhyay 2010); plant remains, streaks of coal, and carbonaceous shale in biozone CF4 of Langpar; pyrolitic organic compounds in biozone CF3.
b. **Evidence of warming of climate**: Gradual increase of calc sediments and limestone content higher up in the upper Maastrichtian succession; sudden disappearance of some species from biozone CF3 (*Pseudotextularia elegans*), change of body plan in some forms leading to the development of new species as *Guembelitria irregularis* near the interface of biozones CF2-CF1 and *Guembelitria langparensis* during CF4 (both are stress-friendly forms); development of morphogroups in some commonly occurring species; sudden increase in the predated

Table 3 Calibration of various K/Pg events with planktonic foraminiferal succession in the Theriaghat Section, Meghalaya

Biozones	Lithic Events/ elemental anomaly	Elemental anomaly	Planktonic foraminiferal composition	Planktonic foraminiferal events	Clay mineral succession	Geochem. anomaly	events signifying increasing temperature	sea event
Parvularugoglobigerina eugubina Zone (= Zone Pα; 3.43Ma)	Thick bedded limestone;		Of 22 species, 9 have FA & 3 have LA; **moderate population**	↓ *P.eugubina* ↑ *P.eugubina*				←
Guembelitria cretacea Zone (= Zone P0; 0.036Ma)	reworked microspherules & phosphosilicate nodules		Of 15 species, 6 have FA & 6 have LA; **poor population**	Irregular *Guembelitria* spp.	Kaolinite & montmorillonite dominated	↑	Greenhouse effects of 3rd phase of Deccan eruption	→
Plummerita hantkeninoides Zone (= Zone CF1; 65.30 – 65.00 Ma)	Yellowish silty mudstone		Of 22 species, 10 have FA & 17 have LA; **moderate population**	↓*P.hantkeninoides* ↑*P.hantkeninoides*		O_2, Al,& Si increase; Na decrease	Thermal radiation from epizonal plume of Abor & 90° E Ridge	←
Pseudoguembelina palpebra Zone (= Zone CF2; 65.45 – 65.30 Ma)	Thin band of Yellowish brown claystone	Au, Pt and Pd anomaly;	Of 12 species, 5 have FA & 6 have LA; **poor population**	Enormous microspherules & phosphosilicate nodules	illite, kaolinite and abundant montmorillonite		Greenhouse effects of 2nd phase of Deccan eruption;	←
Pseudoguembelina hariaensis Zone (= Zone CF3; (66.83 – 65.45Ma)	Thin bedded limestone		Of 9 species, 4 have FA & 3 have LA; **poor population**	↓ *G. gansseri* ↑ *P.hariaensis*		PAH, Ce anomaly	Forest fire *P. elegans* absent; dwarfed & predated forms	→
Racemiguembelina fructicosa Zone (= Zone CF4; 68.33 – 66.83 Ma)	↑ Limestone (lens) millimeter thin limonitic bands Coal streaks carb. Shale	Ir anomaly	Of 30 species, 23 have FA & 3 have LA; **rich population**	↑*Guembelitria langparensis* ↓ammonite ↑*A.mayaroensis* & *R. fructicosa*	illite, illite/ smectite dominant		Greenhouse effects of 1st phase of Deccan eruption	→ ←

Data compiled from Bhandari et al. (1987), Mukhopadhyay (2008, 2009, 2010, 2012a, b, 2013), Shrivastava et al. (2013), Pal et al. (2015a, b, c)

LA last appearance, *FA* First appearance; ↑ appearance; ↓ disappearance; ←/→ Transgression/regression

forms; overall decrease in planktonic foraminiferal contents during biozones CF2-CF1; step-wise disappearance of some species during CF4-CF3; repeated incidents of transgressions and sea level rise during CF3-CF4; unusually high temperature of the formation of some clay minerals in the marine sediments.

c. **Source of heat to increase climatic temperature**: Warming of climate was accentuated by contemporaneous extra-basinal factors including heat radiation from shallow plumes of Abor and Nintyeast Ridge, greenhouse effects of Deccan volcanics, regional forest fire, plate collision and subduction allowing subsurface heat to flow out, outpouring of lavas, showering of enormous microspherule grains and phosphosilicate nodules (Mukhopadhyay 2012b); penetration of sun-rays hindered as shelf water held enormous clay particles in suspension.

d. **Source of flame**: possibly friction of forest woods and friction of rocks during tremor.
e. **Effects**: Biotic effects of very warm climate include disappearance of some forms that re-appeared later during congenial climate; migration of some forms; abrupt increase in the *Guembelitria* population (a stress-friendly taxon), disappearance of ammonite from the basin during biozone CF4 (Mukhopadhyay 2008); inhomogeneous concentration of planktonic foraminiferal population in sediment succession.

Acknowledgements Fieldwork and a part of the laboratory work of this study were carried out under the aegis of IGCP-Project 507; SKM and JPS express sincere thanks to the Chairman INC for IGCP and Director General, Geological Survey of India for funding and logistics. SP and JPS acknowledge CSIR, New Delhi for financial support [Project Grant No. 24 (0315)/11/EMR-II] and Aninda Mazumdar and B. G. Naik (NIO, Goa) for GC-MS analysis of samples.

References

Allamandola LJ, Hudgins DM, Sandford SA (1999) Astrophys J 511:115

Arinobu T, Ishiwatari R, Kaiho K, Lamolda MA (1999) Spike of pyrosynthetic polycyclic hydrocarbons associated with an abrupt decrease in δ_{13}C of a terrestrial biomarker at the Cretaceous-Tertiary boundary at Caravaca, Spain. Geology 27:723–726

Barrera E (1994) Global environmental changes preceding the Cretaceous-Tertiary boundary: early-upper Maastrichtian transition. Geology 22:877–880 (Society of America 106:1254–1266)

Bhandari M, Shukla PM, Pandey J (1987) Iridium enrichment at Cretaceous-Tertiary boundary in Meghalaya. Curr Sci 56:1003–1005

Courtillot V, Jaeger JJ, Yang Z, Feraud G, Hoffman C (1996) The influence of continental flood basalts on mass extinction: where do we stand? Geol Soc Am Spec Pap 307:513–526

Heymann D, Chibante LPF, Brooks PR, Wolbach WS, Smalley RE (1994a) Fullerenes in the K/T boundary layer. Science 265:645–647

Heymann D, Wolbach WS, Chibante LPF, Brooks PR, Smalley RE (1994b) Search for extractable fullerenes in clays from the Cretaceous/Tertiary boundary of the Woodside creek and Flaxbourne river sites, New Zealand. Geochim Cosmochim Acta 58:3531–3534

Heymann D, Chibante LPF, BrooksPR, Wolbach WS, Smit J, Korochantsev A, Nazarov MA, Smalley RE (1996) Fullerenes of possible wildfire origin in Cretaceous-Tertiary boundary sediments. In: Ryder G, Fastovsk D, Garter S (eds) The Cretaceous-Tertiary event and other catastrophes in earth history. Geological Society America Special Paper, vol 307, pp 455–464

Heymann D, Yancey TE, Wolbach WS, Thiemens MH, Johnson EA, Roach D, Moecker S (1998) Geochemical markers of the Cretaceous-Tertiary boundary event at Brazos River, Texas, USA. Geochim Cosmochim Acta 62:173–181

Hoffman C, Feraud G, Courtillot V (2000) 40Ar/39Ar dating of mineral separates and whole rock from the western Ghat lava pile: further constraints on duration and age of Deccan Traps. Earth Planet Sci Lett 180:13–27

Keller G (2001) The end-Cretaceous mass extinction in the marine realm: year 2000 assessment. Planet Space Sci 49:817–830

Killops SD, Massoud MS (1992) Polycyclic aromatic hydrocarbons of pyrolytic origin in ancient sediments: evidence for Jurassic vegetation fires. Org Geochem 18:1–7

Knicker H (1993) Quantative 15°N und 13C-CPMAS-Festkor-per und 15°N Flussigkeits-NMR spectroscopic and pflanzenk-omposten und naturlichen Boden. Dissertation University of Rogensburg, Germany

Lis GP, Mastalerz M, Schimmelmann A, Lewan M, Stankiewicz BA (2005) FTIR absorption indices for thermal maturity in comparison with vitrinite reflectance R_0 in type-II kerogens from Devonian black shales. Org Geochem 36:1533–1552

Ludemann HD, Nimz H (1974) 13C Kernresonanzspektren von ligninen. 2. Buchen-und Fitchen-Bjorkmann-lignin. Die Makromolekulare Chemie 175:2409–2422

Meyers PA, Simoneit BRT (1989) Global comparison of organic matter in sediments across the Cretaceous/Tertiary boundary. Org Geochem 16:641–648

Mita H, Shimoyama A (1999) Characterization of *n*-alkanes, pristine and phytane in the Cretaceous/Tertiary boundary sediments at Kawaruppu, Hokkaido, Japan. Geochem J 33:285–294

Mukhopadhyay SK (2008) Planktonic foraminiferal succession in late Cretaceous to early Palaeocene strata in Meghalaya, India. Lethaia 41:71–84

Mukhopadhyay SK (2009) Convener's report for 2008 on the progress of work in the IGCP Project 507, on 'Palaeoclimate in Asia during the Cretaceous: their variations, causes, and biotic and environmental responses'. IGCP Ind Newslett 29:11–13

Mukhopadhyay SK (2010) Paleoclimates in Asia during the Cretaceous: their variations, causes, and biotic and environmental responses. Final report of the IGCP Project no 507, 2006–2010, pp 1–98; (Circulated through Geological Survey of India portal and available in 'Concluded Projects' of IGCP)

Mukhopadhyay SK (2012a) Morphogroups and small sized tests in *Pseudotextularia elegans* (Rzehak) from the Late Maastrichtian succession of Meghalaya, India as indicators of biotic response to Paleoenvironmental stress. J Asian Earth Sci 48:111–124

Mukhopadhyay SK (2012b) *Guembelitria* (Foraminifera) in the upper Cretaceous-lower Paleocene succession of the Langpar Formation, India, and its paleoenvironmental implication. Geol Soc Ind 79:627–651

Mukhopadhyay SK (2013) Can Late Maastrichtian Planktonic Foraminifera and Paleoclimate help understand the problems of present day global warming? In: Venkatachalapathy R (ed) Earth resources and environment. Research Publishing, Singapore (Ch. 12), pp 193–205

Pal S, Shrivastava JP, Mukhopadhyay SK (2015a) Polycyclic aromatic hydrocarbon compound excursions and K/Pg transition in the late Cretaceous-early Paleogene succession of the Um-Sohryngkew River section, Meghalaya. Curr Sci 109:1140–1150

Pal S, Shrivastava JP, Mukhopadhyay SK (2015b) Physils and organic matter-base palaeoenvironmental records of the K/Pg boundary transition from the late Cretaceous-early Palaeogene succession of the Um Sohryngkew river section of Meghalaya, India. Chemie der Erde—Geochemistry 75(2015c):445–463

Pal S, Shrivastava JP, Mukhopadhyay SK (2015c) Mineral chemistry of clays associated with the late Cretaceous-early Palaeogene succession of the Um Sohryngkew river section of Meghalaya: Palaeoenvironmental inferences and K/Pg transition. J Geol Soc India 86(6):631–647

Shrivastava JP, Mukhopadhyay SK, Pal S (2013) Chemico-mineralogical attributes of clays from the late Cretaceouse early Palaeogene succession of the Um Sohryngkew river section of Meghalaya, India: palaeoenvironmental inferences and the K/Pg boundary. Cret Res 45:247–257

Silverstein RM, Webster FX, Kiemle DJ (2005) Spectrometric identification of organic compounds. Wiley

Tschudy RH, Pillmore CL, Orth CJ, Gilmore JS, Knight JD (1984) Disruption of the terrestrial plant ecosystem at the Cretaceous-Tertiary boundary, Western Interior. Science 225:1030–1032

Venkatesan MI, Dahl J (1989) Organic geochemical evidence for global fires at the Cretaceous/Tertiary boundary. Nature 338:57–60

Von Wehrli FW, Wirthlin T (1976) Interpretation of Carbon-13 NMR spectra. Angew Chem 90:229

Wilson MJ (1987) X-ray powder diffraction methods. In: Wilson MJ (ed) A handbook of determinative methods in clay mineralogy. Blackie, Glasgow, UK, pp 26–98

Wolbach WS, Lewis RS, Anders E (1985) Cretaceous extinctions: evidence for wildfires and search for meteoric material. Science 230:167–170

Wolbach WS, Gilmour I, Anders E, Orth CJ, Brooks RR (1988) Global fire at the Cretaceous-Tertiary boundary. Nature 334:665–669

Wolbach WS, Anders E, Nazarov MA (1990a) Fires at the K-T boundary: carbon at the Sumbar, Turkmenia, site. Geochim Cosmochim Acta 54:1133–1146

Wolbach WS, Gilmour I, Anders E (1990b) Major wildfires at the K-T boundary. In: Sharpton VL, Ward PD (eds) Global catastrophes in Earth history. Geological Society America Special Paper, vol 247, pp 391–400

Yamamoto M, Ficken K, Baas M, Bosch H, Jan Leeuw JW (1996) Molecular palaeontology of the earliest Danian at Geulhemmerberg (The Netherlands). Geol Mijnbouw 75:255–267

Palaeogene–Neogene Tectonics and Continental Aggradational Basins in North-Western India: Implications for Geological Evolution of Thar Desert

Sudesh Kumar Wadhawan

Abstract Thar Desert is characterised by arid-semiarid ecologically fragile environment and occupies a unique tectonic-sedimentary domain in north-western (NW) India. It is confined essentially to West Rajasthan Shelf (WRS). The tectonic disposition and basement configuration of intra-cratonic basins and sedimentary formations here range from the Precambrian Delhi Supergroup in the east, Late Proterozoic to Early Palaeozoic Marwar Supergroup of sedimentary rocks in the middle, to Mesozoic and Cainozoic cover sedimentary formations on the western and north-western fringe areas in western parts of Thar desert in Rajasthan and Haryana in north-western India. It is remarkable that successive geological young age of sedimentary formations is mapped from the highly deformed Precambrian rock formations of the Aravalli Hill ranges on the east to the recent sand covered structurally undisturbed rocky plains on the west in the Thar Desert. Distinctive geomorphic expressions and relative lowering of relief are also deciphered and recorded across an east–west transect along northern parts of the Thar Desert. Geological mapping of Palaeogene–Neogene formations in NW India have helped reconstruction of the palaeo-geographic shore-line limits of the Tertiary sea. Such inferred basement disposition ostensibly had a bearing on the source of evaporate minerals and continuing salinity aspects of the present day inland lakes and playas in the region. Successive northward deepening of the Palaeogene–Neogene sedimentation basin is also inferred from the geological strata logs prepared during the drilling probes for the Potash Mineral Investigation by Geological Survey of India in NW India. It has been inferred that the NE–SW trending Aravalli hill ranges were rejuvenated as a horst bound by the Great Boundary Fault (GBF) in the east and well known Sardarshahar Fault (west of Churu) and a series of westerly down-thrown step faults (called Manpia Faults) on the west thus segmenting the Thar Desert in north-western Rajasthan into several Neogene depressions or the depocenters in-filled with Quaternary continental aggradational deposits of fluvio-lacustrine and aeolian origin. Based on geological field surveys, interpretation of sub-surface data and dug-well inventory, it has been feasible to delineate a series of linear stepped grabben structures hosting a succession of Quaternary deposits separated respectively by denudational relief

S. K. Wadhawan (✉)
D-159, Malviya Nagar, Jaipur 302 017, Rajasthan, India
e-mail: wadhawansk.leo@gmail.com

S. Bajpai et al. (eds.), *The Indian Paleogene*, Society of Earth Scientists Series,
https://doi.org/10.1007/978-3-319-77443-5_5

features. Their stratigraphic succession is established and correlated to build-up a model for improved geological understanding. Present contribution attempts to elucidate role of Palaeogene–Neogene tectonics and its implications for the geological evolution of Thar Desert in NW India.

Keywords West Rajasthan Shelf · Palaeogene–Neogene formations Thar Desert

Introduction

Thar Desert in India is the eastern most extension of the vast dryland belt of Saharo-Arabian Desert. It is dominated by aeolian bed forms: the sand dunes of different dimensions. The Thar Desert is largely located in Rajasthan State, especially between the foothills of the Aravalli ranges in the east and International border with Pakistan in the west. Further west, within Pakistan, the desert extends to the fertile alluvial plains of Indus River. The Indian part of the hot and arid Thar Desert is spread over 19.6 Mha in Rajasthan; 6.6 Mha in Gujarat; 1.5 Mha in Punjab and 1.3 Mha in Harayana (Dhir et al. 1992). Thar Desert is characterised by arid-semiarid ecologically fragile environment and occupies a unique tectonic-sedimentary domain in north-western (NW) India. It is confined essentially to West Rajasthan Shelf (WRS). The tectonic disposition and basement configuration of intra-cratonic basins and sedimentary formations here range from the highly deformed Precambrian Aravalli and Delhi Supergroups in the east, undeformed sedimentary rocks of Late Proterozoic to Early Palaeozoic sedimentary rocks of Marwar Supergroup in the middle, to Mesozoic and Cainozoic cover sedimentary formations on the western and north-western fringe areas in western parts of Rajasthan and Haryana in NW India. It is remarkable that westward successive geological young age of sedimentary formations are mapped from the Aravalli Hill ranges (average 500 m above msl) on the east to the recent sand covered Tertiary rocky plains (80 m above msl) on the west in the Thar Desert. Distinctive geomorphic expressions and relative lowering of relief are also deciphered and recorded in an east-west transect along northern parts of the Thar Desert (Wadhawan 1996, 2010).

Major geological sedimentary basins that have been delineated in the Thar Desert region include: (1) the Nagaur–Ganganagar basin (Marwar Supergroup of rocks) bound by the Aravalli hill ranges in the east and the Delhi–Lahore–Sargoda subsurface ridge/high on the north and northwest; (2) the Jaisalmer basin in the west dominated by the Mesozoic and intermittent Tertiary strata; (3) the Palaeogene Barmer–Sanchor basin and (4) the contiguous Cambay (Khambhat) grabben and (5) the Kachchh basin in Gujarat. Distribution of different rock groups and Quaternary cover sediments in Rajasthan and Gujarat is provided schematically in the lithostratigraphic map (Fig. 1). Palaeogeographical extent of the shallow Tertiary sea is inferred based on occurrences of marine fossiliferous strata and lignite deposits of the marshy/deltaic environment in the region, as it has bearing on the source of evaporite

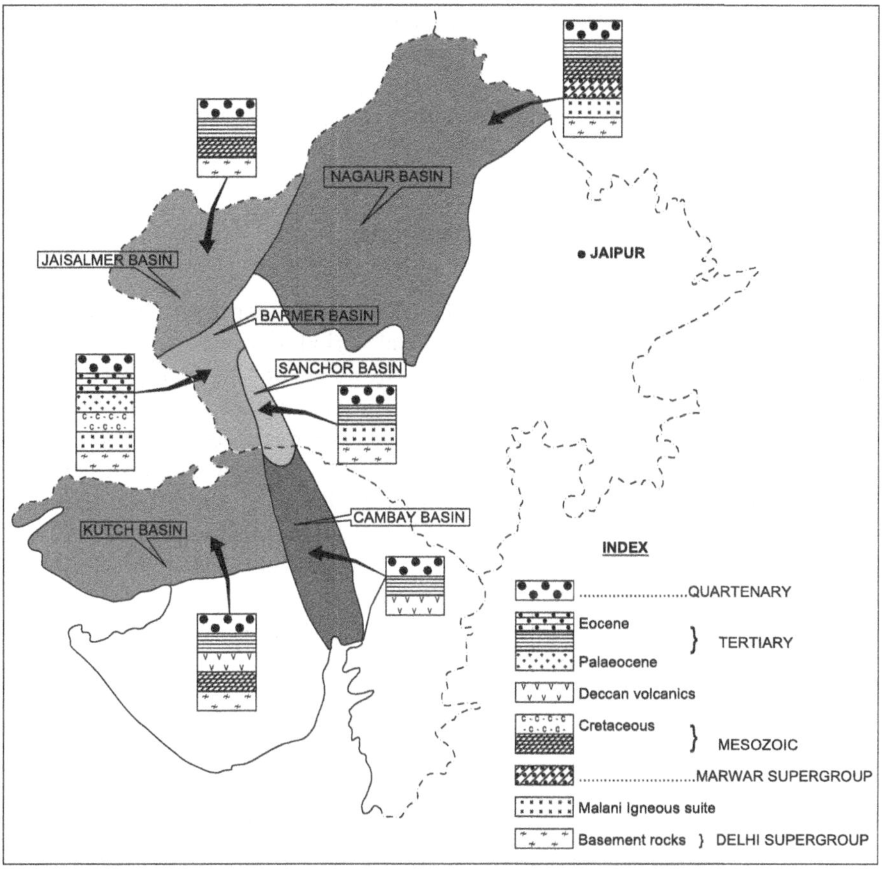

Fig. 1 Tectonically active sedimentary basins of Rajasthan and Gujarat showing generalised lithostratigraphy

minerals and continuing salinity aspects of the present day inland lakes (Wadhawan and Sharma 1997; Roy 1999). As the structural competence or the rheological properties of the constituent rock assemblage in these basins differ considerably from one another, its influence on the mechanism of seismic energy dissipation in the region should have bearing on neotectonic manifestations and related geohazards propensity and their geotechnical remedies in the region (Wadhawan 2015).

A large variety of distinctive dune fields ranging from clustered parabolic to transervese, linear, reticulate, star and barchanoid have been mapped across the vast span of the Thar Desert. The characteristic aeolian landforms of the Thar Desert i.e. the clustered parabolic dunes constitute the present day active desert front that experiences contemporary aeolian dynamism. These older stabilized dune forms have been worked upon by winds episodically that resulted in producing several evolved and superimposed complex dune forms. Morphostratigraphically, episodic nature of multi-phase aeolian deposits are distinguished that rest disconformably over fluvio-aeolian and calcretised colluvio-fluvial sedimentations units in the eastern parts and over a thick pile of polycyclic fluvio-lacustrine sedimentation package in the northern and western parts of the Thar (Wadhawan et al. 1999; Wadhawan 2005).

Tectonic Set-up in Thar Desert Region

Most of the area covered by the present day Thar Desert appears to have remained low lying peneplain throughout the geological period as it was submerged under shallow transgressions of the sea during the Mesozoic and Early Tertiary. Periods of epeirogenic movements in the extra-peninsular India are inferred from the breaks that have been caused to the processes of peneplanation leading to the formation of the base level of erosion or the 'planation surfaces'. Three planation surfaces had been identified and assigned as: 1st during post-Mesozoic; 2nd during Early and Middle Tertiary and 3rd as Early Pleistocene equated to aggradation of the Older Alluvium in the Gangetic plains (Heron 1953). Considerable rise of the Himalayan barrier at about Mid-Miocene should have resulted in establishment of the monsoon conditions over Thar Desert (Ahmed 1969). Ahmad (1986) collated various evidences that suggest that Aravalli hill ranges have risen rapidly as a horst in three phases: (1) an Early–Middle Tertiary uplift by about 1000 m; (2) an Upper Tertiary phase of comparatively slow and discrete peneplain surfaces, and (3) an Early Holocene phase that raised the Guru Shikhar (Mt. Abu) by about 700 m thus placing it at an altitude of 1722 m from apparently submarine situation at the end of Cretaceous or Early Tertiary. It may be mentioned that the tectonic throw had been computed based on the relative relief of different geological formations that were subject to geomorphic planation manifested as the recurring accordant summits and lower elevation of planation surface forming vast pediplain respectively. Rapid uplift of the central Rajasthan as a whole and the Aravalli hill ranges in particular was differential in different parts of the Aravallis and mostly dome-shaped, decreasing in elevation radially from Udaipur. However, it can be interpreted presently as the cymatogenic up-arching. Rapid uplift in the Delhi area during the Late Pleistocene–Holocene rejuvenated a consequent Himalayan stream—the Yamuna. It proceeded to behead the mighty Drishadvati through head-ward erosion. This coupled with westward deviation of the courses of the Sutlej and Beas, the tributaries of the Vedic Sarasvati deprived the perennial water flow leading eventually to its desiccation in the vast sandy tracts of the encroaching Thar Desert (Yash Pal et al. 1980; Bakliwal and Grover 1988; Sridhar et al. 1994; Roy 2003).

Tectonic Lineaments in Thar Desert Region

Although climatic change is responsible for the aeolian episodes, field studies and interpretation of subsurface geological data have now revealed that tectonically active Paleogene–Neogene domain had influenced the successive burial of different sedimentation units of fluvial/fluvio-lacustrine and aeolian generation particularly in the northern and western parts of the Thar Desert. Geological observations support occurrence of these continental scale lineaments were developed as reactivated boundary faults since Paleogene as the confinement of Tertiary basin and the sedimentary deposition record of rocks are also controlled by these bounding structural discontinuities. Some of these lineaments are also active presently, e.g., the Jaisalmer–Barwani and Luni–Sukri (Kachchh). Isopach map of the sub-surface Quaternary and Tertiary formations and palaeo-basin configuration show that the present day Ghagger basin (Nagaur Basin) in northern Rajasthan deepened to abnormal depth and hosted a thick (over 350 m) pile of continental sediments in its northern and northwestern parts (Fig. 1). Northerly flowing structurally controlled palaeo-drainage pattern for the early Quaternary fluvial regime is deduced.

It is inferred that the NE–SW trending Aravalli hill ranges were rejuvenated as a horst bound by the Great Boundary Fault [GBF] in the east and well known Sardarshahar Fault (west of Churu) and a series of westerly down-thrown step faults (called Manpia Faults) on the west thus segmenting the Thar Desert in north-western Rajasthan into several Neogene depressions or the depocenters in-filled with Quaternary continental aggradational deposits of fluvio-lacustrine and aeolian origin (Ahmad and Ahmad 1980; Sen and Sen 1983; Ahmad 1986). Based on geological field surveys, interpretation of sub-surface data and dug-well inventory, it has been feasible to delineate a series of linear stepped grabben structures hosting a succession of Quaternary deposits separated respectively by denudational relief features (Fig. 2). Their stratigraphic succession is established and correlated to build-up a model for improved geological understanding (Wadhawan 1996, 2010).

Interpretation of various geological, geophysical and seismotectonic data in the Thar Desert assist in recognizing that the terrain is traversed by several block faults-manifested asmajor lineaments that have been neotectonically active and shaping the geomorphic and geologic evolution of the dryland environment (Fig. 3; Bakliwal and Ramasamy 1987; Kar 1988a, b; Ramasamy 1999; Virendra Kumar and Wadhawan 2001). Important amongst these are elucidated in the following paragraphs.

Luni–Sukri Lineament

It is a composite lineament system comprising a set of sub-parallel curvilinear lineaments, traversing east–west from Rann of Kachchh through the NE–SW trending Luni river and extending through Aravalli gap and beyond northeast of Ajmer. It delimits a rectilinear fault bound configuration of the Rann of Kachchh. It also

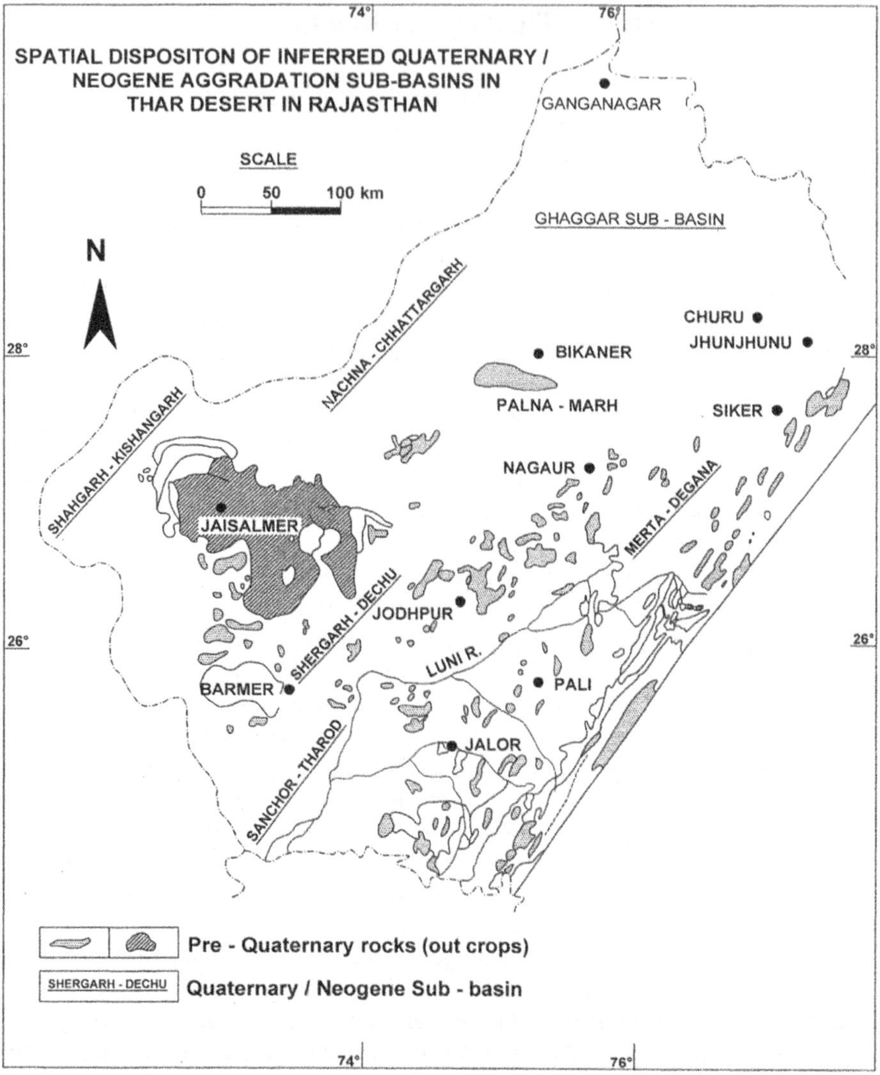

Fig. 2 Spatial disposition of inferred Palaeogene/Neogene aggradation sub-basins in Thar Desert in Rajasthan

defines a sharp contact between the Late Quaternary stabilised parabolic dune fields and the coastal sabkha environment of deposition along the Great Rann of Kachchh that experiences recent reactivation and repeated seismic activity (Ameta et al. 2005). Ramasamy et al. (1991) interpreted a major NE–SW trending Quaternary arch along the Luni–Sukri lineament (Fig. 3). It is also characterised by a graben along its crest

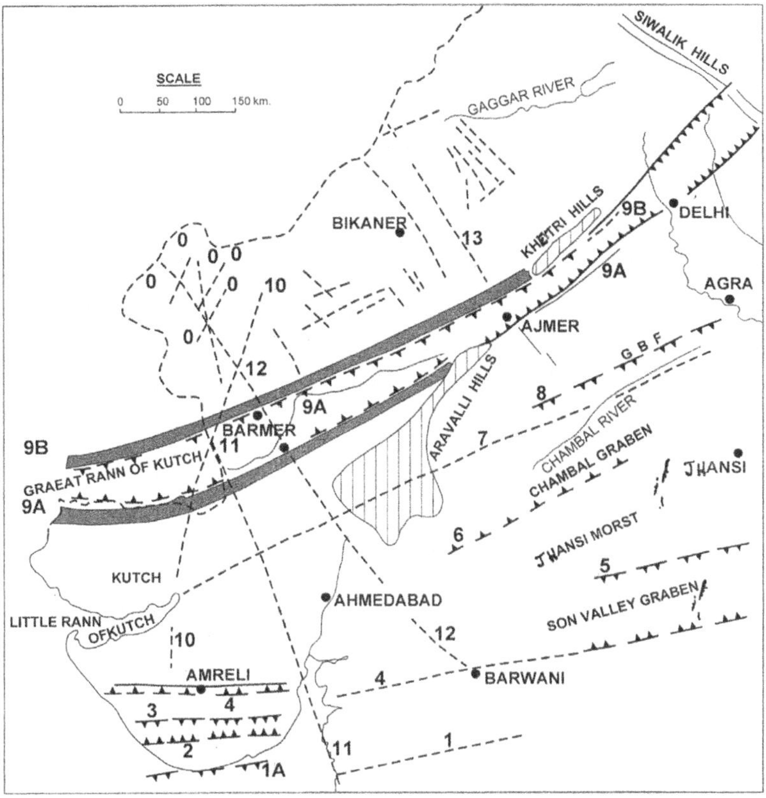

Fig. 3 Major Inferred Tectonic Features and Lineaments with horst and graben structure in the Thar Desert in Rajasthan and Gujarat (*source* Published literature cited in the text). Legend: 1 = Tapti Lineament, 2 = Sawar–Kundla Lineament, 3 = Amreli–Junagarh Lineament, 4 = Bhawnagar–Gondal Lineament, 5 = Asmara Lineament, southern boundary of the Jhansi Horst, 6 = Jalawar Lineament, southern boundary of the Chambal graben, 7 = Chambal–Jamnagar Lineament, 8 = Great Boundary Fault, 9 = Luni–Sukri Lineament, defining a major graben, 10 = Lathi–Rajkot Lineament, 11 = West Coast Lineament, 12 = Jaisalmer–Barwani Lineament, 13 = Raisinghnagar–Tonk Lineament

and its extensions are manifested as tear faults in the Siwaliks towards the NE along the Himalyan foot-hills.

Jaisalmer–Barwani Lineament

It is about 1000 km long tectonic lineament trending NNW–SSE to NW–SE. It is defining the southwestern boundary of the Aravallis northwest of Ahmedabad and show contrasting elevations and fault-line scarps within the Himatnagar Sandstone of Mesozoic age. It abruptly truncates the rocks of the Delhi Supergroup near Vadnagar–Palanpur in north Gujarat plains and marks western margin of the Tertiary basins north of Barmer and its NW extensions have been recorded to coincide with the Jaisalmer–Mari arch in the Afganistan region as well (Dasgupta 1975). Further it coincides with major fault within Mesozoics sedimentary formations of the Jaisalmer area and delimits Gravity High Zone along NE of it and Low Gravity Zone on the SW of it implying thicker fill of sedimentary rocks in the SW. It also coincides with recently reactivated Kanoi fault along which earthquake of intensity 6.3 on Richter scale was recorded on 8th November 1991 (Dharman et al. 1993). Several circular features in the region are interpreted bearing direct linkage with this lineament as surface manifestations of the Tertiary domes and basins (Bakliwal and Ramasamy 1983).

Lathi–Rajkot Lineament

It is estimated to be over 500 km long, N–S aligned curvilinear tectonic lineament. It traverses the Deccan volcanic of Saurashtra (Mesozoic–Cainozoic) and occurs as fault zone within the Mesozoic sedimentary rocks in the Kachchh and Jaisalmer region and cuts through the dune fields of Late Quaternary age thus implying its neotectonic reactivation. Ramasamy (1999) attributed the curvilinear nature of this lineament to be a reflection of the post-collision plate-tectonic phenomenon during Palaeogene–Neogene all along its disposition.

Raisinghnagar–Tonk Lineament

It is about 400 km long, NW–SE trending lineament and demarcates the dune covered pediplain over pre-Aravallis of the Jaipur region from dune free Bhilwara region. It delimits subsidence on north and northeast and confirm relative deepening of the Neogene/Quaternary continental sedimentation in the Ghaggar basin along the northern margin of the Thar Desert.

Sardarshahar Fault

It is aligned along N–S to NNE–SSW with a confirmed down-throw of over 400 m on the west, as recorded in the Quaternary sediments (inferred up to 700 m during subsurface investigations for potash mineralisation in Ghaggar basin). It is inferred to be active fault with frequent mild earthquakes recorded at seismograph observatory located at nearby Churu.

Considering geological, neotectonic–seismic set-up and geothermal gradient characteristics, the Thar Desert area in western India is categorised as the Quaternary tectonic domain undergoing extensional stress fields manifested in a series of normal faults and graben structures with moderate seismic activity and geothermal heat flow attributes. Such neotectonic movements have also been considered responsible for deviation of the river courses originating in the Himalayas such as the Saraswati, Indus, Sutlej and Yamuna, etc., affecting considerably the Mohanjo Daro and Harrapan civilisations and subsequent human settlements in the region (Roy 2003; Sinha-Roy 1986; Dassarma 1986; Valdiya 2002). Post-Tertiary reactivation of Mari–Jaisalmer–Barwani lineament and earthquake ruptures/dislocations along the Kanoi Fault in Jaisalmer area in 1991 confirmed continuing neotectonic activity (Dharman et al. 1993).

In addition to the above lineaments, a wide variety of neotectonically induced geomorphic expressions have been recorded in semi-arid and arid drylands of Rajasthan and Gujarat such as lineament controlled alignment of certain dune fields and abrupt truncation of the different dune morphologic domains, reactivation of Kanoi Fault during the earthquake of 8th November, 1991 in Jaisalmer region (measuring 6.3 on Richter scale), unequal thickness of Quaternary formations in the contiguous blocks in northern and central Thar Desert, distinctive shearing and faulting of Quaternary formations and dislocation of Tertiary and overlying Quaternary strata inferred from subsurface probes in Bikaner–Ganganagar region, preferred and linearity controlled orientation of drainage courses, such as the Luni and Sabarmati river systems, etc. Several faults showing evidences of recent reactivation are manifested as prominent curvilinear tectonic lineaments, for instance, along the Luni–Rupangarh, Nawan–Alwar, Didwana–Dausa, etc. Intersections of these weak zones have created the pull-apart basins occupied presently by the salt lakes of the Didwana, Kuchaman and Sambhar (Sinha-Roy 1986; Dassarma 1986). Further, it is reasonable to infer that the persisting salinity of these lakes could be linked through the tectonic lineaments/shears with the subsurface halite bearing perennial source—the Hanseran Evaporite Sequences in the Bikaner–Nagaur–Ganganagar basin, geologically correlated to the Marwar Supergroup of rocks (Virendra Kumar and Wadhawan, in GEOSAS 2001).

Palaeogene and Neogene/Quaternary Aggradation Basins

It has been postulated that the NE–SW trending Aravalli hill ranges were rejuvenated as a horst bound by the Great Boundary Fault (GBF) in the east and the well known Sardarshahar fault (west of Churu) and a series of westerly down-thrown step faults called the Manpia Faults on the west, segmenting the Thar Desert in western Rajasthan into several Palaeogene–Neogene depressions or depocentres during the that period (Ahmad and Ahmad 1980; Ahmad 1986). Based on detailed field surveys, geological mapping of the Quaternary formations, their nature of surface disposition and inferred sub-surface configuration in arid-semiarid western parts of

India, it is feasible to delineate a few large and broadly linear Neogene sub-basins or tectonic depressions filled-in with aggradational continental Quaternary sediments and separated respectively by the denudational relief features. A series of north-easterly trending linear stepped graben structures have been deciphered showing westerly downthrown blocks respectively (Wadhawan 1991; Wadhawan et al. 1999). It is observed that these structural depressions or the continental depocentres host a succession of Quaternary valley fills comprising fluvio-lacustrine deposits and overlying aeolian sands forming distinctive dune fields in the Thar Desert (Figs. 2 and 3). These Palaeogene/Neogene sub-basins include: the Bikaner–Churu–Ganganagar sub-basin; part of the Ghaggar basin along the northern parts of Thar desert; the Merta–Degana–Chhajoli (Jayal) sub-basin, on the eastern fringe of Thar desert; the Shergarh–Dechu sub-basin in the central parts of Thar Desert; the Shahgarh–Kishangarh sub-basin in Jaisalmer region, etc. (Fig. 2). It is also remarkable that the core part of the different dune fields is largely confined to the Quaternary alluvial basin indicating availability of proximal source of sand for reworking and dune construction. The Quaternary sedimentation history in each continental sub-basin is unique, but broadly comparable and multicyclic and varying in degree of development: starting with coarse colluvium/ill-sorted polymictic fluvial clastic deposits, through playa/lacustrine deposits and culminating with aeolian deposits. It is established that the fluvial and fluvio-lacustrine deposits form the base for aeolian deposition and control the configuration and disposition of older aeolian deposits (dune fields) in the Thar Desert (Wadhawan 2010). Large scale drainage disruption in the Thar Desert is inferred either due to neotectonics or obstruction of rivers by extensive dune fields driven by the southwest winds, for instance present day desiccated Ghaggar River along the northern periphery of the Thar Desert (Sundaram and Rakshit 1994).

Palaeogene rocks in the Jaisalmer Basin of western India are represented by sandstone–shale sequence of the Sanu Formation (Palaeocene). These are capped at places by sand-clay/bauxitic clays and pisolitic laterite occurring as palaeosols. These in turn are overlain by the fossiliferous limestone-bentonitic clays and fullers' earth sequence of the Khuiala Formation (Lower Eocene) and the younger limestone-shale succession of Bandah Formation (Lower to Middle Eocene) (Laul et al. 1984). Calcretised gritty conglomerates, ill-sorted and indurate pebbly-gravelly, ferruginous sandy and clayey fluvial deposits in the area are categorised as the Shumar Formation of Neogene (Plio-Pleistocene)/basal Quaternary age. Geological mapping of Palaeogene–Neogene formations in NW India have helped reconstruction of the palaeo-geographic shore-line limits of the Tertiary sea (Pareek 1984). Such inferred basement disposition ostensibly had a bearing on the source of evaporate minerals and continuing salinity aspects of the present day inland lakes and playas in the region. Successive deepening of the sedimentation basin is also inferred from the geological logs prepared during the drilling probes for the Potash Mineral Investigation by Geological Survey of India in NW India (Wadhawan and Virendra Kumar 1996).

The Quaternary geological maps of the Jaisalmer and Bikaner areas represent the core of Thar Desert in India. Major Quaternary morphostratigraphic units delineated in the area include: the fluvial Shumar Formation, aeolian Shergarh Formation (=Dawa Formation), Ramgarh Formation (fluvio-lacustrine/reworked aeolian) with two Members: playa/inland fluvio-lacustrine facies and as reworked aeolian/fluvio-aeolian sand sheet facies; the younger linear and transverse dune-forming aeolian-deposits as Danana-Mayajlar Formation and the recently mobile spontaneous occurrence of active, simple and compound megabarchanoid dune sands of the Sam Formation (Quaternary Geological Atlas of India, GSI 2014).

The Shumar Formation occurs at the base of the continental Quaternary succession in the area and forms the aggraded and cemented/calcretised sandy plains or in-filled depressions hosting younger fluvio-lacustrine and aeolian sedimentation units. It has been divided into four members namely, delta at the base through gamma, beta to alpha at the top and show variable thickness of development and considerable lateral variations from pebbly sandstone to medium and coarse-grained poorly sorted calcareous sandstones, calcretised pseudo-conglomerates, sandy loam and mottled clays.

Older stabilized dunes-interdune complex designated as the Shergarh Formation (=Dawa Formation) with type sections exposed in the central Thar Desert in India. The dune relief in the Sam-Kinoi area varies from 18 to 27 m, exceptionally attaining heights of 54 m above ground level. More than 0.5 km interdune spacings are recorded in such large dunes. These are modified by superimposed younger generation aeolian accumulations and selective deflation. These stabilised dunes comprise oxidised reddish brown polymodal medium to fine-grained sands extensively impregnated with pedocalcic kankars. The well-sorted older dune sediments have a mean grain size confined between 2.61 and 2.74 phi units and show positive skewness. Luminescence chronology of these dunes indicated that the most dominant aeolian accumulations peaked around 13–14 ka B.P. Ramgarh Formation comprises sand sheets and inland lakes deposits of the clayey silt; playa evaporates and reworked fluvio-aeolian deposits. Fine to coarse sand of quartz and reworked foraminifers and limonite coated granules occur as about 1 m thick top layer as wind erosion lag deposits. Up to 4 m thick fluvial deposits of cross-bedded gravely sands, pebble clasts of grey and pink limestone and quartz occur underlying this unit. Basal unit is up to 4 m thick massive and weekly-consolidated, greenish brown fine sand deposit of well-sorted homogenized quartzose sands of fluvio-aeolian deposits. Intermittent occurrences of thin laminae of silty clays indicate sheet wash and interludes of playa environment of deposition. The Mitha Rann and Khara Rann in the Jaisalmer area are the inland lacustrine basins that persisted as playas through Middle to Upper Holocene.

Closely spaced linear and transverse dune deposits of younger aeolian episode are grouped into the Dhanana–Myajlar Formation. They comprise the younger evolved aeolian landforms or superimposed dunes complex. The Dhanana–Myajlar Formation is the aeolian coeval of the Ramgarh Formation in the area. These longitudinal dunes trend along N 30°–33° E. They appear to have evolved through superimposed aggradation or carved out by the deflationary modifications, respectively. These younger generation aeolian deposits comprise yellowish brown unimodal and bimodal fine sands that are moderately sorted and positively skewed. Their mean grain size varies from 2.51 to 2.91 phi units. Milder pedocalcic pedogenesis is noticed in the form of calc-cemented sand as tiny kankar disseminations in these massive deposits. During ground surveys and mapping of the area, several microliths, leached bone fragments and potsherd were discovered occurring as residual lag deposits on the deflated surfaces of these stabilised dunes. The microliths include flakes, slender blades with cutting and indented edges, fluted cores, etc., representing Mesolithic Culture (8000–7500 year B.C.) that flourished during the subhumid interlude dominated by fluvio-lacustrine environment in the Thar Desert.

Sam Formation constitutes the presently mobile youngest aeolian deposits. Common landforms of these youngest active dune fields are the low clusters of barchans in linear belts and the massive megabarchanoids, vegetation-induced aeolian sandy deposits, shrub coppice and prograding sand sheet. Their dominant linear alignment is along NNE–SSW direction. The modal height of these barchans is 5.5 m and ranges between 3 and 9 m. Similarly the recent mobile dunes and active aeolian sediments in the northern Thar Desert are classified into the Churu Formation.

A large variety of distinctive dune fields ranging from clustered parabolic to transverse, linear, reticulate, star and barchanoid have been mapped across the vast span of the Thar Desert. The characteristic aeolian landforms of the Thar Desert, i.e., the clustered parabolic dunes constitute the present day active desert front that experiences contemporary aeolian dynamism. These older stabilized dune forms have been worked upon by winds episodically that resulted in producing several evolved and superimposed complex dune forms. Morphostratigraphically, episodic nature of multi-phase aeolian deposits are distinguished that rest disconformably over fluvio-aeolian and calcretised colluvio-fluvial sedimentations units in the eastern parts and over a thick pile of polycyclic fluvio-lacustrine sedimentation package in the northern parts of the Thar.

Discussion and Conclusion

Thar Desert is one of the most important and unique geomorphic units in India and covers Marwar Terrain in Rajasthan and Kutch region and adjoining plains of north Gujarat. Palaeogene–Neogene tectonic domain in Thar has been subject to extensional stress fields manifested in series of normal faults and graben/succession of half grabens structure and step faults influenced architecture of continental sedimentation basins. The Indus River appears to be entrenched such as the Ganges in the Himalayan fore-deep region along its southwesterly extension into the Arabian Sea (Radhakrishna and Merh 1999).

The sedimentation basin architecture and origin of Thar Desert are intimately associated with the Cenozoic tectonics in the geological history of the Indian subcontinent when India experienced the beginning of the India–Asia convergence (docking) and rise of the Himalaya. A major regression of similar magnitude occurred during latest middle Eocene (41.3–38.0 Ma) that corresponds to global sea level fall. This ostensibly marked the termination of foraminifer limestone formation in Jaisalmer basin. As a sequel to the great mountain building activity shaping the Himalaya in the northern leading edge, there occurred reactivation of Palaeogene lineaments, block faulting and formation of pull-apart continental sedimentation basins and tilting of blocks in north western parts India. Both tectonic and climatic causes are considered responsible for the aggradations of continental deposits in Thar Desert (DST 2002). It is interpreted that this regression was associated with the global cooling during latest middle Eocene/late Eocene, possibly associated with the nucleation of the Antarctica ice-sheets coupled with the uplift of the Himalaya.

The Paleogene sediments of Western India are well exposed in the Kachchh and Jaisalmer Basins and comprise of fossiliferous limestone sequences. Geological mapping of Palaeogene–Neogene formations in NW India have helped reconstruction of the palaeo-geographic shore-line limits of the Tertiary sea. Such inferred basement disposition ostensibly had a bearing on the source of evaporate minerals and continuing salinity aspects of the present day inland lakes and playas in the region. Successive deepening of the sedimentation basin is also inferred from the geological logs prepared during the drilling probes for the Potash Mineral Investigation by Geological Survey of India in NW India.

Palaeogene rocks in the Jaisalmer Basin of western India are represented by sandstone-shale sequence of the Sanu Formation. These are capped at places by sand-clay/bauxitic clays and pisolitic laterite. These in turn are overlain by the fossiliferous limestone-bentonitic clays and fullers' earth sequence of the Khuiala Formation (Lower Eocene) and the younger limestone-shale succession of Bandah Formation (Lower to Middle Eocene). Calcretised gritty conglomerates, ill-sorted and indurate pebbly-gravelly, ferruginous sandy and clayey fluvial deposits in the area are categorised as the Shumar Formation of Neogene (Plio-Pleistocene)/basal Quaternary age.

It has been inferred that the NE–SW trending Aravalli hill ranges were rejuvenated as a horst bounded by the Great Boundary Fault (GBF) in the east and well known Sardarshahar Fault (west of Churu) and a series of westerly down-thrown step faults (called Manpia Faults) on the west thus segmenting the Thar Desert in northwestern Rajasthan into several Palaeogene–Neogene depressions or the depo-centers in-filled with continental aggradational deposits of fluvio-lacustrine and aeolian origin. Based on geological field surveys, interpretation of sub-surface data and dug-well inventory, it has been feasible to delineate a series of linear stepped grabben structures hosting a succession of Palaeogene–Neogene deposits separated respectively by denudational relief features. Their stratigraphic succession is established and correlated to build-up a model for improved geological understanding.

Strong monsoonal systems in Asia are believed to have originated between 25 and 22 million years (My) ago and are considered to be mainly driven by Tibetan–Himalayan uplift (Awasthi and Ray 2015). It is envisaged that huge volume of sediments supplied to the Himalayan fore-deep basin and also to the continental depocentres in the NW India during late Eocene–Oligocene and later periods were results of large scale weathering in the nascent Himalayas (cf. Awasthi and Ray 2015). Although climatic change is responsible for the aeolian episodes, field studies and interpretation of subsurface geological data have now revealed that tectonically active Paleogene–Neogene domain had influenced the successive burial of different sedimentation units of fluvial/fluvio-lacustrine and aeolian generation particularly in the northern and western parts of the Thar Desert. Isopach map of the sub-surface Quaternary and Terrtiary formations and palaeo-basin configuration show that the present day Ghagger basin in northern Rajasthan deepened to abnormal depth and hosted a thick (over 350 m) pile of continental sediments in its northern and north-western parts. Northerly flowing structurally controlled palaeo-drainage pattern for the early Quaternary fluvial regime is deduced.

Considering presence of a network of intersecting NE–SW, NNW–SSE and N–S tectonic lineaments (in decreasing order of dominance) and resultant pull-apart basins, e.g., the Luni River basin (Ramasamy 1999; Bajpai et al. 2001) and collapse structures created due to dissolution of subsurface strata of Hanseran Group [Marwar Supergroup] evaporites/halite beds, locales were provided for inland drainage terminal sinks/sag ponds or playa depressions with persistent source of salinity. Tertiary lignite depocentres also developed in similar isolated sag ponds in Rajasthan. Further, it is reasonable to infer several NE–SW trending horst and graben structures within Thar Desert which controlled the configuration, spatial disposition and nature of source proximal continental sedimentation basins and that provided scope for intermittent reworking by fluvial and aeolian processes linked to dynamics of the atmospheric monsoonal winds circulation pattern. Neogene/Quaternary sedimentation progressed in marginal fault troughs, grabens and lakes formed by disorganization and desiccation of rivers. It is remarkable that parabolic dunes—the most dominant aeolian bedforms in the Thar Desert are observed to be confined broadly to the fluvial sedimentation base that supported source proximal multiple events of superimposed wind-borne deposition.

Besides, confirmed dislocations of Tertiary and overlying Quaternary strata in Nagaur–Ganganagar and Jaisalmer sub-basins, abrupt truncation of dune fields support contention of neotectonic reactivation of Thar region (Wadhawan 1990). Evidences confirm rejuvenation of pre-existing faults during Neogene times. Tectonically controlled Palaeogene and Neogene fluvial aggradation basins in Thar Desert served favouable sites for aeolian reworking and formation of a variety of dunes depending upon availability of sediments and wind intensity regimes during the Quaternary climatic amelioration. Dunes in Thar were built during late Neogene owing to spurts of intense aeolian activity separated by relative quiescence implying landscape stability and soil formation. Neotectonically active zones in Rajasthan and Kachchh further show focii of major geological events/tectonic movements that have been observed to be shifting westwards at least since the Palaeogene times.

References

Ahmad F (1986) Geological evidence bearing on the origin of the Rajasthan desert (India). Proc Ind Natl Sci Acad 52, A(6):1285–1306

Ahmad F, Ahmad Z (1980) The fan faults of peninsular India and origin of the Himalayas. Tectonophysics 64:97–110

Ahmed E (1969) Origin and geomorphology of the Thar Desert. Ann Arid Zone 8(2):171–180

Ameta SS, Wadhawan SK, Rai DK, Gill PS (2005) Geotechnical micro-zonation of Bhuj town area, Kachchh district in view of 2001 Bhuj earthquake. Geol Surv India Spec Publ 85:267–277

Awasthi N, Ray JS (2015) How old is the Indian monsoon system? Geochemical evidences from the Andaman fore-arc basin. National Conference on Paleogene of the Indian Subcontinent, Lucknow, India, Abstract, p 42

Bajpai VN, Saha Roy TK, Tandon SK (2001) Subsurface sediment accumulation patterns and their relationship with tectonic lineamemnts in semi-arid Luni river basin, Rajasthan, western India. J Arid Environ 48:603–621

Bakliwal PC, Grover AK (1988) Signatures and migration of Saraswati river in Thar Desert, western India. Rec Geol Surv India 116(3–8):77–88

Bakliwal PC, Ramasamy SM (1983) Occurrence of circular features in parts of Thar Desert, Rajasthan. J Geol Soc India 26:225–228

Bakliwal PC, Ramasamy SM (1987) Lineament fabric of Rajasthan and Gujarat, India. Rec Geol Surv India, Jaipur 113(7):54–64

Dasgupta SK (1975) A revision of Mesozoic–Tertiary stratigraphy of Jaisalmer basin, India. J Earth Sci 2(1):1–28

Dassarma DC (1986) Neotectonism in Rajasthan—its manifestations and effects. In: Proceedings of International Symposium on Neotectonics in South Asia. Survey of India, Dehra Dun, pp 282–288

Department of Science and Technology (DST) (2002) Multi-institutional Project Completion Report on Quaternary stratigraphy and palaeo-environmental history of the Thar Desert. DST No. ESS/CA/A3-08/92. New Delhi, 466 pp

Dharman R, Saxena AK, Joshi DD, Mulk Raj (1993) Jaisalmer earthquake of 8th November 1991. Rec Geol Surv India 126(7):107–108 (WR, Jaipur)

Dhir RP, Kar A, Wadhawan SK, Rajaguru SN, Misra VN, Singhvi AK, Sharma SB (1992) Thar Desert in Rajasthan—land, man and environment. Geological Society of India, Bangalore, 191 pp

Geological Survey of India (2014) Quaternary Geological Atlas of India, Hyderabad

Heron AM (1953) Geology of central Rajputana. Mem Geol Surv India 79

Kar A (1988a) Evidence for neotectonism from Indian desert. In: Singh S, Tiwari RC (eds) Geomorphology and environment. Allahabad Geographical Society, Allahabad, pp 300–310

Kar A (1988b) Possible neotectonic activities in the Luni–Jawai plains, Rajasthan. J Geol Soc India 32:522–526

Laul VP, Virendra Kumar, Sahiwala NK, Sen AK, Chakraborty SK (1984) A report on the geological mapping in parts of Jaisalmer and Barmer districts, Rajasthan. Geological Survey of India Unpublished Report, GSI, WR Jaipur, 47 pp

Pareek HS (1984) Pre-quaternary geology and mineral resources of NW Rajasthan, vol 115. Memoir of Geological Survey of India, 99 pp

Radhakrishna BP, Merh SS (eds) (1999) Vedic Saraswati—evolutionary history of a lost river of NW India. Memoir 42, Geological Society of India, 329 pp

Ramasamy SM (1999) Neotectonic controls on the migration of Sarasvati river of the Great Indian Desert. Mem Geol Soc India 42:153–162

Ramasamy SM, Bakliwal PC, Verma RP (1991) Remote sensing and river migration in western India. Int J Remote Sens 12(12):2597–2609

Roy AB (1999) Evolution of saline lakes of Rajasthan. Curr Sci 76(3):290–295

Roy AB (2003) Late Quaternary neotectonic deformations as cause of drainage disorganisation and extinction of the Vedic Sarawati. GSI Proceedings of GEOSAS-IV, pp 182–191

Sen D, Sen S (1983) Post-Neogene tectonism along the Aravalli range, Rajasthan, India. Tectonophysics 93:75–98
Sinha-Roy S (1986) Himalayan collision and indentation of the Aravalli orogen by Bundelkhand wedge: implications for neotectonics in Rajasthan. In: Proceedings of International Symposium on Neotectonics in South Asia. Survey of India, Dehra Dun, pp 18–21
Sridhar V, Chamyal LS, Merh SS (1994) NorthGujarat rivers: remnants of a super fluvial system. J Geol Soc India 44(4):427–434
Sundaram RM, Rakshit P (1994) Occurrence of gypsum deposits at Jamsar and Pallu in Northwest Rajasthan. Ann Arid Zone 33(2):105–108
Valdiya KS (2002) Saraswati: the river that disappeared. University Press, Hyderabad, 116 pp
Virendra Kumar, Wadhawan SK (2001) Neotectonic activities in Thar Desert, India—implications for geological evolution. Proc. GEOSAS, New Delhi, pp 27–31
Wadhawan SK (1990) Quaternary geology, morphostratigraphy and neotectonism in parts of Nagaur district, Rajasthan. Rec Geol Surv India 123(7):53–54
Wadhawan SK (1991) Continental Neogene-Quaternary Stratigraphy in arid-semiarid parts of Rajasthan, India. In: Proceedings of Department of Science & Technology Workshop on Neogene-Quaternary Stratigraphy including the Study of Fluvial and Glacial Systems. University of Delhi, Delhi, pp 101–106
Wadhawan SK (1996) Textural attributes of recent aeolian deposits in different sub-basins of the Thar Desert, India. J Arid Environ 32:59–74
Wadhawan SK (2005) Geological evolution of Thar Desert in Rajasthan and Gujarat, India. GSI Final Report, Jaipur. Extd. Abst. in Rec Geol Surv India 137(7):56–58
Wadhawan SK (2010) Quaternary continental sedimentation basins and dune fields in Thar Desert of NW India. Gondwana Geol Mag (presently J Geosci Res) 12:265–270
Wadhawan SK (2015) Geotechnical applications for geohazards mitigation in dryland environment of Thar Desert, India. J Eng Geol XL(2):48–60 (Proceedings EGNM International Conference on Engineering Geology in New Millennium, IIT, New Delhi)
Wadhawan SK, Sharma HS (1997) Quaternary stratigraphy and morphology of desert ranns and evaporite pans in central Rajasthan, India. Man Environ XXII(2):1–10
Wadhawan SK, Virendra Kumar (1996) Subsurface Quaternary aeolian stratigraphy in the Ghaggar basin, of Thar Desert, India. J Arid Environ 32:37–51
Wadhawan SK, Sareen BK, Pal NK, Raghav KS (1999) Final report on geological and environmental evaluation of Thar Desert, Rajasthan and Gujarat. Final Rep Geol Surv India, Western Region, Jaipur 1–147
Yash Pal BS, Sood RK, Agrawal DP (1980) Remote sensing of the 'lost' Saraswati river. Proc Ind Acad Sci (Earth Planet Sci) 89(3):317–331

Provenance of the Late Paleocene Matanomadh Sandstones, Kachchh, Western India

V. K. Srivastava, B. P. Singh and A. Patra

Abstract The late Paleocene clastic member of the Matanomadh Formation (MF) is an overall sandstone dominated succession wherein sub-ordinate proportion of thinly-laminated silty-mudstones is found lying between the sandstone beds. Provenance of these late Paleocene Matanomadh Sandstones (MS), Kachchh is not known till date and the same has been determined based on petrography and heavy mineral analysis supported by paleocurrent studies. These sandstones reveal an abundance of sub-angular to rounded monocrystalline non-undulatory quartz; and polycrystalline quartz, feldspar and rock fragments occur as minor constituents. These sandstones are classified as quartzose arenite. The rock fragments in these sandstones are dominated by mica-schist, slate, chert and limestones. Q-F-L and Qm-F-Lt diagrams suggest margin of the craton interior to transitional continental stable craton provenance for these sandstones. The paleocurrent measurements of the cross-bedded sandstones suggest NW-SE-directed bipolar and WNW-directed unimodal paleocurrent patterns and suggest a marine-continental transition zone as depositional site for these sandstones where sediments were contributed from both the shallow-marine and continental sources. The heavy mineral assemblages of these sandstones show sub-angular to rounded grains of magnetite, tourmaline, monazite, rutile, kyanite, staurolite and hematite where magnetite is the dominant component; and hence suggest that the heavy minerals might have been supplied from a basic igneous source, low- to medium-grade metamorphic rocks and reworked sedimentary rocks.

Keywords Matanomadh sandstones · Late Paleocene
Kachchh basin · Western India · Heavy minerals · Paleocurrent

---

V. K. Srivastava · B. P. Singh (✉) · A. Patra
Centre of Advanced Study in Geology, Banaras Hindu University, Varanasi 221005, India
e-mail: drbpsingh1960@gmail.com

S. Bajpai et al. (eds.), *The Indian Paleogene*, Society of Earth Scientists Series,
https://doi.org/10.1007/978-3-319-77443-5_6

Introduction

Matanomadh Formation represents the lowermost lithostratigraphic unit of the Tertiary succession of Kachchh and is assigned Paleocene age (Fig. 1). It has been subdivided into a basal lateritic member and an upper clastic member by Saxena (1975). Matanomadh sandstones (MS) whose provenance determination has been attempted in the present study, are exposed along Nakhatrana-Lakhpat road section (N 23° 32′ 43″ E 68° 57′ 04″) and also in the badlands east of the Matanomadh village (N 23° 32′ 06″ E 68° 58′ 16″) in the Kachchh district, Gujarat. Wynne (1872) was the first who studied these Paleocene rocks of Kachchh basin exposed around Matanomadh village and classified them as Sub-Nummulitic Group. Thereafter, Biswas (1965) and Biswas and Raju (1973) proposed a new Time-Stratigraphic classification for the Tertiary sequences of Kachchh and modified it as Madh Series. Recently, Biswas (1992) classified this succession as Matanomadh Formation (MF), after its type area Matanomadh village which is situated at the north-western margin of the Kachchh peninsula (Fig. 1). Lateritic member of the Matanomadh Formation is ~10 m thick and consists of variety of light coloured calcareous and ferruginous mudstones. Laterite and bauxite beds occur in between these mudstones. The clastic member is comprised of a ~17 m thick succession of horizontally-bedded and cross-bedded sandstones, including herringbone cross-bedding and silty-mudstones (Srivastava and Singh 2017) (Fig. 2).

Provenance analysis includes all enquiries that would aid in reconstructing the lithospheric history of the Earth (Basu 2003; Weltje and Eynatten 2004) and can be performed with the help of detrital minerals, including heavy minerals supported by paleocurrent analysis. Provenance studies of sedimentary rocks have chiefly taken into account the mineralogical and/or chemical composition of sandstones and shales (Basu 1976; Dickinson and Suczek 1979; Dickinson et al. 1983; Singh 1996, 2013; Singh et al. 2000; Srivastava and Pandey 2011; Jalal and Ghosh 2012). Provenance study includes the distance and direction of transport, size and setting of the source region, climate and relief in the source area and the specific types of source rocks (Pettijohn et al. 1987). Additionally, heavy minerals have been utilized in provenance studies by numerous workers (e.g. Morton 1985; Basu and Molinaroli 1989; Nechaev and Isphording 1993; Singh et al. 2004; Patra et al. 2014).

Although biostratigraphic aspects of the Paleogene succession in Kachchh are well studied; its sedimentological aspects are least focused. Since little work has been carried out on these sandstones and the determination of provenance for the Matanomadh sandstones has not been attempted until now, authors aim to provide an insight on its plausible provenance. Therefore, an attempt is being made here to illustrate the plausible provenance using light and heavy minerals analysis, rock fragments and paleocurrent analysis as the major tools.

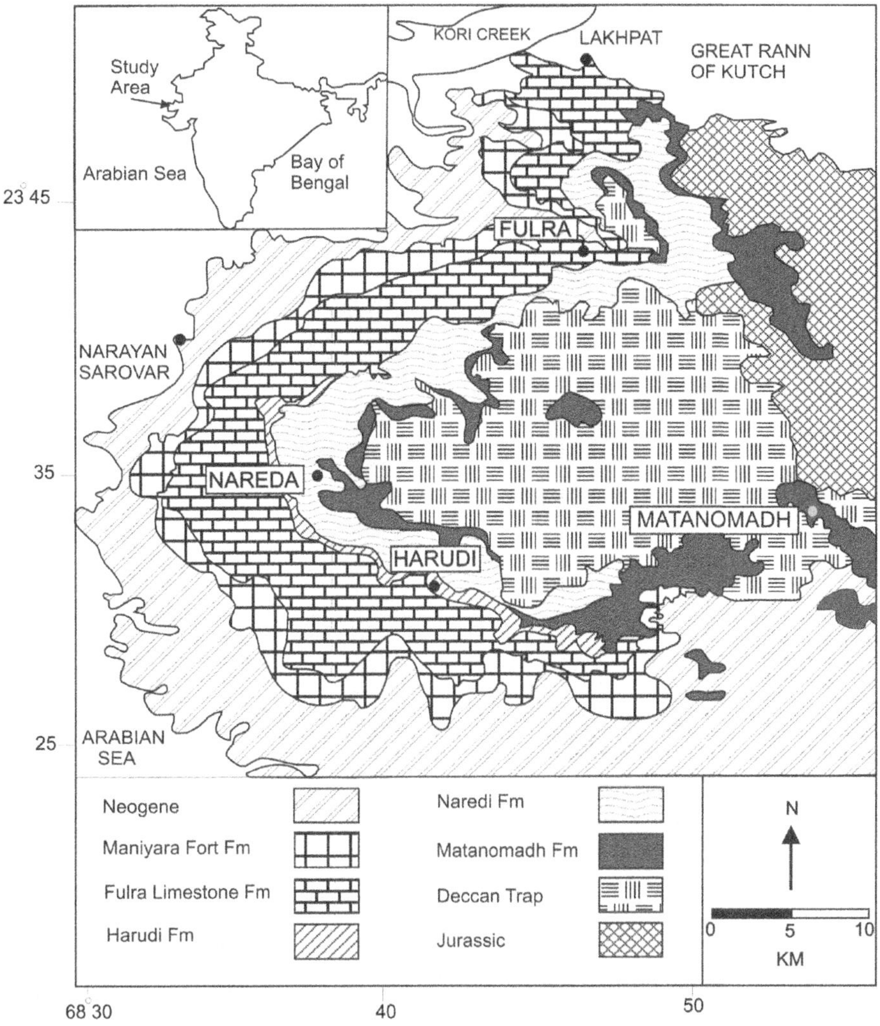

Fig. 1 Geological map showing Paleogene outcrops in the western Kachchh (after Biswas 1992). Here Matanomadh Formation is shown in grey patches. Red circle represents the investigated area. Inset: India in the outline map and arrow indicates the study area

Geological Background

Pericratonic Kachchh basin (KB), extended between latitudes 22° 30′ and 24° 30′N and longitudes 68° and 72°E, evolved due to sequential rifting and repeated movements in relation with the northward drift of Indian plate after the breakup of the Gondwanaland during Late Triassic-Early Jurassic period (~200 ma) (Norton and Sclater 1979; Biswas 1982, 1987). KB is bounded by the Nagar-Parkar fault in

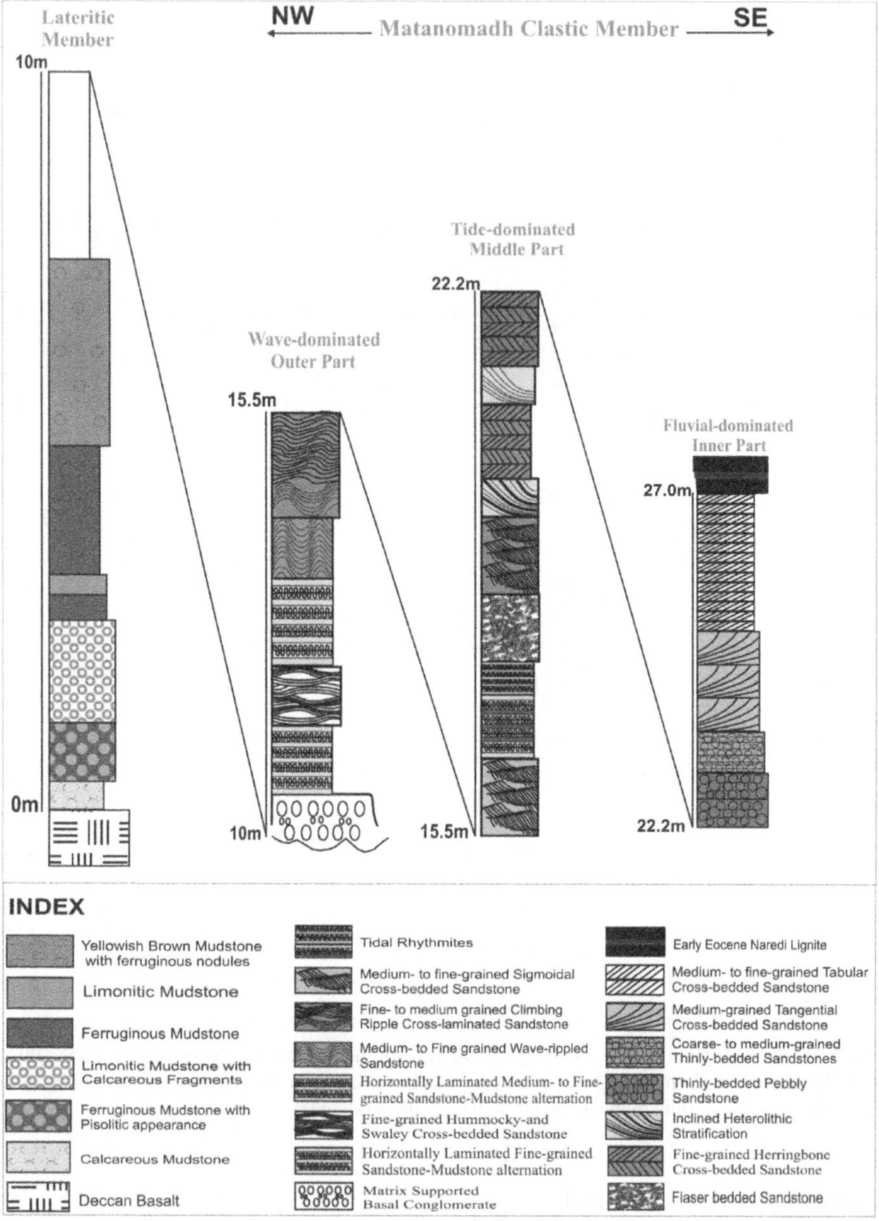

Fig. 2 Lithocolumn exhibiting vertical architecture of different lithofacies in the Paleocene succession of the Matanomadh Formation, Kachchh, including the late Paleocene sandstones of the Matanomadh Clastic Member

the North, Radhanpur-Barmer arch in the east and North Kathiawar fault towards the south (Biswas 1982). The basin has extended far offshore into the Arabian Sea in the west. It has preserved almost a complete sequence from middle-Jurassic to Recent punctuated by several stratigraphic breaks between transgressive cycles. The western marginal Kachchh basin constitutes one of the best developed, undisturbed Mesozoic-Cenozoic sequences in India and provides the repository of geological records related to the Gondwanaland fragmentation (Biswas 2005).

The Tertiary sequences are developed in the western part of this basin and exposed mainly in the narrow coastal plains of Kachchh Mainland and in the peripheral plains of other Mesozoic highlands (Biswas 1992). A complete section of the late Paleocene Matanomadh Clastic Member is outcropped in the western part of the Kachchh Mainland bordering Deccan Basalt (Fig. 1). The ~27 m thick MF represents the oldest lithostratigraphic unit of the Tertiary sequence of Kachchh (Fig. 1) and has been divided into a basal Lateritic Member and an upper Clastic Member (Saxena 1975). The basal lateritic member is about 10.0 m thick and is composed of calcareous mudstone, ferruginous mudstone, bauxite and laterite deposits. Clastic member in the upper part is comprised of ~17 m thick succession of sandstones and sandstone-mudstone alternation that are exposed in the eastern side of the Matanomadh village. The MF is overlain by Eocene lignite-containing Naredi Formation in the western part of the study area (Biswas 1992) where these lignite deposits are being mined by Gujarat Mineral and Development Corporation at Matanomadh and Panandhro localities.

Methodology

Individual facies were identified in the field and underlying and overlying facies were demarcated considering their sedimentologic attributes such as lithology, texture and sedimentary structures and based on that a detailed litholog was prepared. Fresh samples of these late Paleocene sandstones were collected from exposed sections. A total of 12 representative samples were used for thin section preparation for petrographic study out of the total 65 collected samples. Modal analysis was carried out with the help of Image Analysis Software attached with Leica Polarizing Microscope. Heavy minerals were separated from the sandstones using heavy liquid separation method. The sandstone was disaggregated by gently grinding it and the disaggregated fraction was sieved to obtain a size range of 0.0125–0.088 mm. The sieved fraction was then poured in the heavy liquid (Bromoform in the present study having Sp. Gr. 2.89) containing separating funnel. The liquid was stirred with a glass rod for proper mixing and heavy minerals were allowed to settle at the bottom of the funnel for nearly 20 min. This process was repeated three-times and subsequently, the settled heavy minerals were collected on a filter paper and dried at room temperature. Later on, they were washed with acetone to remove coatings. The heavy minerals were then mounted on glass slides with canada balsam. For paleocurrent measurements, the azimuthal data and angle of inclination were recorded for each cross-bed at different locations along and across facies boundaries.

Results

Matanomadh sandstone occurs as planar-bedded, cross-bedded, and ripple cross-laminated and laminated sandstones. Petrography, heavy mineral and paleocurrent analysis of these sandstones was performed and the results are described as under.

Petrography

Matanomadh sandstones are hard and compact and show unimodal to bimodal distribution of the particles (Figs. 3 and 4). Two varieties of the sandstones can be identified as coarse- to medium-grained sandstone and medium- to fine-grained sandstone on the basis of the particle size variation. Coarse to medium framework grains form the thickly- and thinly-bedded sandstones, and cross-bedded sandstones, while medium to fine framework grains form the ripple cross-laminated and laminated sandstones. The coarse- to medium-grained sandstone is moderately sorted and the grains are sub-rounded to rounded, show bimodal distribution (Fig. 3a). A few well rounded grains are also found. In them, quartz forms 74–75% where monocrystalline quartz (~67%) dominates over polycrystalline (8%) and non-undulatory quartz grains dominate over undulatory quartz. Polycrystalline quartz show interlocking texture. Low amount of feldspar (11–12%) is present in these sandstones (Table 1), which are mainly microcline and plagioclase (Fig. 3b, c), including the weathered varieties (Fig. 3d). Rock fragments form a little proportion (2–3%) in the coarse- to medium-grained sandstone. Cement (12–13%), is mainly calcareous and ferruginous in this variety of sandstones. These can be classified as quartzose arenites (following Okada 1971).

In the medium- to fine-grained sandstone, framework grains are mainly sub-angular to sub-rounded (Fig. 4). Quartz forms 76–78% where monocrystalline variety (~74%) dominates over the polycrystalline (~4%) in them (Table 1) (Fig. 4a). Mica is mainly muscovite in these sandstones (Fig. 4b–d). The rock-fragments of mica-schist and chert are observed in addition to few limestone fragments (Fig. 4c, d). Feldspar forms about 5% and rock fragments form 6–8% in this type of sandstone. Cementing material (11–12%) is mainly carbonate and iron-oxide. In some samples only carbonate cement is present, while in others, iron oxide coatings are present over the particles and cement is mainly carbonate. In those sandstones, which contain both the types of cement the carbonate cement looks earlier precipitated than ferruginous cement. The sandstones composition suggests that they belong to quartzose arenite types (after Okada 1971).

Table 1 Modal composition of the Matanomadh sandstones

Constituents (in %)	Coarse- to medium-grained sandstones						Medium- to fine-grained sandstones					
Quartz (Qtz)	73.91	74.55	74.32	74.36	74.13	74.56	78.03	76.82	77.45	75.84	75.79	75.55
Poly. Qtz	8.09	8.22	8.16	7.94	7.78	8.00	3.54	3.31	3.07	3.02	3.08	2.43
Mono. Qtz	65.82	66.33	66.16	66.42	66.40	66.56	74.49	73.51	74.38	72.82	72.71	73.12
Und. Qtz	8.62	7.83	8.22	7.80	8.08	8.16	5.21	5.91	5.46	5.20	5.42	5.32
Non-Und. Qtz	65.29	66.72	66.10	66.56	66.05	66.40	72.82	70.92	71.99	70.64	70.37	70.23
Feldspar	12.01	11.92	12.05	11.47	10.63	10.32	4.43	4.93	4.78	4.02	4.54	4.85
Rock Frag. (m)	–	0.40	0.51	0.53	1.02	0.84	4.26	4.57	4.21	5.94	5.48	6.03
Rock Frag. (s)	1.05	1.02	0.94	0.80	1.24	1.34	2.08	2.29	2.06	2.19	2.45	2.18
Rock Frag. (i)	–	–	–	–	–	–	–	–	–	–	–	–
Carbn. Cement	7.00	6.63	6.76	9.49	9.48	9.32	9.06	9.12	9.27	8.50	8.28	8.36
Ferrug. Cement	6.03	5.48	5.42	3.35	3.50	3.62	2.14	2.27	2.23	3.51	3.46	3.03
Total	100	100	100	100	100	100	100	100	100	100	100	100

Qtz = Quartz; Poly. Qtz = Polycrystalline Quartz; Mono. Qtz = Monocrystalline Quartz; Und. Qtz = Undulatory Quartz; Non-Und. Qtz = Non-undulatory Quartz; Frag. = Fragments; m = metamorphic; s = sedimentary; i = igneous; Carbn. = Carbonate; Ferrug. = Ferruginous

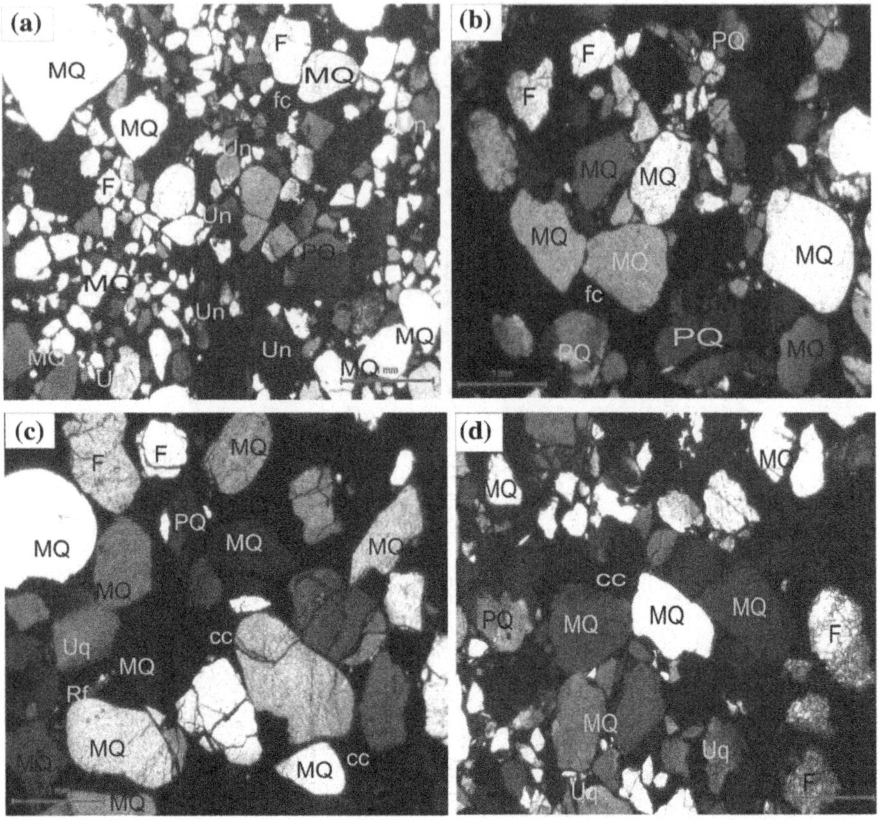

Fig. 3 Photomicrographs of coarse- to medium-grained sandstones. **a** Photomicrograph showing bimodal nature of grain size distribution and dominantly possessing monocrystalline quartz (MQ), polycrystalline quartz (PQ), undulatory quartz (Un), feldspar (F) and chert (Ch) cemented by ferruginous cement (fc). **b** Photomicrograph dominantly possessing sub-angular to sub-rounded monocrystalline quartz (MQ), polycrystalline quartz (PQ), undulatory quartz (Un), feldspar (F) cemented by ferruginous cement (fc). **c** Photomicrograph possessing sub-rounded monocrystalline quartz (MQ), polycrystalline quartz (PQ), undulatory quartz (Un) and feldspar (F) cemented by calcareous cement (cc). **d** Photomicrograph possessing sub-rounded monocrystalline quartz (MQ), polycrystalline quartz (PQ), undulatory quartz (Un) and weathered feldspar (F) cemented by calcareous cement (cc)

In Q-F-L diagram (after Dickinson et al. 1983), the composition of these sandstone units occupy the position within the craton interior (Fig. 5a) (Table 2). Further, they occupy craton interior to transitional continental field in the Qm-F-Lt diagram following Dickinson et al. (1983) (Table 3) (Fig. 5b).

Table 2 Recalculated data with respect to Q-F-L components in the Matanomadh sandstones (after excluding cement proportion)

Constituents (in %)	Coarse- to medium-grained sandstones						Medium- to fine-grained sandstones					
Q	84.98	84.82	84.63	85.31	85.19	85.64	87.87	86.69	87.51	86.19	85.87	85.26
F	13.80	13.56	13.72	13.16	12.21	10.32	4.99	5.56	5.40	4.57	5.15	5.47
L	1.22	1.62	1.65	1.53	2.60	4.04	7.14	7.75	7.09	9.24	8.98	9.27
Total	100	100	100	100	100	100	100	100	100	100	100	100

Q = Total Quartz; F = Feldspar; L = Rock fragments

Table 3 Recalculated data with respect to Qm-F-Lt components in the Matanomadh sandstones

Constituents (in %)	Coarse- to medium-grained sandstones						Medium- to fine-grained sandstones					
Qm	75.65	75.45	75.33	76.20	76.30	76.45	83.88	82.95	84.04	82.76	82.38	82.52
F	13.80	13.56	13.72	13.16	12.21	10.32	4.99	5.56	5.40	4.57	5.15	5.47
Lt	10.55	10.99	10.95	10.64	11.49	11.85	11.13	11.49	10.56	12.67	12.47	12.01
Total	100	100	100	100	100	100	100	100	100	100	100	100

Qm = Quartz monocrystalline; F = Feldspar; Lt = Rock fragments including quartzite fragments (polycrystalline quartz)

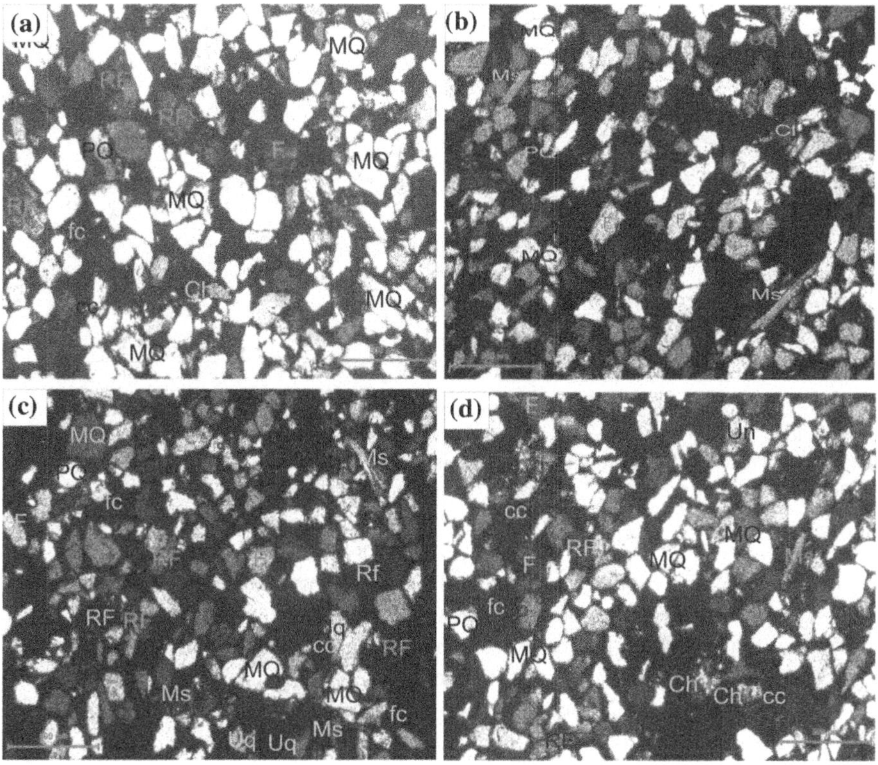

Fig. 4 Photomicrographs of medium- to fine-grained sandstones. **a** Photomicrograph shows sub-angular to sub-rounded monocrystalline quartz (MQ), polycrystalline quartz (PQ), feldspar (F) and rock fragments phyllite (RF)/chert (Ch) cemented by both calcareous and ferruginous cement. **b** Photomicrograph shows sub-angular to sub-rounded monocrystalline quartz (MQ), polycrystalline quartz (PQ), undulatory quartz (Un), feldspar (F) and rock fragments (RF)/calcite (Cl) and muscovite (Ms) cemented by both calcareous and ferruginous cement. **c** Photomicrograph shows sub-angular to sub-rounded monocrystalline quartz (MQ), polycrystalline quartz (PQ), undulatory quartz (Un), feldspar (F) and rock fragments (RF) and muscovite (Ms) cemented by both calcareous and ferruginous cement. **d** Photomicrograph shows sub-angular to sub-rounded monocrystalline quartz (MQ), polycrystalline quartz (PQ), undulatory quartz (Un), feldspar (F) and rock fragments (RF)/chert (Ch) and muscovite (Ms) cemented by both calcareous and ferruginous cement

Heavy Minerals

In these sandstones, both opaque and non-opaque heavy mineral varieties occur where opaque variety dominates over non-opaque. In particular, the proportion of opaque heavy minerals (magnetite) increases up-section and the proportions of non-opaque heavy minerals decrease up-section. Most of the studied heavy minerals are sub-rounded in outline but some of them show sub-angular and rounded forms also.

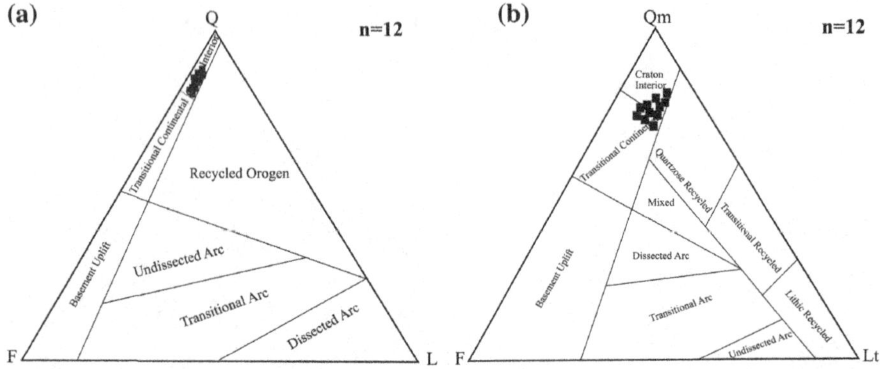

Fig. 5 **a** Q-F-L diagram for the late Paleocene Matanomadh Sandstones, Kachchh basin (after Dickinson et al. 1983). Here n represents the number of samples. **b** Qm-F-Lt diagram for the late Paleocene Matanomadh Sandstones, Kachchh basin (after Dickinson et al. 1983). Here n represents the number of samples

In order of abundance, the heavy minerals suite is comprised of magnetite > tourmaline > kyanite > rutile > monazite > staurolite > hematite (Table 4).

Magnetite ranges from 55 to 66% with sub-angular to sub-rounded nature of the grain corners. It is identified by its black colour and high relief (Fig. 6). Tourmaline occurs in a range of 9–15% and shows sub-rounded to rounded grain boundaries. Two varieties of tourmaline are identified; they are yellow to brown and blue tourmaline. It is identified by its pleochroism, high refractive index, imperfect cleavage and parallel extinction (Fig. 6). Kyanite forms 6–11% of the heavy minerals with sub-rounded shape. It is recognized by its bladed form, colour, oblique extinction and two-sets of mutually perpendicular cleavage pattern (Fig. 6). Rutile occurs in a range of 6–12% with sub-rounded to rounded grain boundaries. It is identified by its characteristic deep red interference colour and pleochroism (Fig. 6). Monazite forming 6–10% of the heavy minerals shows prismatic crystal outline with sub-rounded to rounded grain shape and recognized by its greenish black colour and shining surface (Fig. 6). Staurolite forms 2–4% of the bulk heavy minerals. It shows yellowish colour, angular shape, pleochroism, high refractive index, imperfect cleavage, low birefringence and parallel extinction. It contains numerous inclusions of quartz (Fig. 6). Hematite ranges from 1 to 3% and it shows angular to sub-angular nature of corners. It is identified by its light reddish grain boundaries (Fig. 6).

Dispersal Pattern

Sediments dispersal pattern is mainly controlled by flow directions and hydrodynamic conditions of the depositional processes and depositional milieu. Flow directions in sedimentary successions are commonly associated with primary sedimentary

Table 4 Heavy minerals percentage in the Matanomadh sandstones

Constituents (in %)	Coarse- to medium-grained sandstones						Medium- to fine-grained sandstones					
Magnetite	66	66	65	56	57	57	56	55	55	56	57	56
Tourmaline	10	9	9	13	13	12	11	12	12	14	14	15
Kyanite	7	7	6	11	10	10	9	9	9	8	8	8
Rutile	6	7	7	8	9	10	12	11	12	12	11	11
Monazite	9	9	10	7	6	6	7	7	7	6	7	6
Staurolite	–	–	–	2	3	3	2	3	3	3	3	4
Hematite	2	2	3	3	2	2	3	3	2	1	–	–
Total	100	100	100	100	100	100	100	100	100	100	100	100

Fig. 6 Heavy minerals. **a**, **b** Staurolite; **c**–**e** Rutile; **f**–**j** Monazite; **k**–**n** Kyanite; **o** Hematite; **p**, **q** Magnetite; **r**–**w** Tourmaline

structures (Selley 1980; Pettijohn 1984) and these structures are the only direct source for paleocurrent measurements. The dispersal/paleocurrent pattern measured from the foresets of the Matanomadh cross-bedded sandstones decipher NW-SE, NW and WNW-ESE directed bipolar and unidirectional paleocurrent patterns. The paleocurrent reversal is the characteristic feature of the lower part of the succession which is dominated by sigmoidal- and herringbone cross-beds (Fig. 7a). The NW-directed

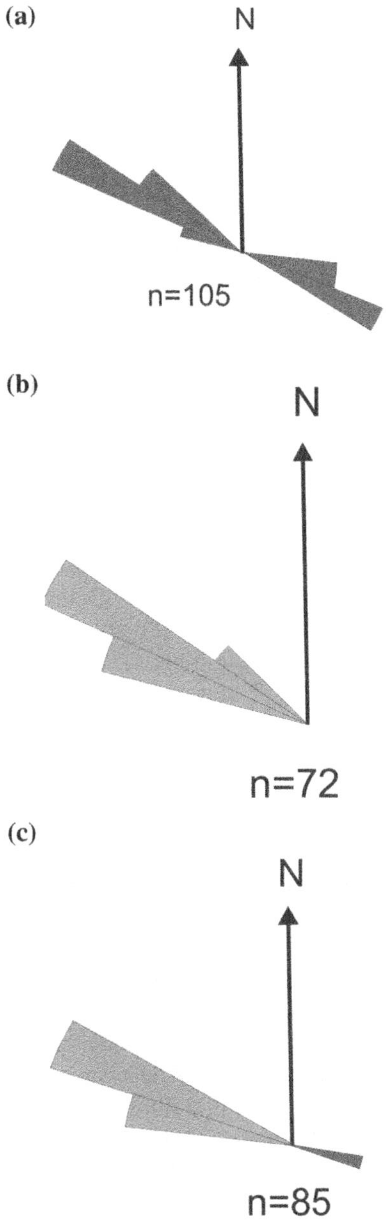

Fig. 7 Paleocurrent diagrams. **a** Paleocurrent diagram for the herringbone- and sigmoidal cross-bedded sandstones. Note here the bipolar dispersal pattern from the tide-dominated lower part of the succession. **b** Paleocurrent diagram for the tangential cross-bedded sandstones. **c** Paleocurrent diagram for the tabular cross-bedded sandstones. Note here the unidirectional dispersal pattern from the fluvial-dominated upper part of the Matanomadh sandstone succession

unidirectional paleocurrent (Fig. 7b) and the WNW-dominated bipolar paleocurrent pattern are observed in the upper part of the succession containing tangential and tabular cross-beds (Fig. 7c).

Discussion

Provenance for the MS is discussed based on the light minerals, rock fragments and heavy minerals analysis. The occurrence of quartz having monocrystalline non-undulatory nature generally points towards sources such as volcanic and hypabasal igneous rocks, fine-grained schists, phyllite and slates, and pre-existing sedimentary rocks (e.g. Blatt et al. 1980). The polycrystalline quartz with interlocking texture is derived from a metamorphic rock, most likely quartzite (e.g. Blatt et al. 1980). Chert in these rocks would have been derived from sedimentary rocks. Plagioclase variety of feldspar might have derived from a basaltic source. The limestone fragments suggest their derivation from a sedimentary source. The metamorphic rocks of the Aravalli hills exposed in the NE must have acted as the chief sources for the MS. Deccan Trap forms the basement of the MF in the Kachchh region. In the Kachchh region, Deccan trap is defined as the outpouring of tholeiitic basaltic lava that overlies the Mesozoic sedimentary rocks of marine and fluvio-deltaic origin (Merh 1995). This must have contributed part of the sediments to the MS.

Kachchh Basin is an east-west oriented pericratonic rift basin at the westernmost periphery of the Indian subcontinent. The Kachchh Rift evolved within the Mid-Proterozoic Aravalli-Delhi fold belt by reactivation of pre-existing faults along NE-SW trend of Delhi fold belt that swings to E-W in Kachchh region (Biswas 2005).Thus, these faults mainly controlled the basin configuration during Late Triassic/Early Jurassic period and the uplifted blocks of the faults containing Mesozoic sequences were weathered and eroded before depositing in the grabens over Deccan basalt and Mesozoic succession during Paleocene epoch (Fig. 8). Most of the quartzose sands are derived from stable craton having low relief while presence of high proportion of lithic fragment suggests upliftment at the source area (Dickinson et al. 1983). The craton interior field in the Q-F-L diagram and craton interior to transitional continental field in the Qm-F-Lt plot occupied by these sandstone units suggest a stable craton or passive margin provenance of a rift basin that had supplied sediments to the depositional basin.

Although it is considered that a part of the heavy minerals is altered during diagenesis (e.g. Garzanti and Vezzoli 2003), the abundance of magnetite among the heavy minerals in the studied sandstones suggests that it was derived from a basic igneous rock such as Deccan basalt. Brown tourmaline indicates low rank metamorphic source, while blue tourmaline indicates a pegmatitic source (Pettijohn 1984). The dominance of the brown tourmaline in the MS suggests its derivation from a low-grade metamorphic rock. The occurrence of kyanite in larger proportion and staurolite in smaller proportion suggest that they were contributed from a low- to medium-grade metamorphic source. Rounded grains of rutile, tourmaline and monazite in the studied sandstones were likely derived either from an acid igneous rock or from reworking of the sediments (Pettijohn 1984).

Bipolar paleocurrent patterns are observed in the depositional basins that are influenced by flood- and ebb-tidal currents in a tide-dominated estuarine or deltaic condition (Selley 1980). According to Dalrymple et al. (1992), an estuary is the

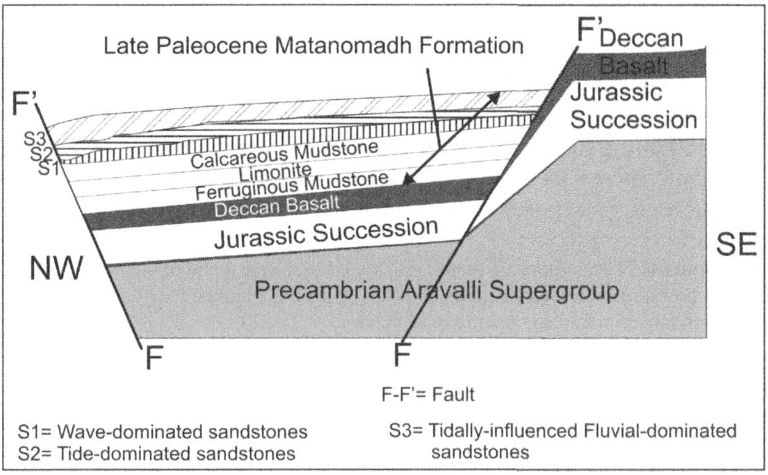

Fig. 8 Cartoon exhibiting vertical distribution of MS lithounits within the late Paleocene half-graven of Kachchh, Gujarat, India and their possible provenance

seaward portion of a drowned valley system which receives sediments from both fluvial and marine sources and which contains facies influenced by marine agents such as tides, waves and fluvial processes. Estuaries may be either tide-dominated or wave-dominated or river-dominated. The bipolar paleocurrent (in parts) suggests that these sandstones were deposited in a wave- and tide-dominated estuary that was later dominated by the unimodal paleocurrent of the fluvial process. The paleocurrent direction for the uppermost tabular cross-bedded sandstone varies slightly than the underlying tangential cross-bedded sandstone, but both are NW-directed. Thus, it is presumed that the NW directed flow was a result of the river current and SE dominated flow was a result of advancing tides. The overall dominance of the WNW paleocurrent direction in the Matanomadh sandstone succession suggests that the main provenance was in the SE direction of the depositional basin. Thus, the Precambrian metamorphic rocks of the basement along with Deccan basalt and Mesozoic sedimentary rocks of the Kachchh basin might have acted as the source during sedimentation of the MS.

Conclusions

Petrographic study of the Matanomadh sandstones shows abundance of sub-angular to rounded monocrystalline non-undulatory quartz in association with minor proportion of feldspar and lithic fragments that classifies them as quartzose arenite variety. The Q-F-L and Qm-F-Lt plots suggest that the sandstones represent craton interior provenance. Abundance of opaque variety of heavy mineral (e.g. magnetite) over the non-opaque tourmaline, kyanite, rutile, monazite and staurolite is observed in these sandstones. The paleocurrents have been bipolar in the lower tide-dominated part of the MS to unidirectional in the upper river-dominated part. The main directions of flow have been NW-SE and WNW, respectively. Petrography and heavy minerals studies along with paleocurrents study suggest that the provenance was dominated by low- to medium-grade metamorphic rocks of the basement i.e. Aravalli metamorphic rocks, volcanic rocks of basaltic composition such as Deccan Trap and older Mesozoic successions exposed in the southeast of the depositional basin.

Acknowledgements The authors are thankful to the Head, Department of Geology, Banaras Hindu University for providing working facilities. Dr. S. K. Ghosh is thanked for going through an earlier version of the manuscript and suggesting modifications.

References

Basu A (1976) Petrology of fluvial sand derived from plutonic source rocks: implications to palaeoclimatic interpretation. J Sediment Petrol 46:694–709

Basu A (2003) A prospective on quantitative provenance analysis. In: Valloni R, Basu A (eds) Quantitative provenance studies in Italy, vol 61. Mem. Deserit. dellacarta Geol. dell Itallia, pp 11–22

Basu A, Molinaroli E (1989) Provenance characteristics of detrital opaque Fe-Ti oxide minerals. J Sediment Petrol 59:922–934

Biswas SK (1965) A new classification of Tertiary rocks of Kachchh, Western India. Bull Geol Min Metall Soc India 35:1–6

Biswas SK (1982) Rift basins in western margin of India with special reference to hydrocarbon prospects. AAPG Bull 66:1497–1513

Biswas SK (1987) Regional tectonic framework, structure and evolution of the western marginal basins of India. Tectonophysics 135:307–327

Biswas SK (1992) Tertiary stratigraphy of Kuth. J Palaeontol Soc India 37:1–29

Biswas SK (2005) A review of structure and tectonics of Kachchh basin, western India, with special reference to earthquakes. Curr Sci 88:1592–1600

Biswas SK, Raju DSN (1973) The rock-stratigraphic classification on the Tertiary sediments of Kachchh. Bull ONGC 10:37–46

Blatt H, Middleton GV, Murray RC (1980) Origin of sedimentary rocks. Prentice Hall, 782 pp

Dalrymple RW, Zaitlin BA, Boyd R (1992) Estuarine facies models: conceptual basis and stratigraphic implications. J Sediment Petrol 62:1130–1146

Dickinson WR, Suczek CA (1979) Plate tectonics and sandstone composition. AAPG Bull 63:2164–2182

Dickinson WR, Beard LS, Brakenridge GR, Erjavek JL, Ferguson RC, Inman KF, Knepp RA, Lindberg FA, Ryberg PT (1983) Provenance of North American Phanerozoic sandstones in relation to tectonic setting. Geol Soc Am Bull 94:222–235

Garzanti E, Vezzoli G (2003) A classification of metamorphic grains in sandstones based on their composition and grade. J Sediment Res 73:830–837

Jalal P, Ghosh SK (2012) Provenance of the Late Neogene Siwalik sandstone, Kumaun Himalayan Foreland Basin: constraints from the metamorphic rank and index of detrital rock fragments. J Earth Syst Sci 121:781–792

Merh SS (1995) Geology of Gujarat. Geol. Soc. India., 220 pp

Morton AC (1985) Heavy minerals in provenance studies. In: Zuffa GG (ed) Provenance of arenites. NATO Advance Science Institute Series, No. 418. D. Reidel, Dordrecht, pp 249–277

Nechaev VP, Isphording WC (1993) Heavy mineral assemblages of continental margins as indicators of plate tectonic environment. J Sediment Petrol 63:1110–1117
Norton IO, Sclater JG (1979) A model for the evolution of the Indian Ocean and the breakup of Gondwanaland. J Geophys Res 84:6803–6830
Okada H (1971) Classification of sandstones: analysis and proposal. J Geol 79:509–525
Patra A, Singh BP, Srivastava VK (2014) Provenance of the late Paleocene sandstones of the Jaisalmer Basin, Western India. J. Geol Soc India 83:657–664
Pettijohn FJ (1984) Sedimentary rocks, 3rd edn. CBS Publishers, 628 pp
Pettijohn FJ, Potter PE, Siever R (1987) Sand and sandstone, 2nd edn. Springer, New York, 553 pp
Saxena RK (1975) Lithostratigraphy of the Matanomadh Formation, Kachchh, India. Palaeobotanist 24:261–262
Selley RC (1980) Ancient sedimentary environments, 2nd edn. Cornell University Press, 287 pp
Singh BP (1996) Murree sedimentation in the northwestern Himalaya. Geol Surv India Spec Publ 21:157–164
Singh BP (2013) Evolution of the Paleogene succession of the western Himalayan foreland basin. Geosci Front 4:199–212
Singh BP, Andotra DS, Kumar R (2000) Provenance of the lower Tertiary mudrocks in the Jammu Sub-Himalayan zone, Jammu and Kashmir State (India), NW Himalaya and its tectonic implications. Geosci J 4:1–9
Singh BP, Pawar JS, Karlupia SK (2004) Dense mineral data from the north-western Himalayan foreland sedimentary rocks and recent river sediments: evaluation of the hinterland. J Asian Earth Sci 23:25–35
Srivastava SK, Pandey N (2011) Search for provenance of Oligocene Barail sandstone in and around Jotsoma, Kohima, Nagaland. J Geol Soc India 77:433–442
Srivastava VK, Singh BP (2017) Shoreface to estuarine sedimentation in the late Paleocene Matanomadh Formation, Kachchh, western India. J Asian Earth Sci 136:1–15
Weltje GJ, Eynatten HV (2004) Quantitative provenance analysis of sediments; review and outlook. Sed Geol 171:1–11
Wynne AB (1872) Geology of Kachchh. Mem Geol Surv India 9:1–289

Palynofacies Study of Lakadong Limestone (Late Paleocene) of Mawsynram Area, Shillong Plateau, India: Implications for Sequence Stratigraphy

Bikash Gogoi, Vandana Prasad, Rahul Garg and Indrabir Singh

Abstract Subtle paleoenvironmental fluctuations are difficult to detect in sedimentologically uniform carbonate rocks, however, study of relative increase and decrease of marine and terrestrial organic matter content provides a useful tool for deciphering the deepening and shallowing cycles in these successions. Lakadong Limestone Member in Khasi and Jaintia hills in South Shillong Plateau is late Paleocene in age and forms part of the carbonate platform. It represents deposition during High Stand Systems Tract. In the present study palynofacies analysis were carried out in two Lakadong Limestone sections i.e., Dohsniang (Kurtinsiang) (KPL) and Laitmowksing (LTL) from Khasi hills for detailed palaeoenvironmental interpretations, for the correlation and for the identification of higher order sea level cycles. Both the sections have been dated as late Paleocene based on the characteristic larger benthic foraminiferal assemblages belonging to the Tethyan Shallow Benthic Zones SBZ 3. For palynofacies analysis various type of organic matter were characterized and counted. The study shows cyclicity pattern in the organic matter types in both the sections which may be linked to the sea level changes of higher order cycles. Based on the variation in the organic matter content both the sections were subdivided into distinct palynofacies units. Each palynofacies unitsstart with high proportion of black oxidized palynomaceral along with dinoflagellate cysts representing a transgressive surface, followed by high quantity of degraded brown and cuticle organic matter from terrestrial source. Each palynofacies unit thus represent progradational deposit of High Stand Systems Tract starting with the deepening facies followed by shallowing facies of more terrestrial origin. Four progradational sequences have been identified in the KPL and three in the LTL section that can be correlated. Palynofacies

B. Gogoi
Department of Geological Sciences, Gauhati University, Guwahati 781014, India

V. Prasad (✉) · R. Garg
Birbal Sahni Institute of Palaeosciences, Lucknow 226007, India
e-mail: prasad.van@gmail.com

I. Singh
17-11-2C Metro City Nishat Ganj, Lucknow 226007, India

S. Bajpai et al. (eds.), *The Indian Paleogene*, Society of Earth Scientists Series,
https://doi.org/10.1007/978-3-319-77443-5_7

study thus offers a logical approach for the study of uniform carbonate facies. Based on the present palynofacies criteria it has been possible to identify higher order sea level cycles in the Lakadong Limestone exposed at KPL and LTL section in Khasi Hills.

Keywords Sedimentary organic matter · Biostratigraphy
Carbonate facies · Palaeoenvironment · High Stand Systems Tract (HST)

Introduction

Palynofacies study helps in the determination of relative increase and decrease of marine and terrestrial organic matter content in the sedimentary successions. It can be an important tool for the documentation of depositional cycles in shallow marine settings to detect subtle change in paleoenvironmetal conditions (Tappan 1980; Parry et al. 1981; Tyson 1987; Gorin and Steffen 1991;Steffen and Gorin 1993; Tyson 1993, 1995, 1996; Bombardiere and Gorin 1998). High rate of supply, rapid burial and good preservation of organic matter makes conditions ideal for application of palynofacies for the recognition of depositional cycles in fine grained shallow marine silciclastic sedimentary sequences (Muller 1959; Combaz 1964; Combaz 1980; Pratt 1984; Tissot and Welte 1984; Evitt 1985). Identification of shallowing-deepening cycles for deciphering paleoenvironmental changes is often difficult in carbonate rocks. The study of relative increase and decrease of marine and terrestrial organic matter content provides a useful tool for deciphering the deepening and shallowing cycles in carbonate rocks. Generally, palynofacies analysis in carbonate rocks is difficult due to low terrigenous sediment supply from land, slow rate of sedimentation and oxidizing environment of deposition. However, during last decade, palynofacies studies have been increasingly applied for the interpretation of paleoenvironment and relative sea level fluctuations in carbonate successions (Gorin and Steffen 1991;Steffen and Gorin 1993; Bombardiere and Gorin 1998; Strohmenger and Strasser 1993; Pittet and Gorin 1997). In early Palaeogene successions, availability of larger benthic foraminiferal biozonation scheme is helpful in determination of a geochronological framework and identification of eustatic sea level cycles.

Early Paleogene is an important time interval due to the incidence of large scale global warming known as the Paleocene Eocene Thermal Maxima and other hyperthermal events (Aubry 1998; Jaramillo and Oboh-Ikuenobe 1999; Dickens 2000; Prasad et al. 2006; Prasad et al. 2013). Most of the studies were carried out in mid and high latitudinal areas (Aubry 1998; Dickens 2000; Jaramillo and Oboh-Ikuenobe 1999), while only few studies are available from low latitudinal regions (Prasad et al. 2006; Prasad et al. 2013). The Indian subcontinent was in equatorial zone during early Paleogene (Kiessling et al. 2003). Shillong plateau, which is the northeastern extension of Indian Craton, shows well developed early Paleogene succession on the southern fringes of Garo-Khasi-Jaintia hills. The succession consists of alternating units of carbonates and terrigenous clastics deposited in tropical climate. Late

Paleocene-early Eocene time period has been considered a significant due to its association with Paleocene Eocene Thermal Maxima (PETM). In order to correctly assess the warming event it is important to understand the pre-PETM environmental conditions. Deposits of Lakadong Limestone Member in East Khasi Hills being late Paleocene in age can provide environmental conditions of pre-PETM. Lakadong Limestone Member is a deposit of High Stand System Tract of 3rd order sea level cycles (Garg and Ateequzzaman 2000; Prasad et al. 2006). In the present study, palynofacies criteria have been applied on the Lakadong Limestone Member (late Thanetian) of KPL and LTL sections of the Mawsynram area, Khasi Hills of Meghalaya to decipher 4th–5th order sea level cycles. The purpose of the present paper is to provide information on the palaeoclimate and palaeoenvironmental changes during late Thanetian, before the onset of the PETM at Palaeocene-Eocene boundary.

Regional Geology

In early Cretaceous separation of India from Antractica-Australia assembly started and formation of east coast of India was initiated. The separation of India took place first in the SW part and gradually moved to the north eastern part. The northeastern part of India was the last to separate from the Antarctica–Australia assembly. In this process a number of sedimentary basins were formed on the east coast of India, namely Cauvary, Pennar, Krishna–Godavari, Mahanadi, Bengal and Assam-Arakan Basin (Singh and Swamy 2006).

Meghalaya is a part of the Assam–Arakan Basin. The Cretaceous-Tertiary succession in Meghalaya is well developed along the southern margin of the Shillong Plateau and succession is exposed in Garo–Khasi–Jaintia hills. The sedimentation in the Meghalaya started in Latest Campanian–Maastrichtian in a rift basin depositing mainly conglomerates and sandstone of Khasi Group. In Palaeocene time the basin became passive margin and sedimentation on inner shelf started with the deposition of fine grained terrigenous clastics and carbonate buildups. The clastic sedimentation with intermittent thin limestone units were deposited in the inner neritic shallow shelf in this region, continued during early Paleocene (Garg and Ateequzzaman 2000). Late Palaeocene witnessed a change in the sedimentation pattern due to reduced supply of terrigenous clastic material from land leading to carbonate sedimentation (Jauhri 1988, 1994; Garg and Ateequzzaman 2000; Jauhri and Agawal 2001; Gogoi 2006). Development of a large carbonate ramp during late Thanetian resulted in the deposition of foraminifera rich carbonate succession, the Lakadong Limestone Member. The larger benthic foraminifera of Lakadong Limestone (Jauhri and Agawal 2001; Gogoi et al. 2009) is well correlated with the Tethyan foraminifera zone of late Thanetian sequence (Serra-Kiel et al. 1998). The deposition of the carbonate ramp was short lived during late Thanetian and the region was converted into shallow marine terrigenous clastic depositional area which continued during early Eocene. Thus, the Paleocene-Eocene transition in this region witnessed change in the depositional regime from carbonate to coal bearing terrigenous clastic deposit. In an earlier study palynofacies analysis were carried out on the early Paleogene succession of Khasi Hills for the identification of 3rd order Systems Tracts and the establishment of sequence stratigraphic framework (Garg and Ateequzzaman 2000; Prasad et al. 2006).

Age of Lakadong Limestone Member in Mawsynram

The Lakadong Limestone Member of Mawsynram yields a good assemblage of larger benthic foraminifera of prominent Tethyan markers. Based on the occurrences of FAD and LAD of foraminiferal species *Miscellanea juliettae*, *Miscellanea yvettae*, *Miscellanea miscella* Gogoi et al. (2009) documented two Taxon Range Zones viz. *Miscellanea juliettae* Taxon Range Zone and *Miscellanea miscella* Taxon Range Zone, and one Concurrent Range Zone (*Miscellanea juliettae–Miscellanea miscella* Concurrent Range Zone) in the KPL and LTL section of Mawsynram. The biozones established were correlated with the standard Shallow Benthic Zone (SBZ) of Serra-Kiel et al. (1998). Both the biozones *Miscellanea juliettae* Taxon Range Zone and *Miscellanea miscella* Taxon Range Zone are placed within the SBZ 3 Zone (Early Thanetian) of Serra-Kiel et al. (1998) which is equivalent to Early P4 Zone of planktonic foraminifera (Berggren et al. 1995) of Thanetian age.

Study Area

Lithostratigraphy and geological framework of Cretaceous-Paleogene successions in Mawsynram and Cherrapunji areas of the Khasi Hills, south of the Shillong Plateau have been studied by several workers (Evans 1932; Medlicott 1869; Wilson and Metre 1953; Nagappa 1959; Biswas 1962; Mathur and Evans 1964; Murthy et al. 1976; Dasgupta 1977; Raja Rao 1981; Garg and Ateequzzaman 2000; Gogoi et al. 2009). The Therria Formation represented by thick terrigenous clastic sediments is overlain by thick carbonate succession of the Sylhet Limestone Formation. The carbonate succession ranges in age from late Palaeocene to middle Eocene and includes three foraminifera rich limestone units interbedded with thick sandstone-shales units.

The present study is confined to the Palaeocene succession in Mawsynram area (Fig. 1). It is represented by calcareous and sandy shales and sandstone followed by carbonate and recognized into distinct stratigraphic units (Table 1). In the region, only the Lakadong Limestone Member and the overlying Lakadong Sandstone Member of the Sylhet Limestone Formation are exposed (Figs. 2 and 3). The Lakadong Limestone Member constitutes the basal most lithounit of Sylhet Limestone Formation and it normally occupies the top of the hillocks in the area. It is light grey to dark grey coloured compact limestone with rich assemblages of foraminifera and calcareous algae. The Lakadong Sandstone Member overlies the Lakadong Limestone Member and constitutes the upper most lithounit of the succession exposed

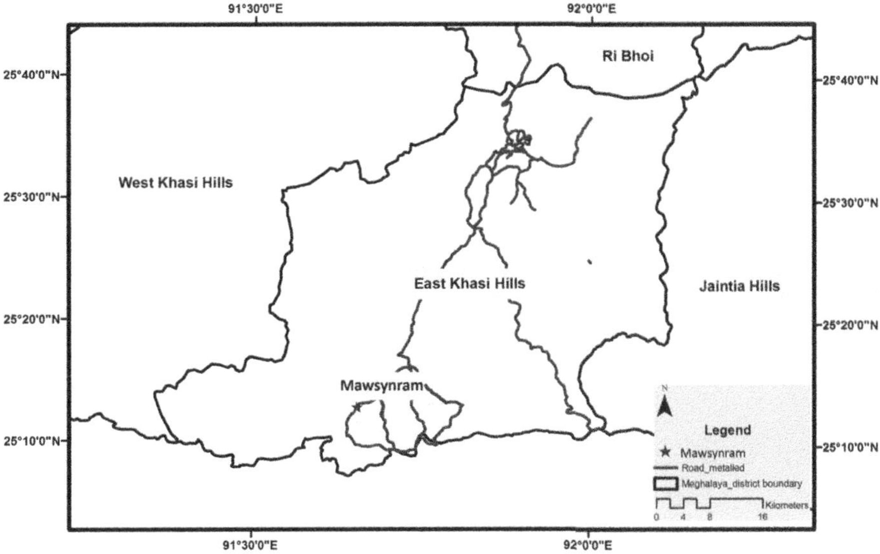

Fig. 1 Location map of the study area

Table 1 Stratigraphic succession observed in Mawsynram area

Formation	Member	Lithology	Age
Sylhet Limestone	Lakadong Sandstone	Fine to medium grained coal bearing sandstone	Thanetian (Late Palaeocene)
	Lakadong Limestone	Dark grey to gray coloured hard and compact algal and foram rich limestone	Thanetian (Late Palaeocene)
Therria	–	Hard, compact, coarse to medium grained sandstone	Selandian (Early Palaeocene)
Langpar	–	Calcareous, fine shales with occasional carbonaceous bands with foraminifera	Late Maastrichtian to Danian (Early Palaeocene)
Mahadek	–	Medium to very coarse grained massive sandstone	Late Campanian to Late Maestrichtian (Late Cretaceous)

in the study area. Towards basinal side, the thickness of the limestone recorded in KPL section is 15.08 m, while in the landward LTL section it is only 9 m thick. The thickness variations of clastic and carbonate facies thought to be related to change in depositional basin configuration of this tectonically undisturbed part.

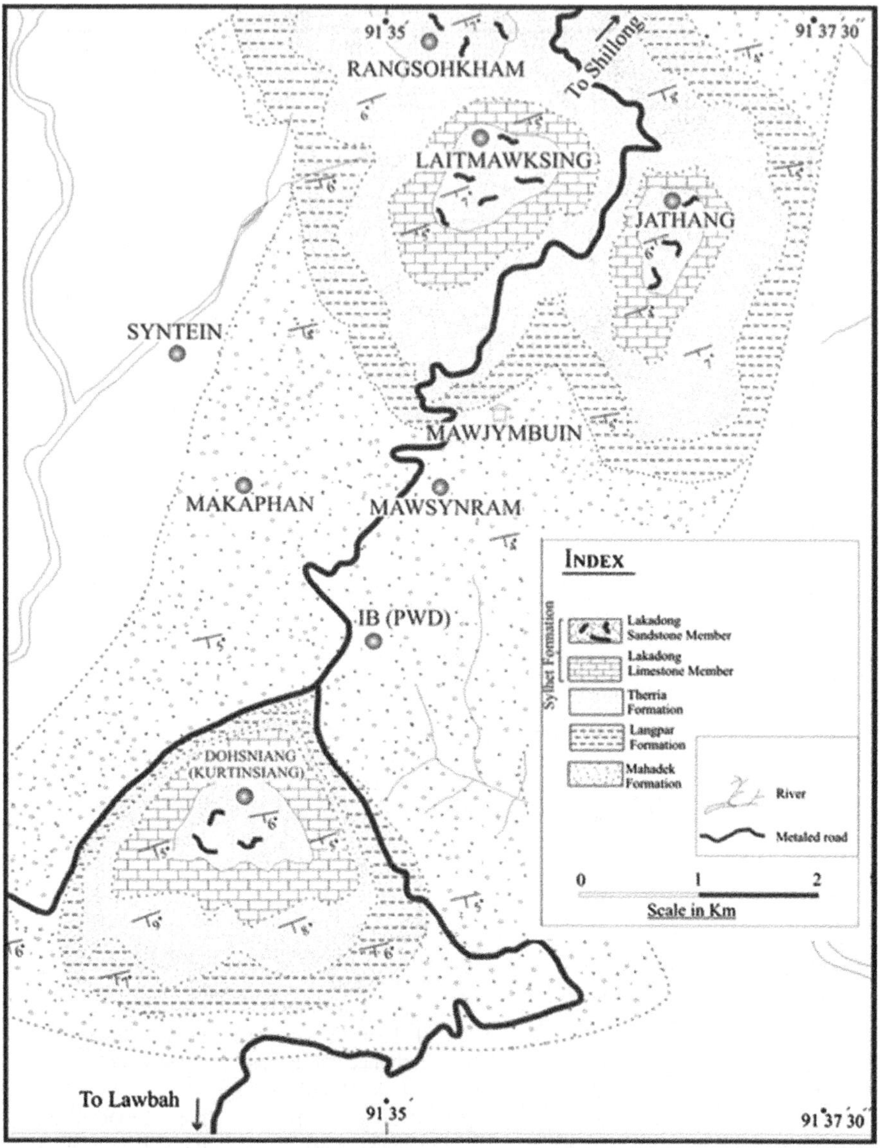

Fig. 2 Geological map of the study area in Mawsynram, East Khasi Hills, Meghalaya (after Gogoi et al. 2009)

Materials and Methods

Selective representative limestone samples and interbedded shale samples were collected for the present study. Samples were thoroughly cleaned by washing under

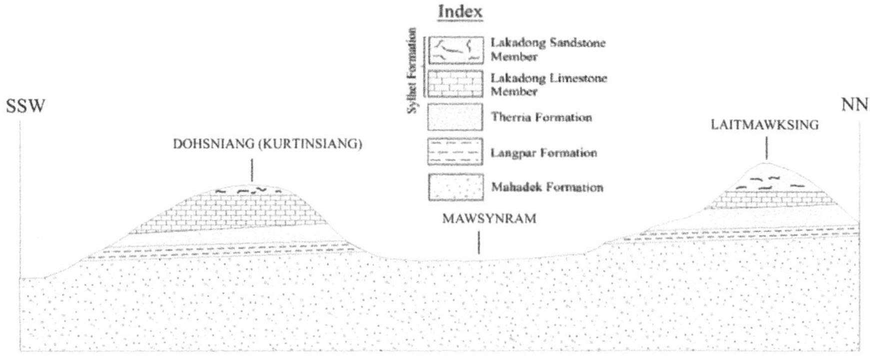

Fig. 3 Schematic cross-section of the study area along NNE-SSW (after Gogoi et al. 2009)

distilled water. Because of the low organic matter content in the carbonate rocks, 100 g of sample were used for maceration. Utmost care was taken to avoid contamination during crushing of limestone samples. Samples were prepared by using standard palynological techniques by digesting all carbonate and silicate mineral constituents were digested using hydrochloric and hydrofluoric acids. Oxidation by nitric acid was done to clean the organic matter. Two sets of slides were prepared; for the study of non-oxidized organic matter analysis one set of slides was prepared prior to HNO_3 treatment. For the study of dinoflagellate cysts and other palynomorphs, slides were prepared after proper oxidization of the maceral with dilute HNO_3. The charcterization of various organic matter types in the palynofacies analysis was carried out under transmitted light microscopy using Olympus Vanox and BHS microscopes with nomarski interference contrast and auto photo attachment. A minimum of 200 counts were made on each slide.

Palynofacies Analysis

The organic matter in the sedimentary succession is composed of both autochthonous and allochthonous particles (Hart 1986; Pocock SAJ 1982). The distribution of allochthonous particles in marine depositional setting is controlled by the physicochemical processes and is a function of distance from the terrestrial source to that of the burial site (Whitaker 1984; Tyson 1987; Huc 1988; Loutit et al. 1988). More the distance from the source, lesser is the concentration of allochthonous organic matter in sediment and vice versa (Boulter and Riddick 1986; Tyson 1987; Bustin 1988; Van der Zwan 1990). On the other hand, the distribution of autochthonous organic matter constituents (dinoflagellate cyst, zooplanktons, marine algae etc.) is controlled by the biological and ecological conditions of the environment i.e. restricted/open marine setting, salinity, pH, nutrient supply and fluctuation in the runoff (Downie et al. 1971; Harland 1973; Whitaker 1984; Summerhayes 1987; Tyson 1987). The relative proportion of allochthonous and autochthonous organic matter in a vertical succession can be utilized to interpret the proximal-distal/shallowing–deepening trend in shallow marine carbonate facies.

There are four different types of palynofacies constituents:

Equidimensional Black oxidized debris: It is also known as inertinite in coal petrology. Completely carbonized or semi-carbonized terrigenous particles, appear as black opaque organic particle with partial or no preserved cellular structure derived prior or during the deposition. These macerals are derived initially from the hard part tissue of plants i.e. lignified vascular tissue. They appear as sharp angular blade or lath shaped particle sometimes irregular in outline and are formed due to mechanical degradation under high oxidizing conditions or due to forest fire (Whitaker 1984; Tyson 1987). They appear as highly sorted in open marine environment and appear as equidimensional. These particles are chemically stable maceral and can withstand high energy oxidizing environment of deposition (Bryant et al. 1988; van der Zwan 1990). Due to their carbonized nature, they are highly buoyant like mica flakes hence can be transported to more distal open marine environment of deposition (Stanley 1986). The open marine carbonate system usually shows enrichment of these organic matter assemblage hence in sequence stratigraphic setup dominance of equidimensional black oxidized debris helps in the detection of maximum flooding/flooding events (Gorin and Steffen 1991).

Brown degraded debris: Dark brown and chocolate coloured plant debris with some cellular structures. They are mostly derived from land plant but due to the fungal and bacterial degradation their cellular structures are obscured. Abundance of brown degraded debris in a shallow marine facies is indicative of proximity of terrestrial source. Fluctuations in the quantity of brown degraded debris are a useful criteria for the interpretation of proximal setting in sea level fluctuation studies. In a carbonate system this group of maceral is interpreted as being transported in a large quantity during maximum progradation phases of High Stand Systems Tracts (Gorin and Steffen 1991).

Structured organic debris: Plant debris with well preserved cellular structure. They are rather rare in carbonate facies due to impoverished clastic supply and oxidizing environmental condition. If present, indicate proximity to the terrestrial source and reducing environmental condition of deposition (Tyson 1987). They include both Woody debris and Cuticle.

Amorphous organic debris: The autochthonous organic matter constituents in a shallow marine environment is represented by amorphous organic matter (derived upon degradation of marine planktons). Amorphous organic matter being highly susceptible to oxidation, does not get preserved in carbonate rocks due to oxidizing environment of deposition of carbonate facies. If present suggest reducing environmental condition of deposition (Tyson 1987).

Dinofalgellate cysts: The dinocyst diversity and abundance indicate more open marine conditions (Wall 1965; Tissot 1979; Smelror and Leereveld 1989; Habib and Miller 1989; Van Pelt and Habib 1988). Under moderate oxidizing depositional environment dinoflagellate cysts get preserved and provide useful information on paleoenvironment.

Pollen and Spore: Pollen and spore derived from the terrestrial source usually provide information of climatic condition on land. In shallow marine depositional environment, pollen and spore help in defining the proximal-distal trend. However, pollen spore are found in very low numbers in carbonate rocks due to low terrigenous supply and oxidizing environment of deposition.

Palynofacies Analysis of Lakadong Limestone

On the basis of relative distribution of various palynofacies constituents, 7 Palyno-Units were identified in the KPL and LTL section, each indicating specific paleoenvironmental conditions.

Palyno-unit A:

1.5 m thick limestone from the base in LTL section and 3.6 m in KPL section is characterised by decreasing trend of Black oxidized debris represented by 16–0.09% in LTL and 38.00–3.52% in KPL section. On the other hand, the Brown degraded debris showed increasing trend, quantitatively represented by 67.05–85.13% in LTL section and 19.89–68.93% in KPL section. (Table 2; Figs. 4 and 5). The percentage of Woody debris varies between 1.64 and 8.02% in LTL section and 1.01–11.01% in KPL section. Cuticle varied between 1.03 and 1.86% in LTL and 0.08–11.86% in KPL section. While pollen and spore are virtually absent in the KPL section, in the LTL section they are represented by 0.07–0.68% *Spinizonocolpites* pollen only. The marine constituents, Dinoflagellate and Acritarchs, were present in low to moderate numbers. The lowest value of Dinoflagellates, 0.04%, was observed in the LTL section and highest 20.09% in the KPL section. Dinoflagellates are represented by *Apectodinium* sp., *Operculodinium* sp., *O. major*, *O. centrocurpum*, *O. israelianum*, *Adnatosphaeridiumvittatum*, *A. multispinozum*, *Spiniferites* sp. and Acritarchs species *Cyclopsiella* sp. and *Collumosphaerafruticosa*. Foraminiferal test lining showed a gradually decreasing trend towards the top of the palyno-unit in both the sections. Lowest value of fungal elements represented by 0.03% was observed in LTL section, the highest of 4.83% in the KPL section.

Table 2 Table showing distribution of palynological organic matters in Lakadong Limestone

Sample No.	Black Oxid. deb.	Brown deg. deb.	Woody deb.	Cuticle	Poll. and Spor.	Foram. Test lining	Dinoflag.	Acritarch	Fun. Ele.
LTL 0	16.91	67.05	8.02	1.09	0.07	2.61	1.18	2.58	1.07
LTL 1	14.09	70.12	6.82	1.53	0.21	1.09	5.31	1.03	0.03
LTL 2	7.82	76.05	4.09	1.03	0.46	1.36	3.08	6.72	0.09
LTL 2B	3.05	75.18	6.22	1.80	0.68	1.04	4.29	5.17	2.58
LTL $2B_1$	6.83	84.07	3.85	1.04	0.41	0.71	1.04	0.35	1.96
LTL 3	2.07	85.13	4.07	1.36	0.29	1.74	1.82	1.01	2.07
LTL 4	0.09	81.01	1.64	1.86	0.12	1.03	5.24	6.01	3.07
LTL 5	16.41	61.83	1.01	1.32	0.38	1.14	8.01	4.92	5.19
LTL 6	23.68	54.83	0.98	4.89	0.71	1.96	9.02	2.65	1.94
LTL 7	19.52	49.46	6.73	3.06	0.42	0.93	1.53	11.72	6.81
LTL 8	15.84	23.64	4.85	1.62	0.73	0.81	14.83	7.05	31.03
LTL 9	13.86	25.08	2.05	4.03	0.61	0.27	4.38	11.25	39.17
LTL 10	22.07	57.19	0.89	0.07	0.57	0.42	1.79	1.93	15.62
LTL 11	19.24	37.85	2.05	1.73	0.31	0.29	1.93	3.58	33.81
LTL 12	33.61	34.02	1.67	4.93	0.08	2.04	1.06	7.29	15.07
LTL 13	29.04	31.18	2.48	1.86	0.91	1.39	1.76	5.84	26.01
LTL 14	21.86	29.06	1.94	4.09	1.47	1.09	3.67	2.04	34.15
LTL 15	26.01	51.72	1.06	0.87	0.91	2.85	1.03	1.31	14.09

(continued)

Table 2 (continued)

Sample No.	Black Oxid. deb.	Brown deg. deb.	Woody deb.	Cuticle	Poll. and Spor.	Foram. Test lining	Dinoflag.	Acritarch	Fun. Ele.
LTL 16	13.63	45.81	1.36	4.16	0.95	1.18	15.29	6.83	11.63
LTL 17	4.32	79.03	2.85	1.74	0.89	1.05	3.72	1.51	4.86
LTL 18	12.19	61.32	6.29	4.01	0.74	0.93	5.08	4.27	6.02
LTL 19	5.07	76.46	2.87	4.91	1.01	0.82	3.25	1.06	4.51
LTL 20	4.47	77.03	0.06	5.35	2.08	0.09	4.61	3.02	3.87
LTL 21	9.91	83.15	4.17	0.74	0.14	0.00	0.38	0.25	1.61
LTL 22	7.03	69.81	5.23	0.00	7.42	0.36	6.73	2.19	1.53
LTL 23	9.16	71.09	3.94	5.34	0.22	0.27	4.61	1.02	4.71
LTL 24	4.53	72.82	5.61	7.04	0.88	0.35	3.05	5.11	0.64
LTL 24A	3.38	77.19	2.73	7.89	0.92	0.41	1.32	6.34	0.26
LTL 25	12.02	58.53	1.07	12.29	0.00	0.31	7.46	7.95	0.44
LTL 26	15.65	64.37	3.51	2.83	0.97	0.25	6.19	5.07	1.54
LTL 27	11.82	74.03	4.81	3.05	0.82	0.43	3.15	0.49	0.84
KPL 1	38.00	31.3	1.01	4.00	0.00	2.08	20.07	2.07	1.06
KPL 2	34.01	22.42	10.08	7.18	0.00	3.06	19.09	2.06	2.08
KPL 3	25.39	19.89	11.01	9.04	0.00	2.64	20.09	9.91	2.02
KPL 4	26.83	29.95	7.61	5.13	0.00	11.27	14.10	4.09	1.05
KPL 5	29.34	32.45	4.27	3.36	0.00	9.82	11.46	6.52	2.84
KPL 6	19.01	47.93	8.38	2.09	0.00	8.76	9.35	3.18	1.15
KPL 7	20.34	34.03	6.97	11.86	0.00	5.74	7.96	9.07	4.83

(continued)

Table 2 (continued)

Sample No.	Black Oxid. deb.	Brown deg. deb.	Woody deb.	Cuticle	Poll. and Spor.	Foram. Test lining	Dinoflag.	Acritarch	Fun. Ele.
KPL 8	17.82	39.92	7.01	8.64	0.00	5.26	10.3	8.63	3.08
KPL 9	27.79	43.91	5.94	7.48	0.00	3.88	6.47	5.08	0.07
KPL 10	32.72	28.06	3.64	4.29	0.00	5.61	12.83	9.17	3.71
KPL 11	31.97	49.02	2.86	3.57	0.00	1.72	7.06	3.17	1.32
KPL 12	21.48	43.19	5.71	9.43	0.00	3.07	11.26	5.04	0.84
KPL 13	24.04	55.41	2.08	7.16	0.00	1.18	5.06	4.17	1.06
KPL 14	22.83	46.81	4.63	6.51	0.00	5.16	6.77	3.18	3.72
KPL 15	24.15	55.02	2.10	4.06	0.00	3.16	3.92	4.84	2.04
KPL 16B	19.89	63.04	2.09	0.11	0.00	2.57	2.38	5.84	4.37
KPL 16	23.84	58.74	3.92	5.73	0.00	2.85	2.61	1.94	1.03
KPL 16T	20.76	68.93	4.82	0.08	0.00	0.19	3.04	2.11	0.93
KPL 18	3.52	62.50	5.50	6.06	0.00	1.01	15.28	5.50	0.98
KPL 19	22.18	60.05	3.29	4.82	0.00	1.03	6.01	1.24	1.71
KPL 20	21.83	36.51	2.97	3.76	0.00	2.94	17.63	9.38	4.91
KPL 21	36.83	29.17	2.09	2.49	0.00	2.28	13.81	8.19	5.08
KPL 22	39.06	26.08	4.14	1.56	0.00	1.98	11.62	8.87	6.3
KPL 23	16.82	61.91	1.69	3.02	0.00	1.97	4.03	5.91	4.77
KPL 24	15.04	67.85	2.06	2.98	0.00	2.79	2.61	2.29	4.07
KPL 25	24.02	62.84	2.71	2.06	0.00	1.79	2.07	3.01	2.39
KPL 26	32.37	47.61	2.05	1.93	0.00	3.86	4.19	2.93	6.02

(continued)

Table 2 (continued)

Sample No.	Black Oxid. deb.	Brown deg. deb.	Woody deb.	Cuticle	Poll. and Spor.	Foram. Test lining	Dinoflag.	Acritarch	Fun. Ele.
KPL 27	25.02	45.38	6.47	5.91	0.00	2.62	3.86	1.97	8.83
KPL 28	23.18	38.86	9.91	6.01	0.00	4.68	6.39	3.64	6.84
KPL 29	19.75	56.28	7.18	3.49	0.00	3.11	4.07	2.85	3.26
KPL 30	17.08	64.51	4.73	5.39	0.00	1.72	5.24	1.96	0.18
KPL 31	15.79	75.06	2.08	1.52	0.00	2.83	1.07	1.02	0.72
KPL 32	12.38	69.05	1.19	4.07	0.00	5.26	2.96	3.68	2.16
KPL 33	14.07	78.11	3.04	0.19	0.00	2.31	0.08	1.42	0.68
KPL 34	11.47	79.54	1.72	0.24	0.00	1.76	0.21	3.06	2.09
KPL 35	8.12	85.06	0.78	0.36	0.00	1.02	0.31	0.64	3.71
KPL 36	39.43	34.71	4.61	12.93	0.00	0.29	5.13	0.25	2.61
KPL 37	25.81	47.11	5.01	10.89	0.00	3.48	5.06	2.07	1.26
KPL 38	12.04	62.19	6.23	11.41	0.00	1.54	4.67	1.94	0.08
KPL 39	11.38	68.94	7.31	10.07	0.00	0.29	2.01	0.06	0.33
KPL 40	11.07	74.05	4.87	6.09	0.00	1.35	1.87	0.31	0.24
KPL 41	6.86	75.13	1.57	14.05	0.00	2.45	0.09	0.51	0.16
KPL 42	19.51	59.01	6.24	12.63	0.00	1.07	1.01	0.62	0.38
KPL 43	6.75	68.04	4.39	16.08	0.00	2.31	1.03	1.67	0.61
KPL 44	14.09	57.23	6.03	18.12	0.00	1.97	0.98	1.01	1.39

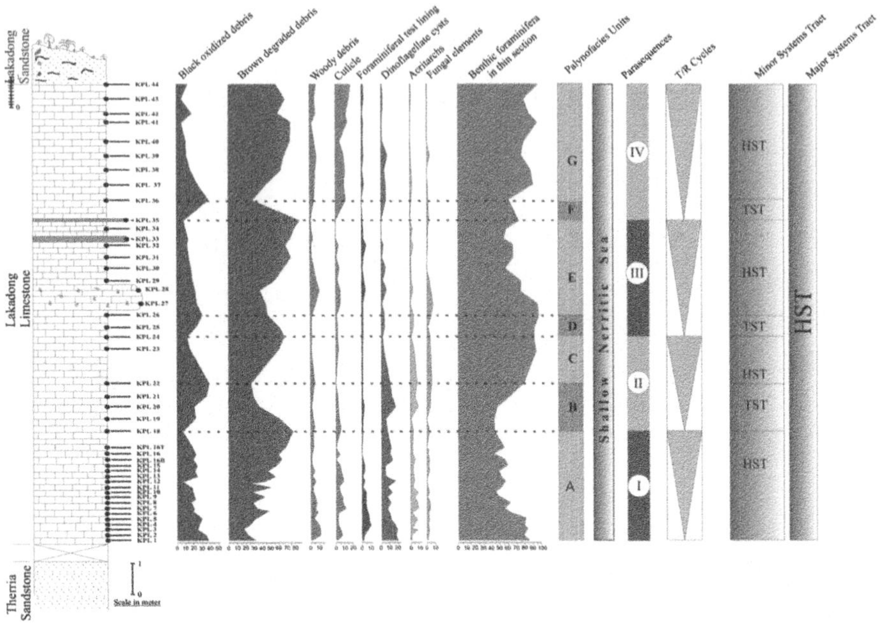

Fig. 4 Palynofacies and benthic foraminiferal distribution pattern in Dohsniang (Kurtinsiang) peak section (KPL) of Lakadong Limestone

Depositional environment: The dominance of Brown degraded terrestrial debris along with increase in cuticle and woody debris revealed a proximal depositional environment for the plyno-unit A.

Palyno-unit B:

This unit is represented by 0.6 m thick limestone sequence in LTL and 1.8 m in KPL section. The unit is characterized by gradual upward increasing trend of Black oxidized debris represented by 0.09–23.68% in LTL section and 3.52–39.06% in KPL section. However, the Brown degraded debris showed decreasing trend, quantitatively represented by 81.01–54.83% in LTL section and 60.05–26.08% in KPL section (Table 2; Figs. 4 and 5). The percentage of Woody debris ranged between 0.98 and 1.01% was observed in LTL section and 2.09–5.50% in KPL section. Percentage of Cuticle varied between 1.32 and 4.89% in LTL section and 1.56–6.06% in KPL section, however, pollen and spore 0.12–0.71% were present in LTL section. The Dinoflagellate constituted up to 5.24% in LTL section and 17.63% in KPL section. Dinoflagellates was represented by *Operculodiniumisraelianum*, *O. centrocurpum*, *O. major*, *Adnatosphaeridium* sp., *Achomosphaera* sp., *Achomosphaeraalcicornu*, *Apectodinium* sp., *Cordosphaeridium* sp., *Hystrichokolpoma* sp., *Polysphaeridiumsbutile* and Acritarch species *Collumosphaera* sp., and *Cyclopsiellavieta*.

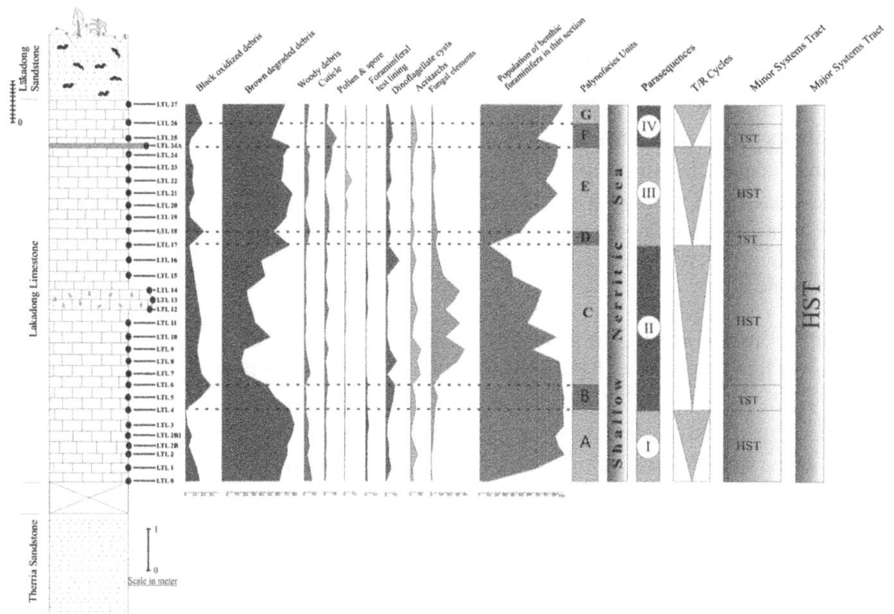

Fig. 5 Palynofacies and benthic foraminiferal distribution pattern in Laitmawksing peak section (LTL) of Lakadong Limestone

Depositional environment: The considerable increased percentage of black oxidized debris along with increased marine component dinocyst and acritarchs and decreased terrestrial component indicated relatively distal environment of deposition of palyno-unit B in carbonate platform.

Palyno-unit C:

This unit in LTL section consisted of 3.2 m thick limestone sequence in LTL section and 0.8 m in KPL section. The unit was represented by decreasing upward trend of Black oxidized debris within the range of 33.61–4.32% in LTL section and 39.06–15.04% in KPL section. The Brown degraded debris showed increasing trend represented by 23.64–79.03% in LTL section and 26.08–67.85% in KPL section (Table 2; Figs. 4 and 5). Percentage of Woody debris ranged between 0.89 and 6.73% in LTL section and 1.69–4.14% in KPL section. However, Cuticle percentage varied between 0.07 and 4.93% in LTL section and 1.56–3.02% in KPL section. Pollen and spore were observed in very few numbers in both the section. Among the marine constituent Dinoflagellatecyst varied between 15.29 and 1.03% in LTL section and 11.62–2.61% in KPL section, Acritarch ranged 11.72–1.31% in LTL section and 8.87–2.29% in KPL section, and both showed decreasing upward trend. The Foraminiferal test lining also showed decrease in numbers in both the section.

Depositional environment: Decreased values of dinocyst, acritrachs and black oxidized debris and the dominance of terrestrial brown degraded terrestrial debris along with increased percentage of cuticle and woody debris revealed a relatively proximal environment of deposition for the plyno-unit C on the carbonate platform.

Palyno-unit D:

This unit consisted of 0.3 m limestone sequence in LTL and 0.7 m in KPL section. The unit was characterized by gradual upward increasing trend of Black oxidized debris 4.32–12.19% in LTL section and 15.04–32.37% in KPL section. The Brown degraded debris showed decreasing trend represented by 79–61% in LTL section and 67.85–47.61% in KPL section (Table 2; Figs. 4 and 5). However, the percentage of Woody debris showed decreasing trend in both the sections. Percentage of Cuticle varied between 1.74 and 4.01% in LTL section and 1.93–2.98% in KPL section, however pollen and spore are observed only in LTL section ranged from 0.74 to 0.89%. The marine constituent Dinoflagellate, Acritarchs showed increased percentage in both the sections within the range of 2.07–5.08%. Dinoflagellatecyst are represented by *Operculodiniumisraelianum*, *O. centrocurpum*, *Spiniferites* sp. and few Acritarch species *Cyclopsiella* sp. *and Collumosphera* sp.

Depositional environment: The considerable increased percentage of black oxidized debris and marine component dinocyst and decreased terrestrial component suggest a distal environment of deposition for palyno-unit D on carbonate platform.

Palyno-unit E:

Palyno-unit E in LTL section comprised of 2.4 m limestone sequence in LTL section and 3.7 m in KPL section. The unit was represented by decreasing upward trend of Black oxidized debris, varied within the range of 12.19–3.38% in LTL section and 32.37–8.12% in KPL section. The Brown degraded debris showed increasing upward trend 8.86–85.06% in KPL section and in LTL section between 61.32 and 83.15% (Table 2; Figs. 4 and 5). Percentage of Woody debris 0.06–6.29% was observed in LTL section and 0.78–9.91% in KPL section. Cuticle percentage varied from 0.00 to 7.89% in LTL section and 0.24–6.01% in KPL section. Pollen and spore was present only in LTL section within the percentage range of 0.14–7.42%. However, considerable decreased dinocyst and acritarchs numbers were observed in this palyno-unit.

Depositional environment: Decreased values of dinocyst, acritrachs and Black oxidized debris and the dominance of terrestrial brown degraded debris along with increased structured and woody debris suggested a relatively proximal environment of deposition for the plyno-unit E on carbonate platform.

Palyno-unit F:

This unit was represented by of a 0.5 m thick limestone sequence in LTL section and 0.78 m thick limestone in KPL section. The unit was characterized by gradual upward increasing trend of Black oxidized debris representing by 3.38–15.65% in LTL section and 8.12–39.43% in KPL section. The Brown degraded debris with strong increasing upward trend in KPL section constituted about 34.71–85.06% while in LTL section it showed a mild increasing trend 58.53–64.37% (Table 2; Figs. 4 and 5). The woody debris varied between the range of 3.51–1.07% in LTL section and 4.61–0.78% in KPL section. Percentage of Cuticle varied from 12.29 to 7.89% in LTL section and 12.93–0.36% in KPL section, however pollen and spore was present only in LTL section 0.92%. The marine constituent Dinoflagellate, Acritarchs and Foraminifera was also observed in both the sections within the ranges of 0.31–7.46, 0.25–7.95 and 0.25–1.02% respectively. Dinoflagellate cysts were represented by *Operculodiniumisraelianum, O. centrocurpum, Spiniferites* sp.

Depositional environment: The increased percentage of black oxidized debris and marine component dinocyst and decreased terrestrial component suggested a relatively distal environment of deposition for palyno-unit F on carbonate platform.

Palyno-unit G:

Palyno-unit G in LTL section is made up of 0.5 m thick limestone sequence in LTL section and 3.7 m thick sequence in KPL section. The unit was represented by decreasing upward trend of black oxidized debris, which varied from 15.65 to 11.82% in LTL section and 39.4–36.75% in KPL section. The Brown degraded debris showed increasing upward trend and is quantitatively represented by 64.37–74.03% in LTL section and 34.71–75.13% in KPL section (Table 2; Figs. 4 and 5). Percentage of Woody debris ranged between 3.51 and 4.81% in LTL section and 1.57–6.24% in KPL section. However, Cuticle varied from 2.83 to 3.05% in LTL section and 6.09–18.12% in KPL section. Pollen and spore was observed only in LTL section within the percentage range of 0.82–0.97%. Among the marine constituent's dinoflagellate cysts, were observed in both the sections within the ranges of 6.19–0.09% and were represented by *Spiniferites* sp.

Depositional environment: Decreased values of dinocyst, acritrachs and Black oxidized debris and the dominance of terrestrial Brown degraded debris along with increased structured organic debris and woody debris suggested relatively proximal environment of deposition for the palyno-unit G on carbonate platform.

The seven Palyno-units of KPL and LTL sections have been grouped into four palynofacies sequences, each corresponds to a parasequence. Each Parasequence starts with a Transgressive Systems Tract represented by distal depositional environment on the carbonate platform setting followed by progradational succession in a proximal depositional setting with more terrestrial influx defining the High Stand Systems Tract. First Parasequence is incomplete and consist of progradational succession only.

Discussion

Indian peninsula during its northward journey towards Eurasia at the time of Paleocene-Eocene transition was located in the equatorial zone. Shallow marine sediments were deposited on the southern flanks of Shillong Plateau during early Paleogene (Evans 1932). Sylhet Limestone Formation comprises three limestone members and two sandstone members of Paleogene age in south Shillong Plateau. It represents several phases of emergence and collapse of carbonate platform conditions and bears the imprint of fluctuating paleoenvironmental conditions and consequent changes in the biota. Lakadong Limestone Member represents the first (oldest) carbonate platform condition after the deposition of sandy facies of Therria Formation (Danian-Selandian) in this region. The overall basin development was tectonically controlled that resulted in reduced terrestrial supply in the basin. Warm and humid tropical climate of late Paleocene resulted in the extensive deposition of a typical foraminifera reef building carbonate platform rich in calcareous algae and benthic foraminifera in Lakadong Limestone Member. Large population of benthic foraminifera namely glomalveolinids, miscellanids and ranikothalids characteristic of SBZ-3 of shallow benthic zone is suggestive of early Thanetian age for Lakadong Limestone Member. Total age span of Lakadong Limestone Member is 56.2–58.23 Ma that also indicates development of carbonate platform condition for a brief time interval (Gogoi et al. 2009). The earlier studies have shown that overlying sandy facies of Sylhet Limestone Formation is Lakadong Sandstone Member, which is a coastal facies. Paleocene Eocene Thermal Maxima (PETM), a globally warm climatic event, has been identified in Lakadong Sandstone Member of Mawsynram Area of Shillong Plateau (Prasad et al. 2006). High terrestrial discharge associated with the excessive warm climatic condition of PETM together with tectonic movement resulted in the collapse of carbonate ramp setting and development of large scale coastal marshes. This suggested predominance of role of climatic conditions in the development of coastal facies in the carbonate ramp in Shillong Plateau. Hence, understanding of the changing palaeoenvironmental conditions during the deposition of Lakadong Limestone Member is significant and can provide clues regarding climatic fluctuation during late Paleocene. Palynofacies can also help in the correlation of coeval carbonate succession, deposited indifferent parts of the carbonate platform (distal-proximal). KPL section has been deposited in a distal platform setting with respect to LTL sections and with the help of palynofacies study both can be correlated. In an earlier study Sequence Stratigraphic framework was provided to Palaeocene-Eocene succession of East Khasi Hills and Lakadong Limestone Member represent the High Stand Systems Tract. In this study palynofacies analysis further divided the High Stand Systems Tract of 3rd order sea level cycles of Lakadong Limestone Member into small scale cycles representing 4th–5th order cycles. The present study demonstrates the high potential of palynofacies analysis in high-resolution stratigraphy and correlation of carbonate succession of late Paleocene. Palynofacies analysis may serve as a correlation tool, since palynomorphs and phytoclasts behave like sediment particles regarding their transport, sorting, distribution and deposition and can complement sedimentological studies. Therefore, palynofacies analysis provides essential information on sedimentary processes and palaeoenvironment within a depositional system. The detailed study of lateral palynofacies patterns enabled a better understanding and precise genetic interpretation of the eustatic history of the carbonate system of Lakadong Limestone. These studies can also help in differentiating the global-eustatic and regional tectonic/climatically controlled sea level fluctuations.

Acknowledgements The BG, VP and RG are grateful to Director, BSIP, Lucknow for carrying out this work. BG sincerely acknowledge the BSIP for financial support under the BSRS scheme.

References

Aubry MP (1998) Early Paleogene calcareous nannoplankton evolution: a tale of climatic amelioration. In: Aubry MP et al (eds) Late Palaeocene-early Eocene climatic and biotic events in the marine and terrestrial records. Columbia University Press

Biswas B (1962) Stratigraphy of the Mahadek, Langpar, Cherra and Tura formations, Assam, India. Bull Geol Min Met Soc Ind 25:1–48

Bombardiere L, Gorin GE (1998) Sedimentary organic matter in condensed sections from distal oxic environments: examples from the Mesozoic of SE France. Sedimentology 45:771–788

Boulter MC, Riddick A (1986) Classification and analysis of palynodebris from the Palaeocene sediments of the Forties Field. Sedimenology 33:871–886

Bryant ID, Kantorowicz JD, Love CF (1988) The origin and recognition of laterally continuous carbonate cemented horizons in the Upper Lias Sands of southern England. Mar Pet Geol 5:108–133

Bustin RM (1988) Sedimentology and characteristics of dispersed organic matter in Tertiary Niger Delta: origin of source rocks in a deltaic environment. AAPG Bull 72(3):227–298

Combaz A (1964) Les palynofacies. Rev Micropaleontol 7:205–218

Combaz A (1980) Les kerogensvus are microscope. In: Durand B (ed) Kerogen: insoluble organic matter from sedimentary rocks. Editions Techmip, Paris, pp 55–111

Das Gupta AB (1977) Geology of Assam Arakan region. Oil commentary 14(17):4–13

Dickens GR (2000) Methane oxidation during the late Palaeocene thermal maximum. Bull Soc Géol Fr 171:37–49

Downie C, Hussain M, Williams GL (1971) Dinoflagellate cyst and acritarch associations in the Paleogene of southeast England. Geosci Man 3:29–35

Evans P (1932) Explanatory notes to accompany a Table showing the Tertiary succession in Assam. Trans Min Geol Inst India. 27:155–260

Evitt WR (1985) Sporopollenin dinoflagellate cysts: their morphology and interpretation. Amer. Assoc. Strati. Palyno. Found., p 333

Garg R, Ateequzzaman K (2000) Dinoflagellate cysts from the Lakadong Sandstone, Cherrapunji area: biostratigraphical and palaeoenvironmental significance and relevance to sea level changes in the Upper Palaeocene of the Khasi Hills, South Shillong Plateau, India. Palaeobotanist 49(3):461–484

Gogoi B (2006) Integrated sedimentological and palaeonotological study of Lakadong Limestone Member of Mawsynram area of East Khasi Hill District, Meghalaya. Unpublished Ph.D. thesis, Dibrugarh University, Dibrugarh, Assam

Gogoi B, Kalita KD, Garg R, Borgohain R (2009) Foraminiferal biostratigraphy and palaeoenvironment of the Lakadong limestone of the Mawsynram area, south Shillong plateau, Meghalaya. J Palaeontol Soc India 54(2):209–224

Gorin G, Steffen D (1991) Organic facies as a tool for recording eustatic variations in marine fine-grained carbonates—example of the Berriasian stratotype at Berrias (Ardeche, SE France). Palaeogeogr Palaeoclimatol Palaeoecol 85:303–320
Habib D, Miller JA (1989) Dinoflagellate species and organic facies evidence of marine transgression and regression in the Atlantic Coastal Plain. Palaeogeogr Palaeoclimatol Palaeoecol 74:23–47
Harland R (1973) Dinoflagellate cysts and acritarchs from the Bearpaw Formatioin (Upper Campanian) of southern Alberta, Canada. Palaeontology 16:665–706
Hart GF (1986) Origin and classification of organic matter in clastic systems. Palynology 10:1–23
Huc AY (1988) Sedimentology of organic matter. In: Frimmel FH, Christman RF (eds) Humic substances and their role in the environment. Wiley, New York, pp 215–243
Jaramillo CA, Oboh-Ikuenobe FE (1999) Sequence stratigraphic interpretations from palynofacies, dinocyst and lithological data of Upper Eocene-Lower Oligocene strata in southern Mississippi and Alabama, U.S. Gulf Coast. Palaeogeol Palaeoclimatol Palaeontol 145:259–302
Jauhri AK (1988) Observations on the foraminiferal fauna of the basal part of Sylhet Limestone Group (Lakadong Limestone) from Therria Ghat, South Shillong Plateau, Meghalaya. In: Maheshwari, HK (eds) Palaeocene of India. Indian Association Palynostra, Lucknow, 4:34–42
Jauhri AK (1994) Carbonate buildup in the Lakadong Formation of the south Shillong plateau, NE India: a micropalaeontological perspective. In: Matteucci, R, Carboni MG, Pignatti JS (eds) Studies on Ecology and Palaeoecology of benthic Communities. Boll Soc Palaeontol Italina 2:157–169
Jauhri AK and Agawal KK (2001) Early Palaeogene in the South Shillong Plateau, NE India: local biostratigraphic signals of global tectonic and ocenic changes, Palaeogeogr Palaeoclimatol Palaeoecol 168:187–203
Kiessling W, Flugel E, Golonka J (2003) Patterns of Phanerozoic carbonate plateform sedimentation. Lethaia 36:195–226
Loutit TS, Hardenbol J, Vail PR (1988) Condensed sections: the key to age determination and correlation of continental margin sequences. In: Wilgus CK, Hastings BS, Kendall CGSC, Posamentier HW, Ross CA, Van Wagoner JC (eds) Sea level changes: an integrated approach. SEPM Spec Publ 42:183–215
Mathur LP, Evans P (1964) Oil in India. Inter Geol Cong. 22nd session, India, p 88
Medlicott HB (1869) Geological sketch of the Shillong plateau in north eastern Bengal. Memoirs Geol Surv India 7:151–207
Muller J (1959) Palynology of recent Orinoco delta and shelf sediments: reports of the Orinoco shelf expedition. Micropaleontology 5:1–32
Murthy MVN, Chakraboty C, Talukdar SC (1976) Stratigraphic revision of the Cretaceous-Tertiary sediments on the Shillong plateau. Rec Geol Surv India 107(2):80–90
Nagappa Y (1959) Foraminiferal biostratigraphy of Cretaceous-Eocene succession in the India-Pakistan-Burma region. Micropaleontol 5(2):145–192
Parry CC, Whitley PKJ, Simpson RDH (1981) Integration of palynological and sedimentological methods in facies analysis of the Brent formation. In: Illing LV, Hobson GD (eds) Petroleum geology of the continental shelf of North-West Europe. Heyden and Son, London, pp 205–215
Pittet B, Gorin G (1997) Distribution of sedimentary organic matter in a carbonate-siliciclastic platform environment: Oxfordian deposits from the Swiss Jura Mountains. Sedimentology 44:915–937
Pocock SAJ (1982) Identification and recording of particulate sedimentary organic matter. In: How to assess maturation and paleotemperatures. SEPM short course 7:7–131
Prasad V, Garg R, Ateequzzaman K, Singh IB, Joachimski MM (2006) Apectodinium acme and palynofacies characteristics in the latest Palaeocene-earliest Eocene of northeastern India: biotic response to the Palaeocene Eocene Thermal Maxima (PETM) in low latitude. J Palaeontol Soc India 51(1):91–107
Prasad V, Singh IB, Bajpai S, Garg R, Thakur B, Singh A, Saravanan N, Kapur VV (2013) Palynofacies and sedimentology-based high-resolution sequence stratigraphy of the lignite-bearing

muddy coastal deposits (early Eocene) in the Vastan Lignite Mine, Gulf of Cambay, India. Facies. https://doi.org/10.1007/s10347-012-0355-8
Pratt LM (1984) Influence of palaeonvironment factors on preservation of organic matter in Middle Cretaceous Greenhorn Formation, Pueblo, Colorado. AAPG Bull 68(9):1146–1159
Raja Rao (1981) Coalfields of India, Bull. G.S.I. 45(1):1–44
Serra-Kiel J, Hottinger L, Caus E, Drobne K, Ferrandez C, Jauhri AK, Less G, Pavlovec R, Pignatti J, Maria Samso J, Schaub H, Sirel E, Strougo A, Tambareau Y, Tosquella J, Zakrevškaya E (1998) Larger foraminiferal biostratigraphy of the Tethyan Paleocene and Eocene. Bull Soc Geol Fr 169(2):281–299
Singh IB, Swamy ASR (2006) Delta sedimentation—East coast of India. Technical Books Publisher and Distributors, Dehradun
Smelror M, Leereveld H (1989) Dinoflagellate and acritarch assemblages from the late Bathonian to early Oxfordian of Montagne Crussol, Rhone valley, southern France. Palynology 13(1):121–141
Stanley DJ (1986) Turbidity current transport of organic-rich sediments, Alpine and Mediterranean examples. In: Meyers PA, Mittere RM (eds) Deep ocean black shales: organic geochemistry and paleoceanography setting. Mar Geol 70:85–101
Steffen D, Gorin G (1993) Palynofacies of the Upper Tithonian-Berriasian deep-sea carbonates in the Vocontian Trough (SE France). Bull Centres Rech Explor-Prod Elf-Aquitaine 17(1):235–247
Strohmenger C, Strasser A (1993) Eustatic controls on the depositional evolution of the Upper Tithonian Berriasian deep water carbonates (Vocontian Trough, SE France). Bull Centres Rech Explor-Prod Elf-Aquitaine 17(1):183–203
Summerhayes CP (1987) Organic-rich Cretaceous sediments from the North Atlantic. In: Brooks J, Fleet AJ (eds) Marine petroleum source rocks. Geol Soc Spec Publ 26:301–316
Tappan H (1980) The paleobiology of plant protests. Freeman, W. H. San Francisco, p 1028
Tissot B (1979) Effect on prolific petroleum source rocks and major coal deposits caused by sealevel changes. Nature (London) 277:462–465
Tissot BP, Welte DH (1984) Petroleum formation and occurrence. Springer, New York, p 538
Tyson RV (1987) The genesis and palynofacies characteristics of marine petroleum source rocks. In: Brooks J, Fleet AJ (eds) Marine petroleum source rocks. Geol Soc Spec Publ 26:47–67
Tyson RV (1993) Palynofacies analysis. In: Jenkins DG (ed) Applied micropaleontology. Kluwer Academic Publishers, Dordrecht, pp 153–191
Tyson RV (1995) Sedimentary organic matter: organic facies and palynofacies. Chapman and Hall, London, p 615
Tyson RV (1996) Sequence-stratigraphical interpretation of organic facies variation in marine siliciclastic systems: general principles and applications to the onshore Kimmeridge Clay Formation, UK. In: Hesselbo SP, Parkinson DN (eds) Sequence stratigraphy in British geology. Geol Soc London Spec Publ 75–96
Van der Zwan CJ (1990) Palynostratigraphy and palynofacies reconstruction of the upper Jurassic to lowermost Cretaceous of the Draugen Field, offshore Mid Norway. Rev Palaeobot Palynol 62:157–186
Van Pelt R, Habib D (1988) Palynology of the Jurassic Twin Creek Limestone. Palynology 12:248
Wall D (1965) Microplankton, pollen and spores from the Lower Jurassic in Britain. Micropalaeontology 11:151–190
Whitaker MF (1984) The usage of palynostratigraphy and palynofacies in definition of Troll Field Geology. In: Offshore Northern Seas-Reduction of uncertainties by innovative reservoir geomodelling. Norsk Petrolumsforening Article G6
Wilson GF, Metre WB (1953) Assam and Arakan. In: Illing, VC (eds) The world's Oil Fields: The Eastern Hemisphere, The Science of Petroleum, Oxford University Press, 6(1):119–123

Nannofossil Imprints of Paleogene Transgressive Events in India

Jyotsana Rai

Abstract The transition from super hot Cretaceous to transient Palaeogene climate and therein reasonably fast northward movement of island Indian plate from equatorial tropics to buckling with Tibetan plate after subducting and engulfing the Tethys Sea and giving rise to lofty Himalayas, affected the oceanic current patterns and both fanua and flora drastically. These rapid tectonic processes from 65 to 50 Ma led to eustatic rise and in its response imprints could be seen in several areas of the Indian subcontinent. The calcareous nannoplankton being marine, tiny (2–10 μm), photosynthetic golden brown algae, and its fossil counterpart the calcareous nannofossils having fast turnover rate during Cenozoic are ideal as biostratigraphic markers. The Palaeogene statotypes are demarcated in Europe and biostratigraphy on varied biotic parameters were erected there but these can be well utilized in India also wherever the imprints of Palaeogene transgressions on Indian craton are recorded. After the global catastrophic late Cretaceous mass extinction around 65 Ma in which over ninetynine percent nannofossil species after gigantism died out. With the dawn of Palaeocene, new Lilliputian species evolved which grew exponentially in number and size during early Palaeogene. Massive release of greenhouse gasses at the beginning of the PETM (Paleocene Eocene Thermal Maximum) occurred and the end of PETM was met with a very large sequestration of carbon dioxide in the form of methane clathrate, coal, and crude oil which reduced the atmospheric carbon dioxide. The beginning of the Eocene Epoch indicated increased amount of oxygen in the earth's atmosphere. At the end of the Middle Eocene Climatic Optimum, cooling and the carbon dioxide drawdown continued through the late Eocene and into the Eocene-Oligocene transition.

Keywords Palaeogene · Transgression · Calcareous nannofossils

---

J. Rai (✉)
Birbal Sahni Institute of Palaeosciences, Lucknow 226007, India
e-mail: jyotsana_rai@yahoo.com

S. Bajpai et al. (eds.), *The Indian Paleogene*, Society of Earth Scientists Series,
https://doi.org/10.1007/978-3-319-77443-5_8

Introduction

Calcareous nannofossils contain the coccoliths, coccospheres of Haptophyte algae (Prymnesiophyceae) and associated nannoliths (enigmatic or unknown provenance). First observed under light microscope with all other microfossils by Ehrenberg (1854) and sketches of both coccolith and discoaster are produced in a celebrated publication "Mikrogeologie" and considered them of inorganic origin. It was Sorby (1861) who realized its organic origin by observing the curvature in coccoliths under the microscope. They are phytoplanktons (photosynthesizing autotrophs) called coccolithophores, which form coccospheres. They biomineralise calcite mineral and after death of population and its postmortem disintegration their calcareous skeletons are found in marine deposits in astronomical numbers. A coccolith is a single disc-like plate which is secreted by the algal organism and held in combination with many more, at times morphologically different shape plates are attached by an organic coating to form the coccosphere. Coccoliths usually get disaggregated and separated upon death and are preserved in the sedimentary record. Coccospheres are usually formed of single type of coccoliths but coccospheres with two or more morphologically different coccoliths are not uncommon as seen in living and culture studies. There are two forms of coccoliths, the holococcoliths which are formed from calcite crystals of identical shape and dimension and the heterococcoliths with modified calcite shape. Calcareous nannofossils are believed to be fossil counterpart of living nannoplanktons and are known to have its ancestry from Late Triassic onwards (Moshkovitz 1982; Jafar 1983b). They are exclusively marine, minute (2–10 μm) phytoplanktons, which are responsible for oxygen budget in the atmosphere and have short stratigraphic and wide palaeogeographic ranges, which make them ideal for biostratigraphy. Quick recovery, acid free preparation technique and precise dating potential, especially in Palaeogene, make them ideal for dating marine sediments under petroleum exploration. The immense biostratigraphic potential of calcareous nannofossils was first realized in two papers of Martini and Bramlette (1963) and Bramlette and Martini (1964). The nannofossil zonal scheme provided by Martini (1971) for land sections and by Okada and Bukry (1980) for open ocean deposits are widely used for Palaeogene biostratigragy the world over (Fig. 1).

Chronological Palaeogene Nannofossil Records from India: The earliest calcareous nannofossils in India were recorded from Maastrichtian-Danian sediments of Khasi Hills, northeast India employing light microscopy by Narasimhan (1963). Rajagopalan (1964) indicated presence of nannofossil bearing horizon from Maastrichtian-Danian age sediments of Pondicherry but provided no illustration. Pant and Mamgain (1969) recorded nannofossils from several Palaeogene sections of Kachchh and illustrated sketches. Mathur (1972) illustrated nannofossils under light microscope of Palaeocene-lower Eocene age from Kuar Bet Hill Point. Bhandari et al. (1977) reported Palaeocene-Eocene age nannofossils employing light microscopy from Ladakh molasse near Kargil area. Nannofossils are also recorded from Upper Cretaceous and Lower Tertiary sediments of Pondicherry, South India (Narasimhan 1975); upper Late Maastrichtian/Danian age Subathu of Dharampur-Simla Himalaya

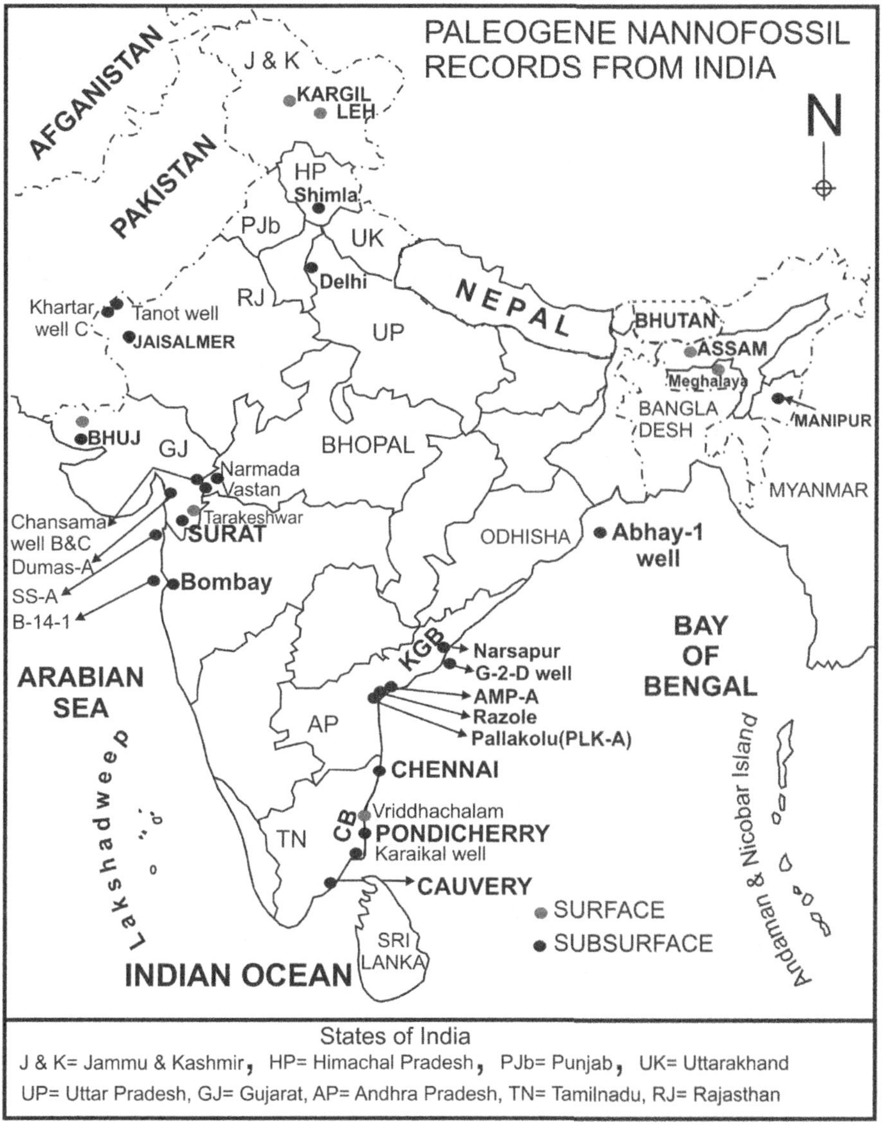

Fig. 1 Palaeogene nannofossil records of both surface (green dots) and subsurface (red dots) occurrences plotted on map of India (color figure online)

(Jafar and Kapoor 1988); K/T boundary species with Early Eocene nannofossils from Subathu Formation, Shimla Himalaya (Jafar and Singh 1992); Paleocene age *Discoaster multiradiatus* Zone from bore hole of Gopurapuram village northeast of Vriddhachalam area (Jain et al. 1983). Nannofossils are also recorded from Early Ypresian age Jafarabad Formation and Lutetian to Priabonian age Belapur Formation of Bombay Offshore Basin (Saxena 1996). K/Pg boundary, Paleocene to Oligocene age nannofossil zones were identified from Amalpuram, Palakollu wells of Krishna-Godavari Basin (Saxena 2000). Precise LPTM boundary was marked in Vastan lignite mine of Cambay basin by nannofossil assemblage substantiated by negative carbon isotope excursion (Samanta et al. 2013a, b). Middle Eocene age nannofossil assemblages were known from several sections of Kachchh (Pant and Mamgain 1969); Sanu (Paleocene) and Bandah (Bartonian) of Kharatar well-C of Jaisalmer (Singh 1998b), Bartonian age nannofossils from Tanot well-1 (Rai et al. 2014). Rakhadi river section (Singh et al. 1980); Berwali stream (Singh 1998a, b); Lakhpat (Singh 1978a, 1980a); Vinjhan-Miani (Singh 1978b, 1980b); Abhay-1 well of Bengal Basin (Singh 1983); RatoNadi section (Singh and Singh 1987, 1991); Rachelo nala section (Rai 1988, 1997, 2007; Jafar and Rai 1994); Babia Hill (Singh and Singh 1986). Nannofossils were encountered from Late Eocene of Broach, (Pant and Mathur 1973); Tarakeshwar area (Singh et al. 1978), near Surat outcrops in Ghalha Nala and Kusumbha Tal area yielded Late Eocene (Priabonian) nannofossils belonging to NP 20 Zone of Martini (1971); ?Lutetian–Priabonian age nannofossils from subsurface Tarapur Shale of Dumas Well-A, Cambay Basin, Surat (Singh and Uddin 2000); Late Eocene of Kopili Formation of Mikir Hills, Assam (Singh 1979); Priabonain age nannofossils from surface sediments representing Rewak Formation exposed in South Garo Hills (Singh et al. 2016); Meghalaya; Lutetian to Priabonian age nannofossils from Chanasma wells of Mehsana block, Cambay Basin (Singh et al. 1998), The early Eocence nannofossils with reworked Cretaceous–Paleocene taxa are also seen near Basgo section, Ladakh area and the work is in progress. Table 1 indicates the illustrated nannofossil taxa employing light microscopy (LM) and under scanning electron microscopy (SEM). A synthesis of published nannofossil data from various basins of India is compiled in Figs. 2 and 3.

Uninterrupted Late Eocene (Priabonian) and Early Oligocene (Rupelian) sediments with boundary of Priabonain/Rupelian could be found in Meghalaya, north-eastern India as Priabonian is hiatus on land in Kachchh and Jaisalmer areas of western India. Rupelian age nannofossils were found in Kachchh, Maniara Fort section, in type section of Maniara Fort Formation (Rai 2002).

Conclusions

Occurrence of nannofossils in Palaeogene sedimentary successions from several sedimentary basins of India is though sparse, but is now well documented. Available dataset demonstrates that nannofossil evidence has proved to be crucial for precise demarcation of time boundaries viz. Maastrichtian-Danian, Danian-Selandian, Thanetian-Ypresian, Ypresian-Lutetian, Lutetian-Bartonian, Bartonian-Priabonian). Although nannofossils are most widely recorded from Eocene contributing to biostratigraphic precision in various basins, there is strong potential of generating nannofossil data also from Danian (early Paleocene), PETM, and Oligocene sequences for detailed biostratigraphic studies as evidenced by the few known records. The proven

Table 1 Known nannofossil records from sedimentary Basins of Indian Subcontinent

Locality	Mode of documentation	Age	Publication
Extra Peninsula Ladakh Molasse Group near Kargil area (Kargil Fm. overlies Ladakh granite)	LM	Palaeocene-Eocene	Bhandari et al. (1977)
Subathu of Dharampur-Shimla	LM	NP3-NP4 Late Danian *Chiasmolithus danicus*/*Ellipsolithus macellus* Zone with reworked elements of Maastrichtian (*Micula mura* Zone)	Jafar and Kapoor (1988)
Koshalia Nala section, Subathu Formation, Shimla Himalaya	LM	NP12 + NP13 *Tribrachiatus orthostylus-Discoaster lodoensis* zones, Late Ypresian with reworked Late Maastrichtian-Danian + NP1-NP4 zones	Jafar and Singh (1992)
Kachchh Basin Kachchh; Denma, Waghopadar, Harudi, Guvar, Dedhapur, Lakhpat, Khari, Sehe, Waghopadar	LM, sketches	Eocene = (Denma, Waghopadar, Harudi, Guvar, Dedhapur, Lakhpat) Oligocene = Khari, Lakhpat, Sehe, Waghopadar	Pant and Mamgain (1969)
Pachham Island (Kuarbet Hill Point)	LM	Palaeocene-Lower Eocene	Mathur (1972)
Lakhpat	LM	Late mid Eocene (NP-16 Zone)	Singh (1978a)
Vinjan Miani	LM	Late mid Eocene (NP-16 Zone)	Singh (1978b)
Lakhpat	LM	Late mid Eocene (NP-16 Zone)	Singh (1980a)
Vinjan Miani	LM	Late mid Eocene (NP-16 Zone)	Singh (1980b)
Berwali Stream Section	LM	Late mid Eocene (NP-16 Zone)	Singh (1998a, b)
Babia Hill	LM	Late mid Eocene (NP-16 Zone)	Singh and Singh (1986)

(continued)

Table 1 (continued)

Locality	Mode of documentation	Age	Publication
Rato Nadi Section	LM	Harudi Formation; Early Bartonian; Early-Late Middle Eocene (NP16/CP14a) Fulra Limestone Formation; Late Eraly-Early Late Bartonian, NP16/CP14a-early NP17/CP14b	Singh and Singh (1991)
Rakhadi River Section	LM	Late mid Eocene (NP-16 Zone)	Singh et al. (1980)
Rato Nadi Section	LM	Late mid Eocene (NP-16 Zone)	Singh and Singh (1987)
Kakdi Nadi (RatoNala Section)	LM + SEM	Late mid Eocene (NP-17 Zone)	Rai (unpublished thesis) (1988)
Rato Nala/Ratchelo Nala Section	LM	NP-17 Bartonian	Jafar and Rai (1994)
Ratchelo Nala Section	LM + SEM	NP-17 Bartonian	Rai (1997)
Kakdi Nadi (RatoNala Section)	LM, SEM	Late mid Eocene (NP-17 Zone)	Rai (2007)
Type section, Maniara Fort Formation (Oligocene), SW Kachchh (=Kutch)	No illustration	Oligocene (Rupelian),	Rai (abstract) (2002)
Kutch Gulf and Continental Shelf	No illustration	Recent + Reworked Cretaceous & Tertiary	Guptha (1979)
Southern Saurastra (B-14-1)		Late Palaeogene	Singh and Ali (abstract) (1983)
Cambay Basin Broach	LM	Late Eocene	Pant and Mathur (1973)
Tarkeshwar	LM	Late Eocene (?NP 19, *Isthmolithus recurvus* Zone)	Singh et al. (1978)
Chansama Wells, Mehsana Block	LM	Chansama Well-B, Kalol & Tarapur Shale, Lutetian to Pribonian (Middle-Late Eocene) Chansama Well-C, Tarapur Shale (Priabonian)	Singh et al. (1998)

(continued)

Table 1 (continued)

Locality	Mode of documentation	Age	Publication
Dumas Well-A, Tarapur Shale, Dadhar, Tarkesvar formations	(No illustrations)	Middle-Late Eocene, Oligocene, Lower Miocene	Singh and Uddin (2000)
Vastan lignite mine	LM + SEM	LPTM	Samanta et al. (2013a, b)
Vastan lignite mine		LPTM	Samanta et al. (2013b)
Bombay Offshore-Tarapur well (B-14-1)	LM	Late Palaeogene	Singh and Ali (1983)
Bombay Offshore well SS-A (Jafrabad, Belapur, Diu, Mahuva, Bombay, Tapti, Bandra & Chinchini formations) All formations nannofossil yielding.	LM	Jafrabad = NP11-?NP15 (Early-Mid.Eocene) Belapur = ?top NP20 (Late Eocene) Diu = top NP 21(Earliest Oligocene) Bombay = Oligocene/Miocene boundary	Saxena (1996)
Mumbai offshore to Saurashtra offshore		Mumbai offshore, Panna clastics, Devgarh, Bassein, Mukta formations = Palaeocene to Oligocene Saurashtra offshore = Early Palaeocene to Late Miocene	Saxena (abstract) (2007)
Jaisalmer Basin Khartar well-C, Sanu, Khuiala & Bandah formations	LM	Sanu = Paleocene, Khuiala = Nannofossil absent Bandah = Bartonian, Middle Eocene	Singh (1998b)
Tanot well-1	LM, SEM	Bartonian	Rai et al. (2014)
Krishna-Godavari Basin Pallakolu-A, intertrappeans, PLK-A well, Razole area, Infratrappeans RZ well	No illustration	Danian (*Cruciplacolithus tenuis*)/Maastrichtian (*M. murus*)	Raju et al. (1995)

(continued)

Table 1 (continued)

Locality	Mode of documentation	Age	Publication
PLK-A well Infra, Inter and Supratrappean	LM	Maastrichtian (*Micula murus* Zone) Danian (NP1/NP2 NP3) & Selandian (?NP5)	Von Salis and Saxena (1998)
Onland to offshore wells (Amalapuram-A, AMP-A; Palakollu-A, PLK-A, G-13-A, GS-5-A, GS-14-A & GS-21-A); Razole, Palakollu, Pasarlapudi & Vadaparru formations	No illustration	PLK-A = K/T to Thanetian/Danian GS-5-A & GS-14-A = Ypresian/Thanetian AMP-A = Ypresian, Earliest to Early Lutetian GS-14-A = Lutetian/Ypresian GS-21-A = Lutetian, Bartonian G-13-A & GS-21-A = Priabonian G-13-A = Early Rupelian, Late Rupelian, Late Chattian GS-21-A = Rupelian AMP-A = Chattian	Saxena (2000)
Cauvery Basin Vridhacahalam, Gopurapuram borehole (VGP), Pallakolai road	LM	*Discoaster multiradiatus* NP 9 Zone, Upper Palaeocene	Jain et al. (1983)
Karaikal Well No. 9 Cutting samples of sediments of few meters above the basement		Middle Paleocene to earliest Miocene (NP4-NP5 to NP11-NP12)	Jafar (1983b)
Pondicherry	(No illustration)	Late Cretaceous-Early Tertiary	Rajagopalan (1964)

(continued)

Table 1 (continued)

Locality	Mode of documentation	Age	Publication
Pondicherry	(No illustration)	Late Cretaceous-Early Tertiary	Narasimhan (1975)
Andaman Island (Middle Andaman)	LM	Late Oligocene	Mathur and Mathur (1980)
Cole Brook, North Passage and Great Nicobar	LM +SEM	Early-Middle Miocene	Wei & Srinivasan (1984)
Islands of Baratang, exuded material from Mud Volcano	LM	Basal Danian marker taxa with Late Cretaceous assemblage	Jafar (1994)
Assam Mikir Hills, Samkherjan area (LowerKopili Formation)	LM	Late Middle Eocene-Early Late Eocene	Singh (1979)
Meghalaya Khasi Hills (Mahadek, Langpar Stage) section	LM	Maastrichtian-Danian	Narasimhan (1963)
Khasi Hills (Langpar Formation)		Maastrichtian-Danian (*Micula prinsii* Zone-NP1 Zone)	Garg (abstract) (2007)
Khasi Hills (Langpar Formation)		Danian (NP1-NP4 and CP1-CP3 zones)	Garg and Jain (1995)
Dilni River	LM	Bartonian Np17-lower NP 18	Rai and Garg (2009)
Rewak Formation, Garo Hills	LM	Late Eocene (Priabonian)	Singh et al. (2016)

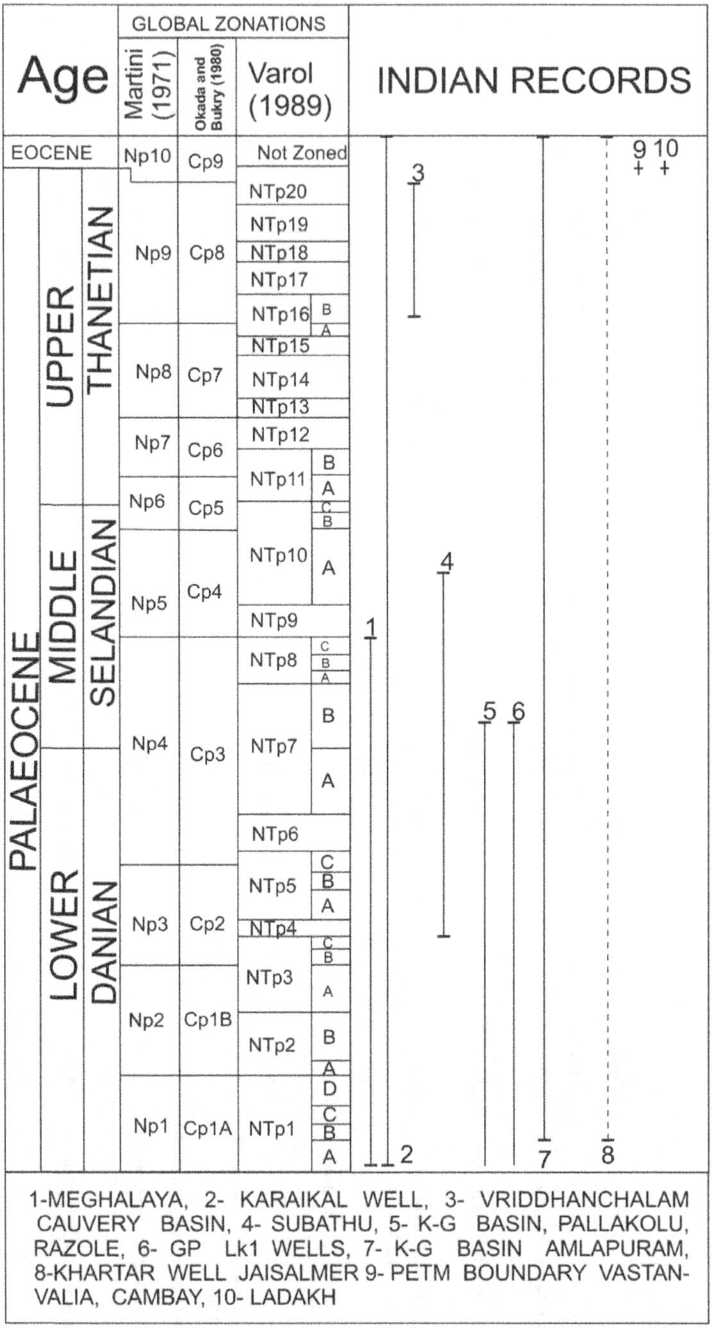

Fig. 2 Palaeocene nannofossil records from various basins of India plotted on nannofossil zones of various scheme

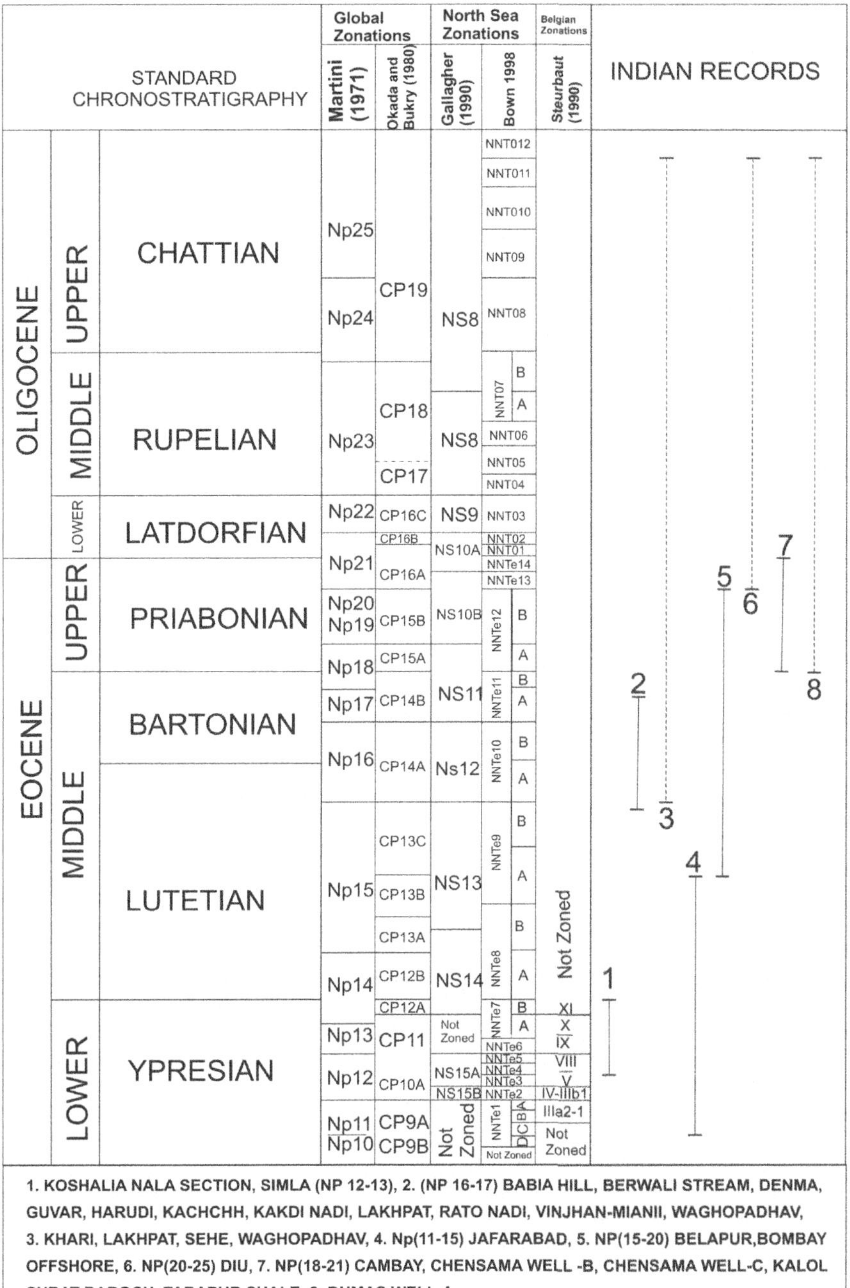

Fig. 3 Eocene-oligocene nannofossil records from various basins of India plotted on nannofossil zones of various scheme

hydrocarbon potential of Palaeogene sequences in western, northeastern and south-eastern sectors of India has provided the impetus for detailed nannofossil studies in view of their value in biostratigraphic precision, especially due to refined bioevents identified by FAD's (First Appearance datum) and LAD'S (Last Appearance datum) of several species which are cosmopolitan markers. In view of the expanded hydro-carbon exploration activities in Indian Subcontinent, biostratigraphic potential of nannofossils is going to be immensely useful in future. Utility of calcareous nanno-fossils in macro-fossil devoid thick Tethyan successions dating has strong potential both for biostratigraphy and for precise timing of India–Asia collision. More detailed and systematic studies of nannofossils are required from Palaeogene successions of various basins of Indian subcontinent with known nannofossil productivity.

Acknowledgements The author is thankful to Dr. Sunil Bajpai, Director BSIP for work facility and constant encouragement. The illustrations prepared by S. Singh and H. Srivastava of Geology Department, Lucknow University, Lucknow are duly acknowledged. BSIP permission Number is BSIP/RDCC/Publication No. 61/2016-17.

References

Bhandari LL, Venkatachala BS, Singh P (1977) Stratigraphy, palaeoecology and paleontology of Ladakh molasse Group in the Kargil area. pp 127–133. In: IV Indian Colloquium on Micropale-ontology and Stratigraphy Dehradun (eds Venkatachala BS, Sastri VV) 1974-75

Bramlette MN, Martini E (1964) The great change in calcareous nannoplankton fossils between the Maestrichtian and Danian. Micropaleontology 10:291–322

Bown PR (1998) Calcareous nannofossil biostratigraphy. In: Bown PR (ed) British Micropalaeon-tological Society Publications Series, Chapman & Hall, Cambridge, pp 1–315

Ehrenberg CG (1854) Mikrogeologie. Das Erden und FelsenschaffendeWirken des unsicht-barkleinen selbstandien Lebens auf der Erde. Leipzig: Voss. 374 + 31 [index to names] + 88 [index to Atlas]

Gallagher LT (1990) Calcareous nannofossil biozonation of the Tertiary of the North Sea Basin. Newsl Stratigr 22:21–41

Garg R (2007) Danian calcareous nannofossils from Khasi Hills, Meghalaya. XXI Indian Collo-quium on Micropaleontology and Stratigraphy, 16–17 Nov 2007, BSIP, Lucknow, p 33

Garg R, Jain KP (1995) Significance of terminal Cretaceous calcareous nannofossil marker *Micula prinsii* at the Cretaceous-Tertiary boundary in the Um Sohrengkew section, Meghalaya, India. Curr Sci 69(12):1012–1017

Guptha MVS (1979) Coccoliths from sediments of the Gulf of Kutch. Ind J Mar Sci 8:85–88

Jafar SA (1983a) Significance of Late Triassic calcareous nannoplankton from Austria and Southern Germany. Neus Jb Geol Paläontol, Abh 166:218–259

Jafar SA (1983b) Middle Paleocene to earliest Miocene calcareous nannoplankton biostratigraphy of well-cutting samples from Karaikal well No. 9 (Cauvery Basin), Southern India. Unpublished Oil and Natural Gas Commission Report, Government of India, Dehradun

Jafar SA (1994) Late Maastrichtian nannofossils from the Lattengebirge (Germany) and the Andaman-Nicobar Island (India)-remarks on events around Cretaceous–Tertiary boundary. Neues Jb Geol Paläontol Abh 191(2):251–269

Jafar SA, Kapoor PN (1988) Late Maestrichtian-Danian nannoplankton from basal Subathu of Dharampur, Simla Himalaya, India-Palaeogeographic Implications. Palaeobotanist 37(1):115–124

Jafar SA, Rai J (1994) Late Middle Eocene (Bartonian) calcareous nannofossils and its bearing on coeval post-trappean transgressive event in Kutch Basin, Western India. Geophytology 24(1):23–42

Jafar SA, Singh OP (1992) K/T boundary species with Early Eocene nannofossils discovered from Subathu Formation, Shimla Himalaya, India. Curr Sci 62(5):409–413

Jain KP, Garg R, Joshi DC (1983) Upper Palaeocene calcareous nannoplankton from Vriddhachalam area, Cauvery Basin, Southern India. Palaeobotanist 31(1):69–75

Martini E (1971) Standard Tertiary and Quaternary calcareous nannoplankton zonation. In: Farinacci A (ed) Proceedings of the second planktonic conference Roma, 1970. Edizioni Technoscienza, Rome, vol 2, pp 739–852

Martini E, Bramlette MN (1963) Calcareous nannoplankton from the experimental Mohole drilling. J Paleontol 37:845–856

Mathur YK (1972) Plant fossils from the Kuar Bet, Pachham Island, Kutch. Curr Sci 41:488–489

Mathur YK, Mathur K (1980) Barail (Laisong) palynofossils and Late Oligocene nannofossils from the Andaman Island, India. Geosci J 1(2):51–66

Moshkovitz S (1982) On the findings of new calcareous nannofossil (*Conusphaera zlambachensis*) and other calcareous organisms in the upper Triassic sediments of Austria. Eclogae Geol Helv 5:611–619

Narasimhan T (1963) Coccolithophorids and related nannoplanktons from the Cretaceous-Tertiary sequence of Khasi Hills, Assam. J Geol Soc India 4:l09–115

Narasimhan T (1975) Nannoplanktons from upper Cretaceous and lower Tertiary formations of Pondicherry, South India, vol II, pp 207–208. In: Venkatachala BS, Sastri VV (eds) Proceedings of 4th Indian Colloquium on Micropaleontology and Stratigraphy. Dehradun 1974–75, Indian Institute of Petroleum Exploration, Oil and Natural Gas Commission, Dehradun

Okada H, Bukry D (1980) Supplementary modification and introduction of code numbers to the low-latitude coccolith biostratigraphic zonation (Bukry, 1973; 1975). Mar Micropalaeontology 5:21–32

Pant SC, Mamgain VB (1969) Fossil nannoplankton from the Indian subcontinent. Rec Geol Surv India 97(2):108–128

Pant SC, Mathur UB (1973) Fossil nannoplankton from Eocene of Broach, Gujarat, India. Rec Geol Surv India 105(2):209–216

Rai J (1988) Calcareous nannoplankton from Eocene of Kutch, Western India. Unpublished PhD thesis, Lucknow University, Lucknow, pp 1–316

Rai J (1997) Scanning-electron microscopic studies of the Late Middle Eocene (Bartonian) calcareous nannofossils from the Kutch Basin, Western India. J Palaeont Soc India 42:147–167

Rai J (2002) Calcareous nnnofossils from the Maniara Fort Formation (Oligocene), SW Kachchh (= Kutch), Western India. In: 9th International Nannoplankton Association Conference, JNR, Parma, Italy, vol 24, issue No. (2), p 153

Rai J (2007) Middle Eocene calcareous nannofossil biostratigraphy and taxonomy of onland Kutch Basin, Western India. Palaeobotanist 56:29–116

Rai J, Garg R (2009) Late Middle Eocene (Bartonian) age calcareous nannofossils from Dilni River Section, Meghalaya, Northeastern India. In: Bhat GM et al (eds) Geoenvironment challenges ahead. McMillan Publication, New Delhi, pp 277–293

Rai J, Singh A, Gulati D (2014) Bartonian age calcareous nannofossil biostratigraphy of Tanot well-1, Jaisalmer Basin and its implications. J Palaeontol Soc India 59(1):29–44

Rajagopalan N (1964) Late Cretaceous and early Tertiary stratigraphy of Pondicherry, South India. Bull Geol Soc India 1(1):10–12

Raju DSN, Jaiprakash BC, Kumar A, Saxena RK, Dave A, Chatterjee TK, Mishra CM (1995) Age of Deccan volcanism across KTB in Krishna-Godavari Basin: New evidences. J Geol Soc India 45(2):229–233

Samanta A, Sarkar A, Rai J, Rathore SS (2013a) Late Paleocene-Eocene carbon isotope stratigraphy from a near—terrestrial tropical section and antiquity of Indian mammals. J Earth Syst Sci 122:63–171

Samanta A, Bera MK, Ghosh R, Bera S, Filley T, Pande K, Rathore SS, Rai J, Sarkar A (2013b) Do the large carbon isotopic excursions in terrestrial organic matter across Paleocene-Eocene boundary in India indicate intensification of tropical precipitation? Palaeogeogr Palaeoclimatol Palaeoecol 87:91–103

Saxena RK (1996) Calcareous nannoplankton biochronostratigraphy of Bombay Offshore Basin. In: Pandey J, Azmi RJ, Bhandari A, Dave A (eds) XV Indian Colloquium on Micropaleontology & Stratigraphy, Dehradun, pp 723–732

Saxena RK (2000) Calcareous nannofossil datum levels in the Paleogene shale sections of Krishna-Godavari Basin, India. In: Proceedings of XVI Indian Colloquium on Micropaleontology and Stratigraphy, NIO, Goa, Bulletin of the Oil and Natural Gas Corporation Limited, vol 37, issue No. 2, pp 161–172

Saxena RK (2007) Nannofossils from Kutch-Saurashtra-Mumbai-Kerala offshore along the west coast of India. XXI Indian Colloquium on Micropalaeontology and Stratigraphy, November 16–17, 2007, BSIP, Lucknow, p 176 (abstract)

Singh P (1978a) A note on the Late Middle Eocene nannofossils from Lakhpat, Kutch. Curr Sci 47(3):87–88

Singh P (1978b) A note on the Late Middle Eocene nannofossils from the Vinjhan-Miani Area, Kutch. Curr Sci 47(2):53–54

Singh P (1979) Calcareous nannoplankton from the Kopili Formation of Mikir Hills, Samkherjan Area, Assam. Bulletin of the Oil Nat Gas Commission, Dehradun 16(2), pp 75–85

Singh P (1980a) Late Middle Eocene calcareous nannoplankton from Lakhpat, Kutch, Western India. Geosci J 1(1):1–14

Singh P (1980b) Late Middle Eocene calcareous nannoplankton and palaeogeographic remarks on Vinjhan-Miani area, Kutch, Gujarat. Geosci J 1(1):15–29

Singh P (1983) Eocene calcareous nannoplankton from the Abhay-l Well, Bengal Basin with palaeoceanographic remarks. In: 1st international conference on Paleoceanography, Zürich, pp 60 (abstract)

Singh P (1998a) Late Middle Eocene calcareous nannoplankton from Berwali River Section, Kachchh, India. Geosci J 9(2):231–246

Singh P (1998b) Calcareous nannoplankton and foraminiferal biostratigraphy of upper cretaceous and paleogene subsurface sequences of Kharatar well–C, Jaiasalmer, Rajasthan. GeoSci J 19(2):145–169

Singh P, Ali QA (1983) Late palaeogene calcareous nannoplankton from the Tarapur Well B-14-1, Bombay Offshore with paleoceanographic remarks. In: 1st international conference paleoceanography. Zürich, p 56 (abstract)

Singh P, Singh MP (1986) Late Middle Eocene calcareous nannoplankton from Babia Hill, Kutch, Gujarat, India. Geosci J 7(2):145–162

Singh MP, Singh P (1987) A note on the geology and micropalaeontology of the RatoNadi Section, Kutch, Gujarat State, India. Geosci J 8(1&2):201–204

Singh P, Singh MP (1991) Nannofloral biostratigraphy of the Late Middle Eocene strata of Kachchh region Gujarat State, India. GeoSci J 12(1):17–51

Singh P, Uddin I (2000) Calcareous nannoplankton from the subsurface of the tertiary sequence of Dumas Well-A, Cambay Basin, Gujarat. In: Proceedings of XVI Indian Colloquium on Micropaleontology and Stratigraphy, NIO, Goa, Bulletin of the Oil and Natural Gas Corporation Limited, vol 37, Issue No. 2, pp 219–224

Singh P, Rastogi KK, Vimal KP (1978) A note on the Late Eocene nannofossils from Tarkeshwar, Gujarat. Curr Sci 47(10):346–347

Singh P, Singh MP, Mathur DN, Srivastava RN (1980) Late Middle Eocene calcareous nannoplankton from Rakhadi River Section, Harudi, Kutch. Curr Sci 49(5):172–176

Singh P, Porwal DK, Uddin I (1998) Microfauna and calcareous nannoplankton from the subsurface tertiary sequence of Chanasma wells B and C, Mehsana Block, Cambay Basin, Gujarat. Geosci J 19(2):87–99

Singh A, Rai J, Garg R (2016) Record of the late eocene (Priabonian) nannofossils from the lower part of the Rewak Formation, Garo Hills, Meghalaya, Northeastern India. Palaeontol Soc India 61(1):91–97
Sorby HC (1861) On the organic origin of the so called "crystalloids" of the chalk. Ann Mag Nat Hist 8(45):193–200
Sterbaut E (1990) Ypresian calcareous nannoplankton biostratigraphy and palaeogeography of the Belgian Basin. Belgische Vereniging voor Geologie 97(1988):51–285
Vandenberghe N, Hilgen FJ, Speijer RP (2012) Chap 28 The paleogene period. In: Gradstein FM, Ogg JG, Schmitz M, Ogg G (eds.) The geologic time scale, 1st edition. Elsevier, Amsterdam, pp. 855–921
Varol O (1989) Palaeocene calcareous nannofossil biostratigraphy. In: Crux JA, van Heck SE (eds) Nannofossils and their applications. British Micropalaeontological Society Publications Series, Ellis Horwood Limited, Chichester, pp 267–310
Von Salis K, Saxena RK (1998) Calcareous nannofossils across the K/T Boundary and the age of the Deccan volcanism in Southern India. J Geol Soc India 51:183–192
Wei KY, Srinivasan MS (1984) Miocene calcareous nannofossils from Colebrook, North Passage and Great Nicobar Islands, Northearn Indian Ocean. Revista Esp de Micropal 16:345–366

Implication of the Occurrence of Minute Biotic Bodies on the Conjoined *Nummulites* aff. *Nummulites acutus* (Sowerby) in the subsurface Eocene of Cauvery Basin, India

Sanjay Kumar Mukhopadhyay

Abstract Conjoined and aberrant adult tests of *Nummulites* aff. *Nummulites acutus* (Sowerby) from the subsurface Middle Eocene sediments south of Pondicherry in Cauvery Basin, India show in situ development of minute biotic bodies as rare and unusual incidence. Since usual test of extinct as well as modern nummulitids hardly shows such development, these bodies are studied to know their nature of relation with the host adults. Being indigenous, intact and regular in shape, these forms have diverse mode of occurrence from cluster of embryo-like tiny featureless globular bodies to discrete lenticular bodies with granulated surface and as minute lenticular juvenile tests. The rudimentary and pre-mature morphologic features in the juveniles appear septal filaments, involute spiral sheets, planispiral whorls, quadrangular to rectangular chambers, narrow alar prolongations, marginal cord and septa, whose mature counterparts typify adult *Nummulites*; therefore, these are the offspring of the adult *Nummulites* aff. *acutus*. Their remarkably small size compared to the size of the host test indicates young age whereas their embedded emplacement at crucial morphologic parts of the host test points for a genetic relationship between adult and 'baby'. With minute initial chamber (<1 μm) followed by several whorls, the offspring resemble microspheric juveniles. Their persistent development only in the 'conjoined' and 'separated' adult tests indicate that test abnormality is linked with their origin. An entity of conjoined *Nummulites* aff. *Nummulites acutus* comprises two megalospheric adult individuals of the species with fused apertures, and resembles 'plastogamy' in modern smaller benthic foraminifera, in which two adult individuals of a species unite for sexual reproduction. As in the latter, the two individuals in conjoined *Nummulites* are fused using secreted lime deposit as adhesive. Since modern larger foraminifera including nummulitids are not known to develop conjoined entity, taking a cue from plastogamy, sexual reproduction is held responsible for the development of the offspring in *Nummulites* aff. *Nummulites acutus*. This is supported by the occurrence of minute bodies inside the cavities, pores, and fused apertures and along the fused contact zone, and resemblance of a studied offspring

S. K. Mukhopadhyay (✉)
Plot 2, Nabaroon Co-Operative Housing Society, Santoshree Palli, Thakurpukur 700063, Kolkata, India
e-mail: crazytintin65@hotmail.com

S. Bajpai et al. (eds.), *The Indian Paleogene*, Society of Earth Scientists Series,
https://doi.org/10.1007/978-3-319-77443-5_9

with the offspring of modern nummulitids. The study indicates that sexual reproduction by adult individuals was prevalent in extinct *Nummulites* aff. *Nummulites acutus* to generate microspheric forms. This particular life cycle is supposed to prevail in the species towards the end of the Middle Eocene.

Keywords *Nummulites* aff. *N. acutus* · Adult-Juvenile association
Conjoined tests · Sexual reproduction · Middle eocene · Cauvery basin

Introduction

Nummulites Lamarck is the oldest known and most widely studied foraminifera that reached acme in population growth and species diversity during Eocene. Being the most characteristic marine organism of the Paleogene Tethys, the corresponding deposits are commonly referred to as 'nummulitique'. As index and marker for the Paleogene succession, its species are immensely useful for biostratigraphy, basin correlation, age assignment and paleoenvironmental reconstruction of the host successions. Nevertheless, the mechanism of proliferation of the species of extinct *Nummulites* is not well known. The cause of development of dimorphic forms in *Nummulites* (de La Harpe 1879; Munier-Chalmas 1880) was explained by Lister (1895) as alternations of sexual and asexual reproduction after taking a cue from the life cycle of modern *Polystomella crispa*. Laboratory culture on some modern species of nummulitids including *Paleonummulites venosus* provided crucial information on their life cycle and reproduction that were useful to conjecture life-cycle of their extinct counterparts. Röttger et al. (1998) during laboratory culture of a reproductive agamont of modern *Nummulites venosus* (=*Paleonummulites venosus*) documented release of flagellated gametes with mean diameter of proloculus 113 µm, which corresponds with a megalosphere of megalospheric form. This find endorsed Lister's (1895) view on the life cycle of *Nummulites*. Laboratory culture also recorded gametogamy type of sexual reproduction in the megalospheric *Heterostegina depressa* (Biekart et al. 1985; Röttger et al. 1986, 1989); however, megalospheric adult individuals of *Paleonummulites venosus* have not been cultured for adult sexual reproduction. Gamontogamy (=plastogamy of Schaudinn 1895) type of sexual reproduction with attendant offspring as occurring in some modern smaller benthic foraminifera (Myers 1938; cf. Preobrazhenskaya and Tarasova 2004) has not been recorded in modern nummulitids. Mukhopadhyay (2003a, 2007 and 2016) from the Middle Eocene of Cambay basin, Gujarat recovered plastogamic paired tests and conjoined tests in some extinct species including *Nummulites boninensis*, *Nummulites* aff. *Nummulites discorbinus*, *Nummulites* cf. *Nummulites pengaroensis*, *Nummulites fabianii* and *Rotalia* cf. *Rotalia trochidiformis*; and in *Pellatispira aff. crassicolumnata*; the assemblage included juvenile bearing broken test of *Nummulites* cf. *Nummulites beaumonti*. Since no indigenous juvenile was found attached on the entire adult megalospheric conjoined tests of an extinct species of *Nummulites*, the adult-juvenile association in the broken test may be conjectured as plastogamic reproduction. In the absence of conjoined entity in modern larger foraminifera, fossilized conjoined tests are of growing interest to know the cause of conjugation and to understand the nature of association between the adults and juveniles in the anomalous tests.

Nummulites aff. *Nummulites acutus* as 'conjoined tests' of two adult individuals and as 'separated' individuals that once remained in conjoined state is recovered from the Middle Eocene drill core sediments of the Cauvery Basin, India (Fig. 1) as rare occurrence and conjoined specimens attached with minute bodies are reported for the first time. Present study aims to know the nature and cause of relation existing between the attached adults, between the adults and associated tiny forms, and among the tiny forms. The relationship between adult and tiny forms, lacking in a normal adult test of *Nummulites*, was explored in a fossilized conjoined entity and presumed as 'parent-daughter association' (Mukhopadhyay 2015). The incidence of sexual reproduction in extinct *Nummulites* is explained in detail in this paper with the documentation of juvenile-adult association in conjoined tests and 'separated' tests of *Nummulites* aff. *Nummulites acutus*.

The studied core contains other larger foraminifera including *Nummulites* sp., *Assilina* sp., *Operculina* sp., *Discocyclina javana* and *Pellatispira* sp. (Plate 1, Fig. 8), of which the co-occurrence of *Assilina* and *Pellatispira* assigns middle Eocene (Bartonian Stage) age to the host sediments since similar association contains *Orbulinoides beckmanni* in the Amravati Formation in Bharuch District Gujarat (Mukhopadhyay 2003b) and the middle Eocene occurrence of *Pellatispira* was reported by many (Romero et al. 1999; Hottinger et al. 2001; and references in Mukhopadhyay, 2003b). Furon and Lemoine (1939) and Furon (1941) recognized *Nummulites* bearing middle Eocene rocks in the subsurface of Pondicherry area at about 150 m depth and Govindan (2013) from the boreholes of southern Cauvery basin reported *Assilina-Alveolina* bearing larger foraminiferal assemblage containing *Nummulites vrendenbergi*, which according to Racey (1995) is a synonym of *Nummulites acutus*. This concept is largely followed in this study while noting that the studied specimens bear close affinity with *N. acutus*. The studied conjoined tests and separated tests of *Nummulites* aff. *N. acutus* are distinct from previously reported fused and paired tests of *Nummulites* (Mukhopadhyay 2003a, 2007) in the clear development of fused apertures and in the occurrence of attached minute forms that are construed as juveniles. *Nummulites* aff. *N. acutus* is described to adjudge the role of morphologic features of its anomalous forms in the development of the attached juvenile forms.

Material and Methods

A drill core sample about 10 cm long and 8 cm diameter from the Paleogene strata of the Cauvery Basin was received from the Director, Palaeontology Division-II, Geological Survey of India, Kolkata to investigate foraminiferal remains and to determine the age of the sediments though the stratigraphic level, depth of occurrence and pre-

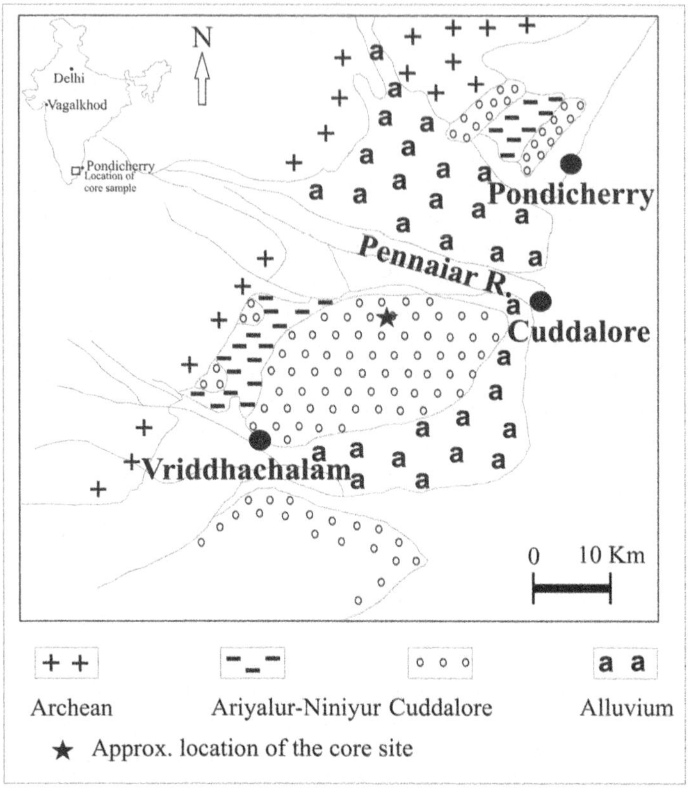

Fig. 1 Geology of a part of the Cauvery Basin, SW of Pondicherry (Geological map after Krishnan 1960). The drill core is from an uncertain location between Cuddalore and Vriddhachalam

cise location of the studied core were not known. The core comprising grey shaly marl with 1.5 cm thick yellowish marlstone intercalation towards the upper part was longitudinally halved with core splitter; one half was used for this study and the counterpart was preserved for reference. About 200 g shaly marl was processed for planktonic foraminifera following standard procedure (Mukhopadhyay 2003a, b) and about 0.3 g finest residues (250+ mesh with 63 μm pore size) were scanned under stereo zoom binocular microscope; however, age diagnostic forms were not found. Coarse residues (80+ mesh) contain resembling larger foraminifera as occurring in the thin sections of the core, and these include *Nummulites* aff. *Nummulites acutus*, *Nummulites* sp., *Assilina* sp., *Pellatispira* sp., *Operculina* sp. and *Discocyclina javana*. The species named first outnumbered the rest; it was picked to include 23 megalospheric adult tests, 6 distorted tests and 8 conjoined tests. The latter two categories are anomalous tests that contain tiny biotic forms on the test surfaces as well as inside some crucial morphologic features. The anomalous tests were studied externally and no attempt was made to study the internal morphology through oriented

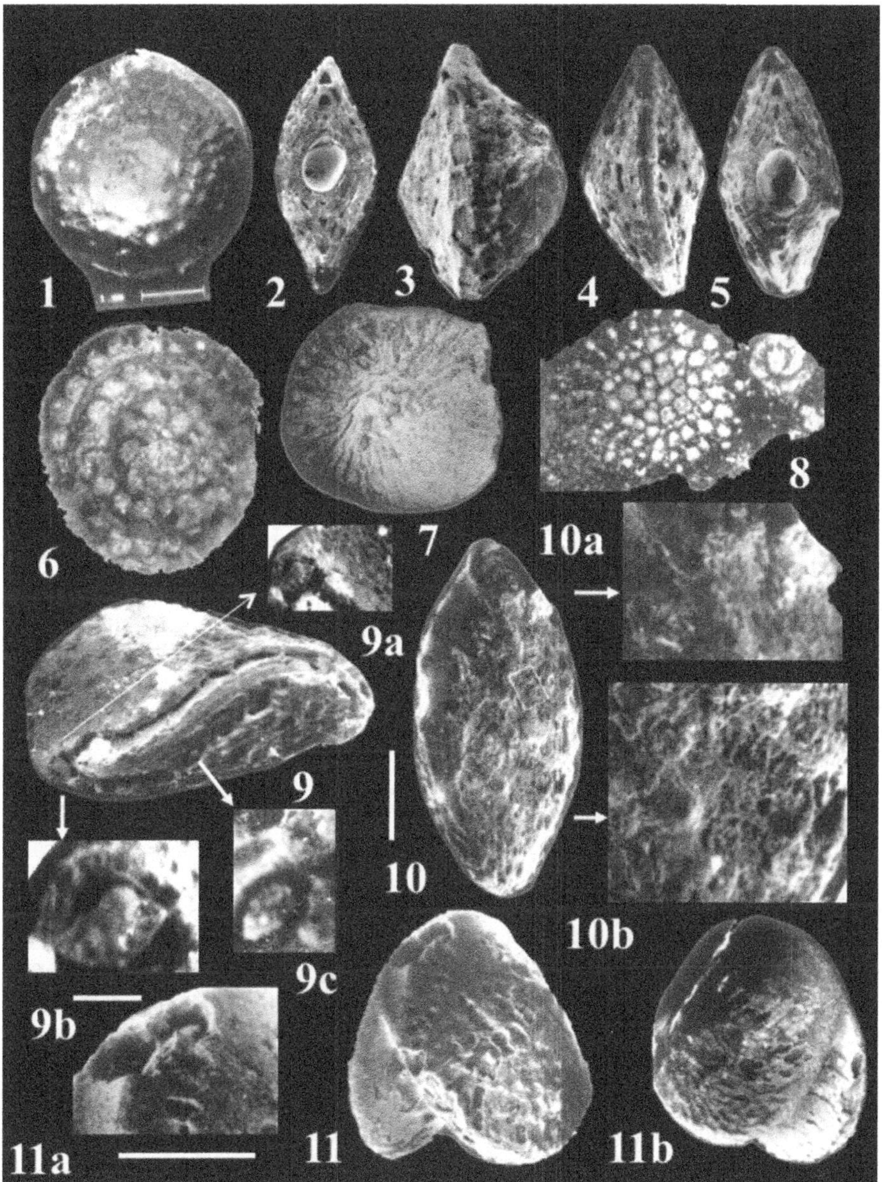
1
2
3
4
5
6
7
8
9a
9
9b
9c
10a
10
10b
11a
11
11b

◂**Plate 1** *Nummulites* sp. aff. *Nummulites acutus* (Sowerby) as megalospheric (A-form) normal tests, and juvenile bearing anomalous conjoined tests and 'separated' tests; 1 mm bar in 1 is valid for 2–8, 11, 11b; 1 mm bar in 10 is valid for 9 and 11; 50 μm bar in 9a stands for 9b, 10a, 10b, and bar is 500 μm in 11a. All measurements are approximate. **1**. External polar view showing spirally arranged granules and radial septal filaments; **2, 5**. Axial sections showing bilaterally symmetric test with acute periphery, large proloculus and often with flat polar knob; **3**. Strongly asymmetric aberrant test with highly convex right side showing smooth wide polar area, surface cavities and asymmetric aperture; **4**. External peripheral view showing swerved peripheral margin; **6**. Near equatorial section showing about five whorls in tight spire; **7**. *Nummulites* sp. showing rare development of basal peripheral aperture in external polar view; **8**. Thin section containing pre-mature *Nummulites* sp. (belonging to ? *Nummulites* sp.aff. *N. acutus*) in association with *Pellatispira* sp., both in oblique sections; **9**. Strongly asymmetric aberrant 'separated test' showing thick sinuous marginal cord, surface undulations and tiny juveniles within cavities and on test surface; weakly convex left side showing surface features in strong relief whereas strongly convex right side with broad polar area is near smooth containing patchy lime deposit; **9a**. oblique section of a juvenile occurring inside the apertural cavity of the adult test, whose enlarged view **(9b)** shows massive, globular central (early) part followed by incomplete whorls containing numerous quadrangular to sub rectangular chambers, near straight septa that are often distally weakly curved and granulated surface; **9c**. lenticular granulated juvenile with tapered peripheral margin occupying a chamber cavity of the adult test; **10**. Unequally biconvex 'separated' test containing large sub circular aperture bordered by thick rim; left side showing smooth surface (spiral sheet) with faint radial septal filaments; broadly convex bio-eroded right side containing discontinuous septal filaments, few subdued granules, patchy lime deposit and numerous juveniles; **10a**. part of aperture showing filamental lime deposit containing thread-like ultra structure and a globular body at the north central margin; **10b**. part of the right surface showing numerous juveniles in oblique and tangential sections; a sub circular juvenile contains minute initial chamber followed by 2–3 whorls, faint septa and quadrangular chambers; **11**. Oblique view of conjoined tests showing fused apertures, wedge shaped contact zone containing fused septal filament, and numerous irregular cavities on the surface; **11a**. fused apertures of the conjoined tests showing marginal cord of previous whorl associated with tiny globular bodies having thread like feature; **11b**. Dorsal view of the conjoined tests showing granulated septal filaments and cavities, some of which are occupied by tiny juvenile bodies

sections because these unusual rare forms occur in few numbers. Some premature *Nummulites* tests appearing in the washed residues (150+ mesh) were avoided due to uncertain relation with *Nummulites* aff. *Nummulites acutus*. The adult individuals that constitute the conjoined entities are provisionally recognized as *Nummulites* aff. *Nummulites acutus* considering resemblance in test size, external morphological features, overall appearance resembling as undistorted normal adult tests and co-occurrence of solitary tests and conjoined tests. All the recovered individuals of the species are megalospheric forms; no microspheric adult individual was found. A conjoined specimen of *Nummulites* sp. collected from the rubble of the Eocene rocks occurring at the base of exposures near Vagalkhod village in Bharuch District, Gujarat provided additional information on the association of minute forms with conjoined tests.

In order to dissociate matrix material and tiny forms from the attached conjoined and separated tests, the specimens were mildly boiled in distilled water dissolved with sodium hydroxide for about one hour (cf. Glaessner 1963) and then cleansed with tap water in 330 mesh sieve (=35 μm pore size). With this process only few detrital grains could be detached; however, minute forms were not found. The oven-dried specimens were placed in ultrasonic vibrator in three spells each for about thirty seconds when more detrital clay and silt were separated. Gold-coated anomalous specimens in scanning electron microscope (SEM) showed the occurrence of minute forms on the test surfaces. Due to rare occurrence, the minute body bearing anomalous tests were carefully processed and examined. Since the minute forms are randomly orientated, their size as measured from attached linear scale in the SEM photographs is tentative. External features of some grown up minute bodies were studied from their in situ random occurrences. The firmly attached bodies being minute, it is extremely difficult to prepare thin section of an individual, though some grown up forms exhibit recognizable internal morphologic features due to partial erosion. Mild hand rubbing and brushing of the test surfaces containing minute bodies with fine carborundum powder under the stereo zoom binocular microscope helped exposing internal morphologic features more clearly. Selected parts of SEM photographs were studied in 'hpscanjet 2400'scanner. An interested part marked under SEM was zoomed in until the visibility of the subjects began to appear clearly. Contrast and brightness were calibrated on grey scale background. Subsequently, the objects were again zoomed in to get highly pixilated pictures which became sharp showing clearer internal morphology; then these were photographed. The preservation of the aberrant specimens and minute forms remains largely satisfactory. All the studied material and illustrated specimens are preserved in the Micropaleontology Laboratory of the Palaeontology Division-I, Central Headquarters of the Geological Survey of India, Kolkata.

Systematic Paleontology

Order Foraminiferida

Family NUMMULITIDAE de Blainville, 1825

Subfamily NUMMULITINAE de Blainville, 1827

Genus *Nummulites* Lamarck, 1801

Type species *Camerina laevigatus* Bruguiere 1792

Nummulites aff. *Nummulites acutus* (Sowerby)

Plate 1, Figs. 1–6, 9–11; Plate 2, 1–5

aff. 1840 *Nummularia acutus* Sowerby, p. 329, pl. 24, Figs. 13–13a

aff. 1982 *Nummulites acutus* (Sowerby); Samanta, pp. 27–32, Figs. 3–5

Megalospheric (solitary) tests (Plate 1, Figs. 1–6, 9–10):

Test medium, sub circular in polar view, lenticular biconvex in peripheral view, bilaterally symmetrical, peripheral margin acute with blunt tip, test thickest along polar diameter; granules circularly arranged, coarser in polar area than peripheral part; septal sutures radial, radiating septal filaments often granulated; trace of marginal cord distinct in peripheral part; spire regular tight with 4–6 whorls; chambers quadrangular in early whorls, later slightly longer than high; septa straight near base, distally gently curved; alar prolongations thin, lateral wall thick, pillars emerging

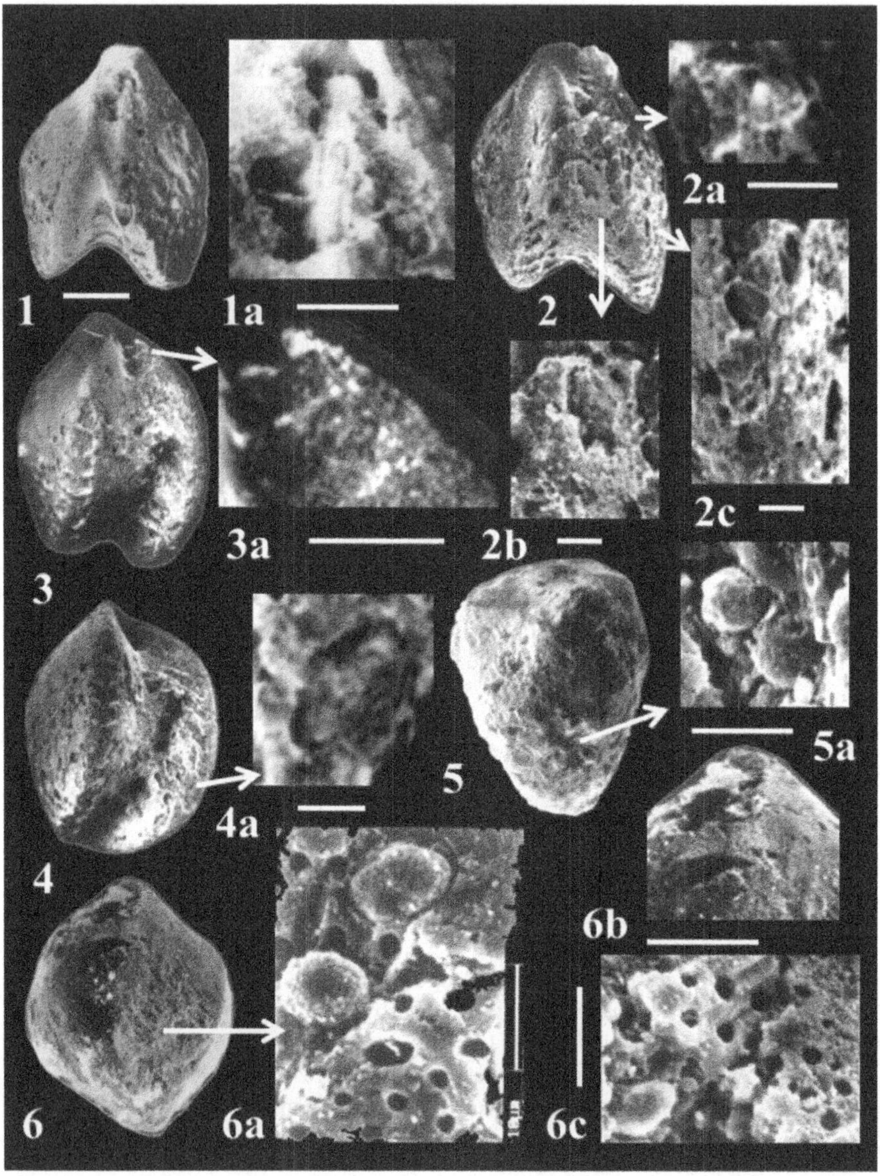
1
1a
2
2a
2b
2c
3
3a
4
4a
5
5a
6
6a
6b
6c
10µm

◂**Plate 2** Juvenile bearing conjoined tests of *Nummulites* aff. *Nummulites acutus* (Sowerby); 1 mm bar in 1 is valid for 2–6; bar in 1a and 3a is 500 μm; 250 μm in 2a, 2b, 2c, 4a; 100 μm in 6b; 50 μm in 5b; 10 μm in 6c. **1**. Conjoined tests showing fused apertural area fused and furrowed contact zone and merged septal filaments within the contact zone; two individuals containing surface cavities fuse at low angle; **1a**. Fused apertural area enlarged to show tiny globular bodies having thread-like feature (?fossilized flagella) are attached with porous marginal cord; **2**. Highly porous and cavernous conjoined tests containing several minute juveniles (2a, 2b, 2c) attached on the surfaces; the adult individuals fuse with peripheral margins at low angle; **2a**. tangential section of an underdeveloped juvenile *Nummulites* showing massive initial part followed by 'v' shaped chambers with alar prolongation, spiral sheets and faintly developed septal filaments; **2b**. underdeveloped juvenile showing partly broken granulated involute spiral sheet; **2c**. Part of peripheral margin of the right test showing several juveniles in oblique sections; **3**. Conjoined tests showing huge biogenic cavity along the contact zone; the two tests fuse at the apertures maintaining equatorial planes in almost parallel position; **3a**. tangential section of a juvenile *Nummulites* occurring beside the fused apertures, it shows 'v' shaped chambers, involute whorl walls, alar prolongations and septal filaments; **4**. Unequally biconvex conjoined tests with more inflated right test and fused apertures asymmetrically placed on the right test; biogenic cavities retain consumed septal filaments and test surface; **4a**. Transverse section of an underdeveloped juvenile showing imperfect involute spiral sheets and alar prolongations; **5**. Conjoin tests comprising two adult tests of unequal size with right test much smaller than the left test and the contact zone is highly narrowed; surface contains imperfect juveniles; **5a**. sub spherical granulated lenticular bodies (offspring) around fused apertures; hole in a body may represent ?predation activity; **6**. Loose specimen of conjoined tests of *Nummulites* sp. collected from the rubble of Eocene exposures near Vagalkhod village in Bharuch District Gujarat, showing fused anterior and posterior ends and wide contact zone; **6a**. Part of the fused contact zone showing strong pores on the surface and sub spherical granulated lenticular bodies; **6b**. Fused apertures retaining lime deposits while minute bodies at the apertural border; **6c**. Part of the fused contact zone showing cluster of tiny globular bodies resembling gel of 'fish egg' (NE corner) and discretely occupying pores; globular tiny offspring of varied size in association with lenticular bodies

from lateral walls; megalosphere 570–600 μm; test diameter 2.5–3.8 mm, test thickness 1.2–1.8 mm.

Remarks: The adult tests closely agree with *Nummulites acutus* (Sowerby) in test size, thickness, acute periphery, decreasing size of granules from polar to peripheral parts, radiating and often granulated septal filaments, tight regular spire, moderately thick marginal cord and quadrangular equatorial chambers whose length increases in successive whorls, and septa upright near base and weakly curved distally. Differences with the typical occur in larger megalosphere, lesser whorls and occasional development of pseudo polar knob. Considering overall characters the described forms are considered to have close affinity with *Nummulites acutus* (Sowerby). Minor variation occurs in test symmetry, peripheral undulations and thickness of the peripheral marginal cord. Samanta (1982) and Racey (1995) described microspheric forms of *Nummulites acutus* (Sowerby). The studied species is distinct in the development of anomalous 'conjoined tests' and 'separated' tests that contain minute biotic bodies and the present study is concerned with this hitherto unreported association.

Conjoined tests (Plate 1, Figs. 11, 11a; Plate 2, Figs. 1–5):

A conjoined entity comprises two megalospheric adult individuals of the species that remain fused at the apertures. The sides of the individuals that face each other merge with the development of a furrowed contact zone produced after fusion of the outer spiral sheets of both the tests. The tests adhered by either secreted lime deposit (cf. Myers 1936) or assimilation and fusion. The two peripheral margins may converge and fuse towards the apertural end or remain almost parallel in close proximity; accompanying marginal cord contains distinctly developed canal orifices. The fused sides occasionally retain granules in low relief and merged septal filaments; the free sides show granules and septal filaments in usual relief. The tests and the contact zone contain numerous pits, pores and cavities some of which may be biogenic as occupied by minute forms. Fused apertures and fused contact zones may contain minute biotic forms of varied shapes and size; fused spiral sheets may retain merged septal filaments. Due to involvement in fusion an adult individual in conjoined tests entity shows partial development of characteristic features that make it difficult to ascertain the species as *Nummulites* aff. *Nummulites acutus*; its recognition up to species level is tentative as based on comparisons of shape and size of tests, distinctive external morphological features on the free sides, overall appearance and association with the typical tests of the species.

'Separated' tests with attached minute biotic bodies (Plate 1, Figs. 9, 10):

The 'separated' tests are individuals that were detached from once conjoined state. In the studied material only two adult individuals of differing shape represent this category. With strong surface undulations and swerving peripheral marginal cord, the nearly planoconvex form (Plate 1, Fig. 9) is highly distorted; patchy lime deposit, discontinuous septal filaments, cavities, asymmetric aperture and scattered granules in low relief are signatures of involvement of the convex side in fusion; the opposite side with weak undulation, low convexity and strong relief of the granules, septal filaments and polar knob indicates that the surface remained free during fusion. The test contains minute biotic forms inside the aperture (Plate 1, Fig. 9a) and in the chamber cavity (Plate 1, Fig. 9b). The other separated biconvex test (Plate 1, Fig. 10) shows smooth granulated left side and a rough right side. Development of numerous minute biotic bodies in random orientations (Plate 1, Fig. 10b), remnant septal filaments, subdued granules and patchy lime deposit on the right side indicate the involvement of the side in fusion. Well developed semicircular aperture with bordering thick rim on the left wall contains apertural cavity filled with lime deposit containing dispersed subglobular minute bodies of varied size and filamental lime having ultrafine thread-like mesh (Plate 1, Fig. 10a) possibly representing mingled protoplasmic material in fossilized state.

Distorted test without minute biotic bodies (Plate 1, Fig. 3):

The strongly asymmetric (unequally biconvex) test with flat tipped peripheral margin has low convexity on the left side and strong convexity on the right side; the former side shows broadly angular polar margin with surface features in usual relief whereas the latter surface is distinct with broadly curved polar margin, scattered granules and septal filaments, and sub triangular peripheral aperture. Due to partial erosion at the peripheral part, the outer spiral sheet exposes straight and wide peripheral marginal cord containing several sub parallel canals. The form resembles the distorted separated test (Plate 1, Fig. 9) in overall appearance, but differs in the lack of surface undulations, lime deposit and minute biotic forms. Absence of minute forms suggests that the test might represent a stage of plastogamic fusion prior to mingling of protoplasmic material when it was separated.

Minute Bodies—Categories and Nature

The anomalous tests contain globular and lenticular minute forms, which are difficult to differentiate from tiny clay particles of resembling size which also scatter on the test surfaces. The clay particles usually have straight sides, angular periphery and dark featureless surface whereas the minute globular and lenticular forms possess near smooth to granulated surface, rounded periphery and apparently bilateral to radial symmetry; larger biotic bodies exhibit some internal morphologic features. The tiny biotic forms normally occur discretely and in cluster within the apertural cavities and along the contact zones showing gradation in size and features and remain firmly attached on the host tests in diverse orientations. These are devoid of signatures of transportation and weathering. Lenticular tiny forms that associate the adult tests seemingly made up of similar test material and contain resembling morphologic features; their persistent association and good preservation indicate in situ and indigenous nature. A minute form is provisionally referred to as 'body' when only surface granulations are either in rudimentary state or with distinct outlines whereas a grown up form showing internal morphological features is referred to as 'juvenile'.

Based on size and development of features the attached forms are divided into four categories—globular bodies, lenticular bodies, underdeveloped juveniles and developed juveniles.

Globular bodies (Plate 1, Fig. 10a, Plate 2, Figs. 1a, 6c)

These are the tiniest (<1–4 μm) biotic forms whose coarser representatives acquire subspherical to lenticular shape having bilateral and radial symmetry. In all the anomalous host tests these forms resemble in appearance and mode of occurrence; coarser forms are discretely dispersed in and around the fused apertures; finest forms often cluster on the contact zone to appear as 'gel of fish-eggs'. A sub spherical tiny form (<1 μm) appears featureless; however, one or two ultrafine thread-like feature may emerge from the body to resemble fossilized 'flagella'. The tiny form has similar size as the initial chamber of a juvenile (Plate 1, Fig. 10b); so, the body is presumed to be single chambered. Slightly larger sub globular bodies (1–3 μm) occur discretely to occupy pores and cavities; these apparently multi chambered bodies might have originated as the tiniest bodies. Modern flagellated gametes of *Paleonummulites venosus* illustrated by Röttger et al. (1998, Plate 1, Figs. 1–3) bear close resemblance with the studied globular bodies.

Lenticular bodies (Plate 1, Fig. 9c; Plate 2, Figs. 5a, 6a, 6c)

These forms (5 to 8 μm) with sub circular to sub elliptical outline, bluntly acute peripheral margin and granulated surface occur discretely and conspicuously along the contact zones and inside the chamber cavities. Being 5 to 10 times larger than a globular body, these multichambered forms occur in different size with developed surface features; in polar view these appear as miniature *Nummulites* aff. *Nummulites acutus*. Modern lenticular propagules of *Allogromia* sp. (Goldstein and Alve 2011) bear overall similarity with the studied ones.

Underdeveloped juveniles (Plate 2, Figs. 2b, 4a)

Incompletely developed lenticular forms (>50 to 150 μm) may discretely scatter on the surface of anomalous tests and are distinguished by imperfect development of surface and internal morphologic features. Externally, the surfaces may show granules, and internally few septal filaments, one or two spiral sheets accompanying alar prolongations and lateral walls may appear as involutions. These forms might have developed from lenticular bodies to represent more advanced growth.

Developed juveniles (Plate 1, Figs. 9a–b, 10b; Plate 2, Figs. 2a, 2c, 3a)

With large size ranging from 70 to 500 μm, an individual juvenile is distinct in external and internal features. Its lenticular test has sub circular peripheral outline, tapered peripheral margin, granulated surface, regular spire, tight whorls and thick whorl wall, radial septa and slightly longer than high chambers; axial and transverse sections have 'v' shaped chambers in equatorial part with thin alar prolongations and moderately thick lateral walls in involute coiling; tangential and oblique sections contain granulated radial septal filaments. Although marginal cord is not well recognized in axial section, apparently massive wall appears in the place of the marginal cord. Most of these features usually characterize typical *Nummulites*; while the external features resemble those in the associated adult test, some are characteristic of *Nummulites* aff. *N. acutus*. A horizontal section of a form with 70 μm diameter shows a spire of 2–3 whorls and the initial chamber having <1 μm size; the offspring resembles microspheric juvenile. With the occurrence of external and internal features of the minute forms in the adults, these are considered as offspring of adult *Nummulites* aff. *N. acutus*. These forms discretely occur in and around merged apertures, along fused contact zones and on the test surfaces. Some of the developed juveniles closely resemble the fossilized juveniles of *Nummulites* (Mukhopadhyay 2007, Plate 1, Fig. 12).

Nature of the forms: Contained morphologic features and mode of occurrence of the minute forms are compared with those of the anomalous adult hosts and normal adult tests of the species (Table 1). The comparison reveals that the minute biotic bodies and lenticular bodies resemble the offspring of modern foraminifera while the juveniles are comparable with the minuscule *Nummulites*. Based on morphologic distinctions, mode of occurrence and size variations these minute forms may represent various growth stages in the development of the offspring. The apparently featureless minutest globular bodies containing fossilized 'flagella type' features may represent early embryonic forms, the minute lenticular bodies containing rudimentary surface features may correspond to advanced embryonic forms; the underdeveloped juveniles with immature internal morphologic features akin to nepionic growth stage whereas developed juveniles containing recognizable morphologic features may correspond to neanic stage. The forms show gradation in size and morphologic features indicating in situ growth. The early embryonic forms occur within or adjacent to the pores along the contact zone and around the fused apertural cavities, and the advanced embryonic forms are dispersed stand alone within the contact zone and within the coarse cavities; the underdeveloped and developed juveniles are discretely occurring on the surface and inside cavities of the conjoined tests. The peripheral location of some juveniles suggests a posture just ready to leave the parent tests. Co-occurrence of diverse fossilized offspring in a conjoined entity indicates their multiphase reproduction while their growth occurred at the expense of granules, septal filaments and spiral sheets of the parent tests.

Possible Cause of Development of the Minute Forms

Nummulites aff. *N. acutus* is conspicuous in the development of rare conjoined tests entities and 'separated' tests with attached minute forms. A conjoined entity reflects a state of fusion of two adult megalospheric individuals of the species; fusion primarily involves apertures and outer spiral sheets, though septal filaments often merge. The individuals remain attached on one of their sides by using lime material as adhesive while their other sides remain free. Although fused and merged tests with random orientations of the participating individuals are known in modern and fossil foraminifera to form due to environmental stress, the purpose of formation of this atypical architecture with invariable fused apertures is not well known. The style of conjoining is analogous to that of modern plastogamy (Schaudinn 1895), in which two adult megalospheric individuals of a species of modern smaller benthic foraminifera unite for sexual reproduction and produce juveniles. However, modern larger foraminifera neither conjoin for sexual reproduction nor contain juvenile. The studied conjoined entities contain minute biotic forms that have remarkably small size compared to the size of the host test, gradation in size and morphological features, similarity in appearance in all the conjoined hosts, and similar test material as of the hosts; moreover, the conjoined entities are generally absent with exotic material including propagules of diverse taxonomic group. Their intact tests and lack of signature of transportation and erosion suggest their indigenous, in situ and integral nature. Some grown up forms contain premature morphological features that characterize adult *Nummulites*; some features including acute peripheral margin, surface granulations, radial septal filaments and overall straight septa closely agree with those of *Nummulites* aff. *N. acutus*; these suggest a possible hereditary relationship between the adults (=parents) and the associated minute bodies (=offspring).

The occurrence of the offspring in and around the crucial morphologic parts of the hosts suggests protection and nurturing of the offspring by the adults. Such occurrence may have a biological explanation. Fusion of apertures in a conjoined entity

Table 1 Similarity and differences between normal tests, separated tests, conjoined tests

Nummulites sp. aff. *Nummulites acutus* (Sowerby)		Megalospheric adult tests				Attached minute bodies (=offspring)	
Sl. No.	Type morphologic features	Normal solitary test	Distorted solitary test	'Separated' solitary test	Conjoined tests of two adult individuals	Globular-lenticular bodies	Underdeveloped and developed juveniles
1.	Test shape, symmetry and size; d/t	Lenticular, biconvex; diameter 3000–3500 μm; 2.2	Lenticular, unequally biconvex; diameter 3000–3800 μm; 1.6–1.9	Lenticular, unequally biconvex; diameter 3800–4000 μm; 1.8–2.29	Lenticular, to sub triangular; diameter 3800–4000 μm; 1.3–1.35	Subspherical lenticular, biconvex; radial to apparently bilateral; <1 to 4 μm: 1.25–1.5	Lenticular, apparently bilateral; 5–8 μm; 2
2.	Polar view	Almost circular; polar part broadly rounded	Lenticular, polar part broadly rounded	Often strongly undulatory; polar part broadly rounded	Subcircular, subtriangular	Subcircular	Subcircular
3.	Peripheral view	Lenticular, peripheral margin acute with blunt tip	Lenticular to irregularly lenticular, periphery undulatory, acute to appressed with tip subrounded	Lenticular to irregularly lenticular, peripheral margin acute, tip subrounded	–	Subcircular to lenticular. periphery acute with subrounded tip	Subelliptical, margin bluntly acute

(continued)

Table 1 (continued)

Nummulites sp. aff. *Nummulites acutus* (Sowerby)		Megalospheric adult tests				Attached minute bodies (=offspring)	
Sl. No.	Type morphologic features	Normal solitary test	Distorted solitary test	'Separated' solitary test	Conjoined tests of two adult individuals	Globular-lenticular bodies	Underdeveloped and developed juveniles
4.	Surface features-polar knob, granules, septal filaments, suture, surface cavities	Occasionally polar bulge, dense granules spirally arranged, granulated septal filaments radial often branching, spiral suture weak	Granules and septal filaments may differ in concentration in two sides of a test; often with cavities	Granules, septal filaments and pores differ in density and intactness in two sides; often cavities on surface	Density of granules, septal filaments and cavities differ in two sides of a test; these are subdued on attached side	Often with thread-like features, surface granulations and predator cavity	Surface with numerous granules, septal filaments granulated
5.	Internal features in equatorial section–spire, whorl, septa, chambers; marginal cord, proloculus	Planispiral, tight, 4–6 whorls, thick marginal cord, quadrangular chambers about 16 in 3rd whorl, septa upright at base, distally weakly curved; large proloculus (570–600 μm)	Not observed	Not observed	Not observed	Not observed	Planispiral, tight, involute, marginal cord and initial whorls indistinct; later whorls 3–4, about 14 quadrangular chambers in 3rd whorl, septa straight to weakly curved; spiral wall thin to moderate, tiny (<1 μm) initial chamber

(continued)

Table 1 (continued)

Nummulites sp. aff. *Nummulites acutus* (Sowerby)		Megalospheric adult tests				Attached minute bodies (=offspring)	
Sl. No.	Type morphologic features	Normal solitary test	Distorted solitary test	'Separated' solitary test	Conjoined tests of two adult individuals	Globular-lenticular bodies	Underdeveloped and developed juveniles
6.	Internal features in axial/tangential section–alar prolongation, wall, pillars, marginal cord, polar characters	Whorl wall involute, moderately thick, marginal cord moderate, thin alar prolongation, pillars distinct, polar area often bulged	Peripheral marginal cord thick with canal orifices	Peripheral marginal cord thick with canal orifices	Peripheral marginal cord thick with canal orifices	Not developed	Whorl wall involute, moderately thick; initial chamber tiny to indistinct; large number of involute chambers with thin alar prolongations, granulated septal filaments
7.	Mode of occurrence	Individual solitary test	Individual solitary test	Individual solitary test	Two megalospheric adult individuals in fused state with fused apertures and outer spiral sheets	Discretely or in cluster inside fused apertures, pits, cavities and on test surfaces of conjoined tests and separated tests	Discretely inside fused apertures, cavities and on test surfaces of conjoined entities and 'separated' tests

(continued)

Table 1 (continued)

Nummulites sp. aff. *Nummulites acutus* (Sowerby)		Megalospheric adult tests				Attached minute bodies (=offspring)	
Sl. No.	Type morphologic features	Normal solitary test	Distorted solitary test	'Separated' solitary test	Conjoined tests of two adult individuals	Globular-lenticular bodies	Underdeveloped and developed juveniles
8.	Distinguishing characters	Spirally arranged granules, acute peripheral margin, radial to faintly branching granulated septal filaments, thick marginal cord and large proloculus	Swerving peripheral margin, test asymmetry, differences in surface features on the two surfaces of a test	Similar as distorted tests, but differ in containing minute biotic bodies, numerous cavities, subdued surface features, presence of lime deposit and bioerosion	Fusing of individuals, apertures, peripheral margins, outer spiral sheets and contact zone; attached minute biotic bodies in diverse development	Minute biotic bodies show gradation in growth and development	Underdeveloped and developed biotic bodies resemble microspheric *Nummulites*
9.	Previous instance of conjoined tests and separated tests		conjoined tests and separated tests were found in *Nummulites boninensis*, *Nummulites* aff. *Nummulites discorbinus*, *Nummulites* cf. *Nummulites pengaroensis* and *Nummulites fabianii* from the middle Eocene of Cambay basin, Gujarat			Attached minute bodies occur in a broken test of *Nummulites* cf. *Nummulites beaumonti* from the middle Eocene of Cambay basin, Gujarat	

(continued)

Table 1 (continued)

Nummulites sp. aff. *Nummulites acutus* (Sowerby)		Megalospheric adult tests				Attached minute bodies (=offspring)	
Sl. No.	Type morphologic features	Normal solitary test	Distorted solitary test	'Separated' solitary test	Conjoined tests of two adult individuals	Globular-lenticular bodies	Underdeveloped and developed juveniles
10.	Inferences	1. Minutest globular bodies are single chambered forms; earliest chamber in the multi-chambered lenticular and juvenile forms has almost same size 2. Globular-lenticular bodies and underdeveloped-developed juveniles occur only in anomalous conjoined tests and separated tests. These do not occur in normal solitary adult tests 3. Close similarity between normal adult tests and minute bodies in respect of external and internal morphologic features points for genetic relation in between 4. A conjoined tests entity of *Nummulites* sp. aff. *Nummulites acutus* represents adult plastogamy; separated tests are individuals dissociated after plastogamy whereas minute bodies attached with the hosts largely represent microspheric offspring originated during sexual reproduction 5. In the association of minute bodies with conjoined tests, the adult individuals represent parents and the minute bodies as daughters 6. Some cavities, pits and diminution of surface features in conjoined tests are due to consumption by the offspring during life tenure 7. Conjoined tests with associated offspring point for gamontogamy part of the life cycle in extinct *Nummulites* sp. aff. *Nummulites acutus* 8. Conjoined tests with attached juveniles are not known in modern nummulitids					

facilitated successful mingling of protoplasmic material during sexual reproduction between the adult pair when mingling is considered prerequisite for the genesis of offspring. Occasional lime deposit in and around the apertures might represent the fossilized protoplasmic material (Plate 1, Fig. 10a, Plate 2, Fig. 6b). Wide dispersal of the minute bodies on conjoined tests is closely comparable with similar dispersal of gametes on parent individuals during reproduction of modern smaller benthic foraminifera. Two megalospheric adults of *Nummulites* aff. *N. acutus* developed conjoined entity by fusion of their apertures and outer spiral sheets; at times, the individuals twisted their tests for the purpose. The occurrence of remnant lime deposit within the fused apertures, development of minute bodies and microspheric juveniles inside reproductively crucial morphological features and consumption of some morphologic features of conjoined tests indicate that the offspring that originated during sexual reproduction were nurtured by the adults in protected parts within the tests prior to their release in the sea.

Discussion

Foraminiferal life cycle basically comprises alternation of sexual and asexual reproductions when microspheric and megalospheric forms developed respectively as recorded during the culture of smaller benthic foraminifera. Based on this, similar life cycle was postulated for *Nummulites* because its species contain dimorphic forms. Laboratory culture of some nummulitids confirmed development of megalospheric juvenile during schizogony (Röttger et al. 1998). Among the two most prevalent types of sexual reproductions known in foraminifera, gametogamy has been observed during the laboratory culture of *Heterostegina depressa* (Röttger and Spindler 1976). Gamontogamy or plastogamy type of sexual reproduction has not been recorded in any living larger foraminifera though the process occurs in some modern smaller benthic foraminifera (Myers 1940; Grell and Bardele 1977; Grell 1979). The find of conjoined tests of some extinct species of *Nummulites* (Mukhopadhyay 2003a, 2007) opens a question whether plastogamy was in vogue in the extinct species of *Nummulites*. As commonly seen in some modern smaller benthic foraminifera, two megalospheric adult individuals of a species voluntarily unite for sexual reproduction and develop plastogamic paired tests and offspring (cf. Preobrazhenskaya and Tarasova 2004). Almost analogous pairing of two adult megalospheric individuals of extinct *Nummulites* aff. *N. acutus* is preserved in the fossilized conjoined entity; this and earlier records of conjoined entities from the Middle Eocene of Cambay basin, Gujarat indicate that the biological phenomenon of conjoining was more common than rare.

Merging of two adult microspheric individuals of extinct *Nummulites millecaput* was explained as plastogamy by Kecskemčti (1962); however, in the revised life cycle model of foraminifera by Goldstein (1997, 1999) two microspheric adult individuals of a species do not unite for sexual reproduction. Pavlovec (1976) while illustrating almost identical merged tests of two adult microspheric individuals of the same species explained that environmental stress could be the cause of fusion. Wöger et al. (2012) opined that the individuals in conjoined *Nummulites* continued to grow even after fusion to acquire new structure of the test. In the studied conjoined specimens no new chamber was found to develop on the fused apertures. Ferrández Cañadell et al. (2012) with the help of μ-CT scan and three dimensional models observed that individuals shared a common test after a period of independent growth of 30–40 chambers. With almost equal number of chambers, a megalospheric adult individual of *Nummulites* aff. *Nummulites acutus* completed nearly three initial whorls to attain sub-adult stage; with the addition of chambers and whorls the individual became adult and ready to conjoin for sexual reproduction. Fusion of chambers of the successive 4th, 5th and 6th whorls would have severely assimilated the tests of the participating individuals; however such fusion is not developed in the studied conjoined specimens which show no signature of fusion of the inner whorls; conjoining occurred by the fusion of the apertures and outermost spiral sheets by using lime material as adhesive, and after sexual reproduction the participating individuals got separated. Incidence of forced fusion of two megalospheric adult individuals of *Nummulites aspermontis* after surviving predation by a metazoan was described from the Lower Eocene (Ferrández Cañadell et al. 2014); the specimens differ from the studied conjoined tests primarily in the fusion of inner whorls and in the post-fusion growth of tests. The studied conjoined tests are distinct in fused apertures and fused outer spiral sheets, in the development of juveniles and diverse cavities; the feature mentioned last served as passage for the exit of the juveniles and as locales for their sheltering and nurturing; some cavities represent hollow spaces due to consumption of the test material by the growing juveniles prior to their leaving the parent tests. Ferrández Cañadell et al. (2014) while commenting on Mukhopadhyay's (2003a, b, 2007) conjoined specimens possibly overlooked these distinctive features; however, on a later occasion Ferrández Cañadell examined the fresh SEM photographs of the same conjoined specimens and opined (pers. communication) that 'the specimens are very interesting'. The specimens actually contain minute biotic forms on the recrystallized surface.

According to Glaessner (1963), an individual develop plastogamy at adult stage by establishing body contact with another adult; during plastogamy an individual attains complete growth and development and becomes ready for union. Debenay (2012) opined that the flagellated gametes during plastogamy fuse in a space developed by partially dissolved apertural faces of the gamonts. Tiny globular bodies containing flagella-like ultra-structure are preserved inside the fused apertures of conjoined *Nummulites* aff. *N. acutus*, but their occurrence in fused state is not found. In order to facilitate mingling of protoplasmic material the two adult individuals held themselves intimately by fusing the outer spiral sheets and bringing their apertures in closest proximity for fusion; individuals often twisted their tests for the purpose of intimate association (Mukhopadhyay 2003a). Voluntary twisting of tests for the purpose of conjoining results in forms that have inflated aperture bordered with thick rim and asymmetric aperture and thus this distortion is different from the distortion developed due to environmental stress (Toler and Hallock 1998). *Nummulites* aff. *N. acutus* and other co-occurring larger foraminifera generally have normal tests; aberrations developed in a few adult tests of the species. Environmental stress would have affected the majority in the whole population.

The juveniles in the studied conjoined tests are indigenous, in situ and intact unlike some transported modern juveniles (cf. Alve and Goldstein 2002). Randomly oriented offspring of various growth stages may appear as having different parental sources; at times they may be confused as new species. That all the categories of minute forms had same parental source is evident from their co-occurrence, gradation of size and morphology among forms of different growth stages occurring in a conjoined tests entity, miniature morphologic features in some resemble mature counterparts characterizing the host adults, similar test material comprising both host and attached forms, and similar growth pattern of the minute forms in different conjoined specimens and separated tests. While gradation among the minute forms indicates in situ growth and development, commonness of internal morphologic features in juvenile and adult forms points for hereditary relationship in between.

Intact association of juvenile and adult is not recorded in modern larger foraminifera; their occurrence in fossilized conjoined tests and 'separated' tests of *Nummulites* aff. *N. acutus* points for a link between adult conjugation, sexual reproduction and offspring development. The grown up juveniles in this association being of microspheric generation, sexual reproduction by the adults seemingly played a vital role in the development and propagation of microspheric forms. This find also endorses the 'alternation of generations' hypothesis in extinct *Nummulites* whereby microspheric and megalospheric generations developed. As yet, complete life cycle has been studied in 30 species of foraminifera (Goldstein 1999) of which some belonging to suborder Rotaliina, superfamily Nummulitacea and family Nummulitidae show gamontogamy or plastogamy type of sexual reproduction with the development of juveniles in their life cycle (Myers 1940; Grell and Bardele 1977; Grell 1979). Fossil Plastogamy was recorded in smaller foraminifera (Le Calvez 1950) and extinct larger foraminifera (in *Nummulites*, Mukhopadhyay 2003a, 2007); juvenile bearing broken adult test of middle Eocene *Nummulites* suggests that the juveniles were plastogamic origin. Present study of juvenile bearing conjoined tests and separated tests of extinct *Nummulites* aff. *N. acutus* provides unequivocal evidence of parent-daughter relation in the species. Since the species belongs to the same race of family-superfamily-order and as the development of adult-juvenile association is restricted to conjoined tests, it is likely that the species followed gamontogamy or plastogamy type of sexual reproduction during the alternation of asexual and sexual life cycles.

Acknowledgements The work was carried out under Project No. 2003-2004/RP/CHQ/CGL/2002/006 of Central Headquarters of Geological Survey of India (GSI). The author expresses sincere thanks to P. K. Basu, the then Director, Palaeontology Division-II, Geological Survey of India, Kolkata for providing the drill core sample, T. C. Lahiri, the then Director of Palaeontology Division-I of Central Headquarters, GSI for facilities in the micropaleontology laboratory and S. Shome, Senior Geologist for help in Scanning Electron Microscopy. Thanks are due to the Organizing Committee of the Paleogene Conference, Lucknow, especially S. C. Tripathi Convener, GSI, for providing an opportunity to present the paper in the Conference. The author is grateful to Barun K. Sengupta (Louisiana State University, USA), György Less (University of

Miskolc, Hungary) and Andrea Benedetti (Università di Roma, Italy) for encouraging comments on the photographs of juvenile bearing conjoined tests of *Nummulites* and author's concept of plastogamy in extinct *Nummulites*. The author also thanks Carles Ferràndez Cañadell (University of Barcelona, Spain) and Cesare Andrea Papazzoni (University of Modena and Reggio Emilia, Italy) for suggestions to improve the SEM photography of the conjoined tests.

References

Alve E, Goldstein T (2002) Resting stage in benthic foraminiferal propagules: a key feature for dispersal? Evidence from two shallow-water species. J Micropalaeontology 21:95–96

Biekart JW, Bor T, Röttger R, Drooger CW, Meulenkamp JE (1985) Megalospheric *Heterostegina depressa* from Hawaii in sediments and laboratory cultures. Pal Proc 88(1):1–20

Debenay J-P (2012) A guide to 1,000 foraminifera from southwestern Pacific: New Caledonia: IRD Éditions. Publications Scientifiques du Muséum, Paris, p 378p

de La Harpe P (1879) Etude sur les *Nummulites* du Comté de Nice. Bull Soc Vaudoise Sci Nat 16(2):201–243

de Sowerby J (1840) Systematic list of organic remains. Appendix to Grant C. W., Memoir to illustrate a geological map of Cutch. Trans Geol Soc London, Ser 2, vol 5, pp 317–329

Ferrández Cañadell C, Briguglio A, Wöger J, Wolfgring E, Hohenegger J (2012) Conjoined Foraminifera: Test fusion in adult *Nummulites*. In: Liao J-C et al (eds) Libro de Resúmenes de las XXVIII Jornadas de la Sociedad Española de Paleontología, Homenaje a Guillem Colom Casanovas (1900–1993). Valencia-Sóller, Universitat de València, Valencia, Sociedad Española de Paleontología, Madrid, pp 57–58

Ferrández Cañadell C, Briguglio A, Hohenegger J, Wöger J (2014) Test fusion in adult foraminifera: a review with new observations of an Early Eocene *Nummulites* specimen. J Foramin Res 44(3):316–324

Furon R, Lemoine P (1939) Sur la presence du Nummulitique a Pondicherry. C R Acad Sci 207:1424–1426

Furon R (1941) Geologie de l'Inde Orientale Francaise: France, Bur. Études Géol. Et Minières Coloniales, Publication No. 17, pp 1–26

Glaessner MF (1963) Principles of micropalaeontology, Harper Publishing Co., New York, 297p

Goldstein ST (1997) Gametogenesis and antiquity of reproductive pattern in the Foraminiferida. J Foramin Res 27:319–327

Goldstein ST (1999) Foraminifera: a biological overview. In: Sen Gupta BK (ed) Modern foraminifera. Kluwer Academic Publishers, Dordrecht, pp 37–55

Goldstein ST, Alve E (2011) Experimental assembly of foraminiferal community from coastal propagule banks. Mar Ecol Prog Ser 437:1–11

Govindan A (2013) Larger foraminiferal biostratigraphy of early Paleogene sections in India. J Geol Soc 1:24–45 (special publication)

Grell KG (1979) Cytogenetic systems and evolution in Foraminifera. J Foramin Res 9(1):1–13

Grell KG, Bardele CF (1977) Light and electron microscopical studies of the foraminiferan *Rotaliella heterocariotica*. In: Abstracts of the fifth international congress of protozoology, New York, p 369

Hottinger L, Romero J, Caus E (2001) Architecture and revision of the Pellatispirines, planispiral canaliferous foraminifera from the Late Eocene Tethys. Micropaleontology 47(supplement 2):35–77

Kecskemčti T (1962) Patologikus Jelenségek Nummuliteszeken. Bull Hung Geol Soc 92:209–219

Krishnan MS (1960) Geology of India and Burma. Higginbothams Pvt. Ltd., Madras, p 536

Le Calvez J (1950) Recherchessur les Foraminifères. 2. Place de la mèioseetsexualit. Arch Zool Exp Gèn 87:211–243

Lister J (1895) Contributions to the life history of the Foraminifera. PhilosTrans Roy Soc London Ser B 186:401–453
Mukhopadhyay SK (2003a) Plastogamy and its early morphological indication in *Nummulites boninensis* Hanzawa from the Middle Eocene of Cambay Basin, India. Rev Paléobiol 22:231–242
Mukhopadhyay SK (2003b) Earliest *Pellatispira* Boussac from the Middle Eocene of India: morphological speciality of the ancestral stock. J Asian Earth Sci 22:209–225
Mukhopadhyay SK (2007) Conjoined tests of *Nummulites* from the Paleogene of the Cambay Basin, India and their possible origin. J Foramin Res 37:41–45
Mukhopadhyay SK (2015) Implication of the occurrence of minute biotic bodies on conjoined tests of *Nummulites* aff. *Nummulites acutus* (Sowerby) in the subsurface Eocene of Cauvery Basin, India. Abstract of the National Conference on Paleogene of the Indian Subcontinent, Lucknow, 23–24 Apr 2015, pp 63–64
Mukhopadhyay SK (2016) Morphogroups, test shape variability and test asymmetry in *Pellatispira crassicolumnata* Umbgrove from the Eocene of Surat-Bharuch (India). Neues Jb Geol Paläontol Abh 281(3):313–326
Munier-Chalmas E (1880) Sur le dimorphisme des *Nummulites*. Bull Soc Geol France 3(8):200–301
Myers EH (1936) The life-cycle of *Spirillina vivipara* Ehrenberg, with notes on morphogenesis, systematics, and distribution of the foraminifera. J Roy Microscopical Soc London 56:120–146
Myers EH (1938) The present state of our knowledge concerning the life cycle of the Foraminifera. In: Proceedings of US National Academy of Science, vol 24, pp 10–17
Myers EN (1940) Observations on the origin and fate of Flagellated Gametes in multiple tests of *Discorbis* (Foraminifera). J Mar Biol Ass U K 24(1):201–226
Pavlovec R (1976) Patologija numulitin: Hie Pathology of Nummulitins. Geologija Razprave in Porocila 19:83–106
Preobrazhenskaya TV, Tarasova TS (2004) Sexual reproduction of the foraminiferan *Planoglbratella opercularis* (d'Orbigny, 1839) in nature. Russ J Mar Biol 30(5):323–327
Racey A (1995) Lithostratigraphy and larger foraminiferal (nummulitid) biostratigraphy of the Tertiary of northern Oman. Micropaleontology 41(supplement):1–123
Romero J, Hottinger L, Caus E (1999) Early appearance of larger foraminifera supposedly characteristic for late Eocene in the Igualada Basin, Spain. Rev Esp de Paleontol 14(1):79–92
Röttger R, Spindler M (1976) Development of *Heterostegina depressa* individuals foraminifera, Nummulitidae) in laboratory cultures. In: Schafer CT, Pelletier BR (eds) First international symposium on benthonic foraminifera of continental margins. Halifax, NS. Maritime Sediments Special Publication, vol 1, pp 81–87
Röttger R, Fladung M, Schmaljohann R, Spindler M, Zakharias H (1986) A new hypothesis: the so called megalospheric schizonts of the larger foraminifera *Heterostegina depressa* d'Orbigny, 1826, is a separate species. J Foramin Res 16:141–149
Röttger R, Schmaljohann R, Zakharias H (1989) Endoreplication of zygotic nuclei in the larger foraminifer *Heterostegina depressa* (Nummulitidae). Eur J Protistology 25:60–66
Röttger R, Dettmering C, Kruger R, Schmaljohann R, Hohenegger J (1998) Gametes in nummulitids (Foraminifera). J Foramin Res 28:345–348
Samanta BK (1982) A restudy of two Middle Eocene species of *Nummulites* (Foraminiferida) from Cutch, Gujarat. Bull Ind Geol Assoc 15:21–50
Schaudinn F (1895) Plastogamie bei Foraminiferen. Sitzungsber Gesselschaft naturforch: Freunde Berlin 10:179–190
Toler SK, Hallock P (1998) Shell malformation in stressed *Amphistegina* populations: relation to biomineralization and paleoenvironmental potential. Mar Micropaleontol 34:107–115
Wöger J, Wolfgring E, Briguglio A, Ferrandez Canadell C, Hohenegger J (2012) The investigation of conjoined tests in larger benthic Foraminifera; state of the art. Ber der Geologischen Bundesanst 94:20

Palaeocene Faunal Events and Fossil Records of Andaman Islands, India

Tarun Koley, Amitava Lahiri, K. M. Wanjarwadkar and C. S. Anju

Abstract Two distinct foraminiferal assemblages, viz., (a) larger benthic foraminifera viz; *Ranikothalia* sp., *Miscellanea* sp., *Daviesina* sp. and *Discocyclina* spp. and (b) planktonic foraminifera viz; *Globigerina daubjergensis, Globorotalia fringa* and *Chiloguembelina* sp. cf. C. *wilcoxiensis* represent two prominent Palaeocene faunal events and marks the basal and top part of the Palaeocene sequence in Andaman Islands. Deep water depositional environment beyond the influence of clastic supply prevailed during Lower Palaeocene and shallow marine high energy condition during Upper Palaeocene period. Process of accretion started during later part of Palaeocene as evidenced in the North and Middle Andaman but not witnessed in South Andaman. Non-availability of the Palaeocene sequence in South Andaman maybe due of presence of barrier causing hindrance to sediment supply towards South.

Keywords Andaman islands · Palaeocene · Larger foraminifera · Ranikot sea

---

T. Koley (✉) · A. Lahiri · C. S. Anju
Palaeontology Division, Eastern Region, DK-6, Sector-2, Salt Lake, Kolkata 700091, India
e-mail: tarunkoley@rediffmail.com

K. M. Wanjarwadkar
Department of Geology, Government Institute of Science, Nipat Niranjan Nagar, Caves Road, Aurangabad 431004, India

Present Address
T. Koley
Palaeontology Division, Southern Region, GSI, Hyderabad 500068, India

Present Address
A. Lahiri
3A, Aashiyana Apartment, Balia, Garia Station Road, Kolkata 700084, India

Present Address
C. S. Anju
DGCO, GSI, Pushpa Bhavan, New Delhi 110062, India

S. Bajpai et al. (eds.), *The Indian Paleogene*, Society of Earth Scientists Series,
https://doi.org/10.1007/978-3-319-77443-5_10

Introduction

Marine Palaeocene sequences are described form numerous localities in India, though some of which are substantiated by undoubted palaeontological evidences (Samanta 1974). Among the Indian occurrences, Palaeocene sequences of Andaman Islands, due to its crucial geographic position and tectonic setup (Fig. 1), deserve special attention for reconstruction of palaeogeography and interpreting the tectonic history of the islands.

Andaman and Nicobar Islands are part of the tectonic belt which is extending from Arakan Yoma to the far East (Karunakaran et al. 1964) and are situated on the outer arc ridge (Fig. 1) of the northern segment of the Sunda Subduction Complex (Weeks et al. 1967; Curray 2005). There was an long standing dispute on the development of Palaeocene sequence in this belt and extension of the Palaeocene Sea beyond Burma toward south. Palaeocene sequence is well developed in Burma, towards north of the Andaman Islands (Nagappa 1959). Towards south, in the islands of the Far East, Middle Eocene sediments unconformably overlie the Cretaceous rocks (Adams 1965, 1970; Nagappa 1959; Glaessner 1959). Hence, there is a gap in sedimentation during Palaeocene—Early Eocene can be inferred.

Prior to 1964, geological investigations by pioneer workers like Oldham (1885), Rink (1870), Tipper (1911) and Gee (1926) considered the thick pile of sedimentary sequence of Andaman Islands as Eocene. Existence of Palaeocene sequence in the Andaman Islands was first mentioned by Chatterjee (1964), though the palaeontological evidences provided by him were not much supportive and firm. The claim of presence of Palaeocene sequence in Andaman Island was based simply on the identification of *Distichoplax biserialis* Dietrich and *Rotalia trochiformis* in stray boulders (Chatterjee 1964). This report of occurrence of Palaeocene rocks was undermined by Samanta (1974) on the ground that *Distichoplax biserialis* is long ranging from Upper Cretaceous to Upper Eocene. Similar conclusion regarding the status of this species of dascycladacean algae was drawn by Mishra and Kumar (1988). However, Samanta (1974) did not make any comment on the report of *Rotalia trochiformis*, a well known Palaeocene foraminifera. It must be mentioned that none of the taxa i.e. *Rotaliatrochi formis* or *Distichoplax biserialis* Dietrich has been illustrated by Chatterjee (1964). In addition to this, records of definite Palaeocene foraminiferal assemblage are available from few localities in Andaman Islands. Limited occurrence of in situ fossiliferous horizon poses hindrance for proper understanding of the Palaeocene events and deciphering the Cretaceous-Palaeocene boundary in the Andaman Islands.

The well illustrated assemblage of Early Palaeocene planktonic foraminifera (Kumar and Soodan 1976) is registered from the area viz. (a) between Mayabunder and Webi and (b) 8 km east of Webi, Middle Andaman. This assemblage is comparable with that of *Globorotalia uncinata* Zone of Bolli (1957) and is equivalent to Zone P-2 of Blow and Berggren (1969). Five species of *Daviesina* (Table 1) along with fossil algae from the limestone of Tugapur, Middle Andaman were reported by Kundal and Wanjarwadkar (2002) indicating Upper Palaeocene age. Later, Koley

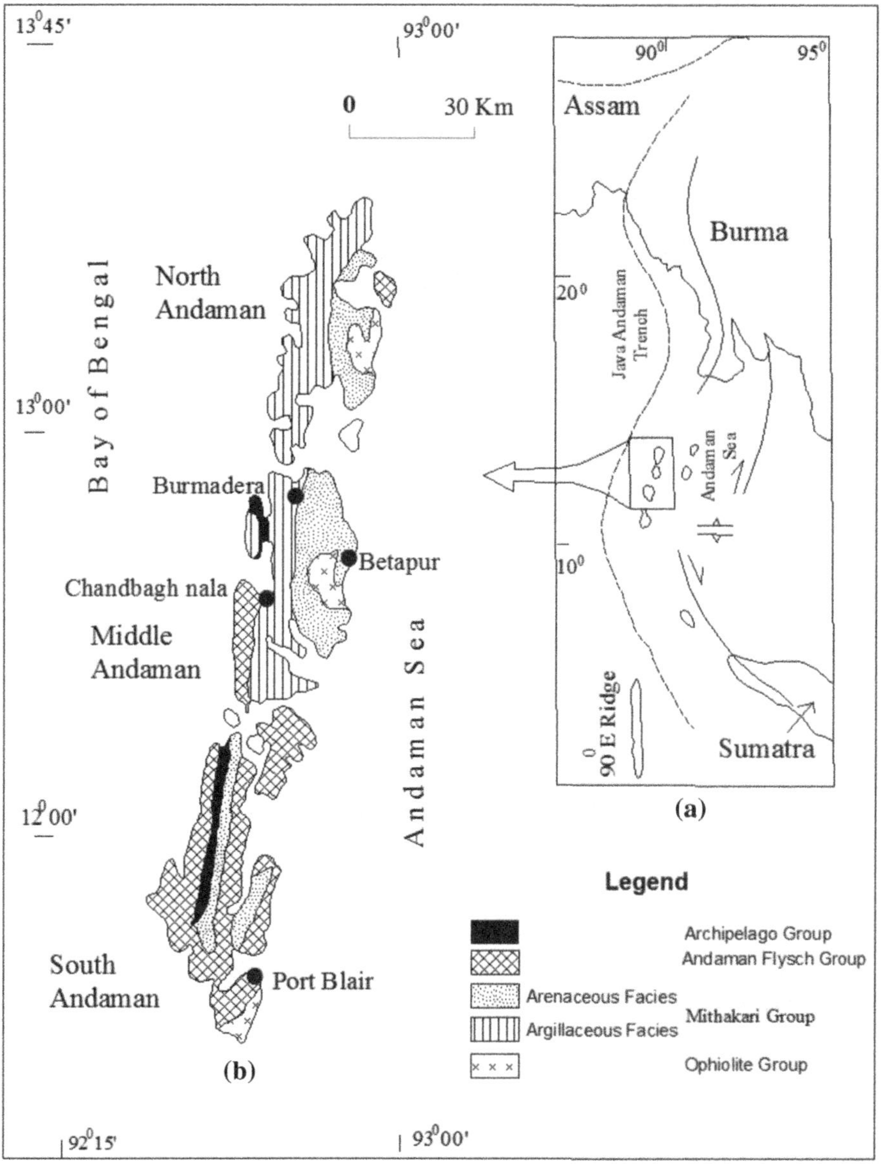

Fig. 1 **a** Location map with tectonic elements of the region. **b** General geological map of South, Middle and North Andaman (modified after Bandopadhyay 2005)

and Wanjarwadkar (2013) indicated the interbedded nature of the Tugapur limestone and reported *Ranikothalia* and *Discocyclina* (Table 1) indicating Upper Palaeocene age of the limestone.

Table 1 Available records of foraminifers of palaeocene affinity from Andaman Islands

Authors	Fossil records	Locality	Remarks
Present work	*Ranikothalia* sp., *Miscellanea* sp., *Daviesina* sp., *Discocyclina ramaroi, Discocyclina haynesi Globigerina daubjergensis, Globorotalia fringa, Globorotalia compressa* and *Chiloguembelina* sp. cf. C. *wilcoxiensis*	Limestone exposed in melange zone at Betapur coast Shale from Chandhbagh nala, Middle Andaman	Taxa identified from random thin section Taxa identified from discrete forms
Koley and Wanjarwadkar (2013)	*Ranikothalia sidensis,Ranikothalia* cf. *tobleri, Ranikothalia* cf. *nuttalli, Discocyclina* sp.	Bedded limestone from Tugapur and limestone boulder in conglomerate	Identified from random thin sections
Bandopadhyay (2011)	*Assilina* sp.	Rampur and Kalipur coast	Three oblique sections are provided
Kundal and Wanjarwadkar (2002)	*Daviesina danieli* Smout, D. *garumensis Tambareau*, D. *langhami* Tambareau, D. *ruida* Tambareau	Burmadera Limestone	–
Roy et al. (1988)	*Globigerina linaperta*, G. *triloculinoides, Globigerina* sp., G. cf. *daubjergensis*, G. cf. *velascoensis, Globigerinelloides* sp., *Globorotalia compressa*	Pelagic sediments from South Andaman	Taxa are identified from random thin section
Kumar and Soodan (1976)	*Globoconusa daubjergensis* Bronninmann, *Subbotina triloculinoides* Plummer, *Globorotalia angulata* White, *G. compressa*, Plummer, *G. imitata* Subbotina, *G. perclara* Loeblich and Tappan, *G. pseudobuloides* Plummer, *G. uncinata* Bolli, *G. varianta* Subbotina	Shale exposed east of Tugapur and Mayabunder	

(continued)

Table 1 (continued)

Authors	Fossil records	Locality	Remarks
Karunakaran et al. (1964)	*Distichoplax biserialis*	Limestone boulder from Conglomerate	This species of algae is no longer index of Palaeocene
Chatterjee (1964)	*Distichoplax biserialis, Rotalia trochiformis*	Limestone boulder from Chiriyatapu Conglomerate	Taxa are neither described nor illustrated This algal species is no longer index of Palaeocene

Records of Palaeocene fossils are rare from North Andaman except the *Assilina* reported by Bandopadhyay (2011) from Kalipur and Rampur coast. However, the illustrations raise some doubts in identification. Moreover, the present authors studied the section and reported undoubted Middle-Upper Eocene Nummulitids and Discocyclinids (Koley and Anju 2013).

Palaeocene fauna from the Tertiary sequence of South Andaman were reported by Roy et al. (1988) from the pelagic sediments associated with the pillow basalt (Table 1). Their report of Palaeocene planktonic foraminiferal assemblage including *Globigerina eugubina*, *G.linaperta*, *G. velascoensis*, *G. trilocunoides*, *G.* cf. *daubjergensis*, *Globorotalia compressa* are identified in the random thin sections. Moreover, typical Cretaceous planktonic foraminiferal genera like *Globotruncna, Reugobgobligerina, Heterohelix, Pseudotextularia* are associated with the Palaeocene forms in their assemblage. Causes for the anomalous mixing of the distinct assemblages are not explained. As sanctity of the identifications of the planktonic foraminifers from random thin sections is always questionable, records of definite Palaeocene fossils are not available.

General consensus is that the trench outer arc setting with active subduction prevailed during the Cretaceous time onwards (Pal et al. 2003; Bandopadhyay 2005) in Andaman and Nicobar Islands. In a subduction zone setting the sediments deposited on the oceanic crust is finally accreted to the continental mass by the process of underplating and tectonically to the toe of the prism (Wakita and Metcalfe 2005). Fossil records of the marine sequence are a necessity to reconstruct the history of the oceanic crust from its initiation and to subduction in the trench. Scarcity of the Cretaceous–Palaeogene fossil records from Andaman leaves limited scope to reconstruct its tectonic history with high degree of confidence. In this work, attempt has been made to establish the limits the Palaeocene sequence of Andaman Island based on the Palaeocene faunal events and to discuss its implications on the palaeogeographic reconstruction and tectonic events of the area.

Material and Method

Identification of fossils is based on the examination of rock sections and processed samples of shales from the argillaceous sequence and limestone exposure which is observed to occur as huge boulders within the melange zone exposed in Betapur coastal section. Planktonic foraminifera are ill preserved and identification of the planktonic foraminifers is based on the SEM images. The cream colour limestone from the melange zone is very hard and porcellanetic as found in the Tugapur and Burmadera area, Middle Andaman. Larger benthic foraminifers are completely fused with the matrix and impossible to extracts. Therefore, the larger benthic foraminifera were identified from random thin sections.

Geological Set Up

Ophiolite Group comprising ultramafics, dacite, basalt, plagiogranite and pelagic sediments is the oldest unit of the Andaman Islands. Thick Tertiary sediments comprising dark grey shale, sandstone, conglomerate and limestone unconformably overlie the Ophiolite Group (Fig. 1b).

In Middle Andaman ophiolites are exposed as dismembered bodies and are mainly concentrated in the eastern part of the Island (Fig. 1b). However, pillowed basalt with chert and limestone are observed in the western part of the Island. Betapur–Pancahwati coast exposes the melange zone consisting of assorted blocks and boulders of basalt, chert limestone, greywacke, shale in a fine grained shale matrix. Western part of the island exposes dark grey shale with interlayered thin sandstone and occasional limestone. Limestone band with linear isolated out crops is observed toward the top of the argillaceous facies. Ultimately the argillaceous sequence is overlain by anarenaceous sequence of polymictic conglomerate, coarse to fine grained sandstone and shale.

Similar geological setup is continuing from Middle Andaman towards north to North Andaman (Fig. 1b). The limestone facies is not exposed in the North Andaman. The sedimentary sequence displays two distinct facies viz. lower argillaceous facies and upper arenaceous facies. The argillaceous facies comprises of dominantly shale with minor interlayered sandstone and the arenaceous facies comprises polymictic conglomerate, gritty sandstone, coarse to fine sandstone, calcareous sandstone, siltstone and interlayered thin shale. Dismembered ophiolite bodies occur mainly within arenaceous facies. The melange zone is exposed in the Rampur coast.

Ophiolite slices are mainly occurring along the eastern coast of the South Andaman Island (Fig. 1b). Chert beds are associated with the ophiolite and as dismembered bodies within the polymictic conglomerate, gritty sandstone of Namunagarh and Hopetown formations. Interestingly, the argillaceous facies (=Lipa Black Shale Formation) is not developed in South Andaman Islands. Thick sequence of con-

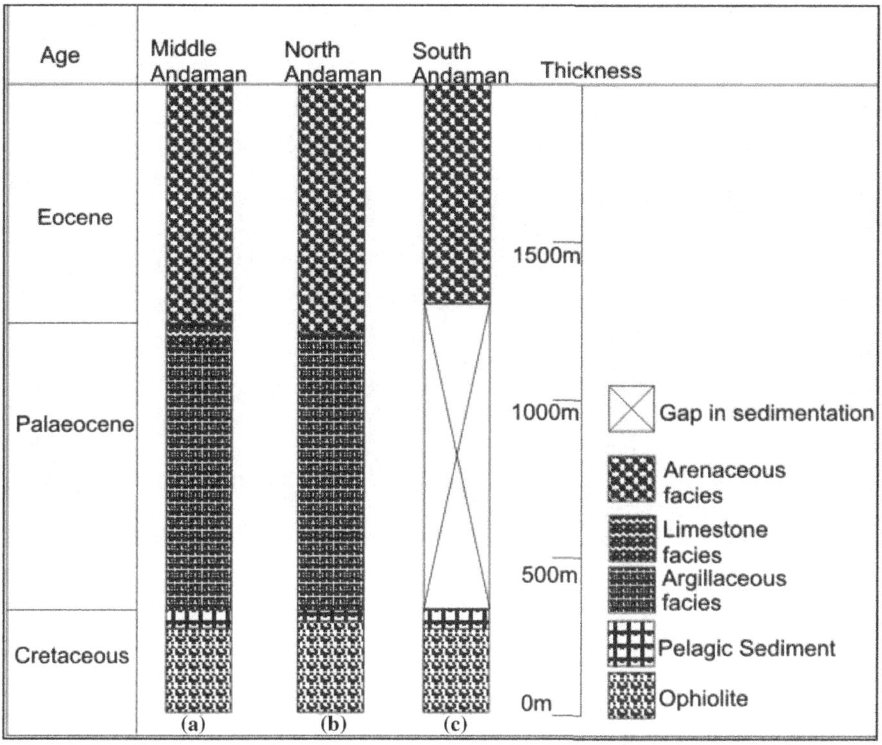

Fig. 2 Temporal distribution of Cretaceous- Eocene litho-facies in Middle, North and south Anaman

glomerate, coarse grained sandstone and shale is overlain by thick turbidite sequence of Andaman Flysch Group of Oligocene age (Fig. 2).

Observations

Two distinct foraminiferal assemblage have been identified from Middle Andaman—(a) planktonic foraminifera assemblage comprising *Globigerina daubjergensis* (Fig. 3a–c), *Globorotalia fringa* (Fig. 3d–e) and *Chiloguembelina* sp. cf. *C. wilcoxiensis* (Fig. 3f–g) assemblage from the argillaceous facies and (b) larger foraminifera assemblage comprising *Ranikothalia* sp. (Fig. 4), *Miscellanea* sp. (Figs. 4 and 5), *Daviesina* sp. (Figs. 4 and 5) and *Discocyclina* sp. (Figs. 4 and 5) from the arenaceous facies.

Limestone containing Upper Palaeocene larger benthic foraminifera are recorded from the sediments of younger stratigraphic level (Koley and Wanjarwadkar 2013). Larger foraminifera indentified are from limestone occurring in the arenaceous

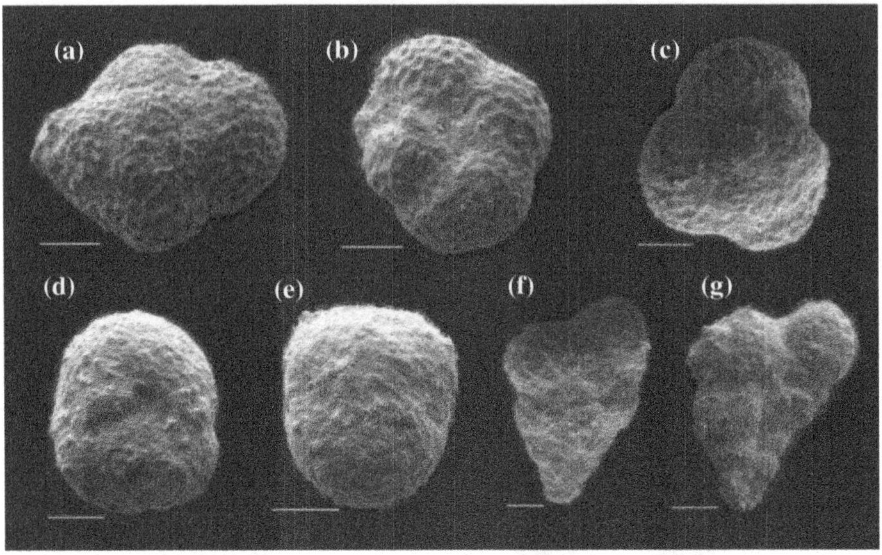

Fig. 3 Planktonic foraminifera from shale exposed in western part of Middle Andaman. Scale bar represent 60μm

facies of Mithakhari Group exposed in the melange zone of Betapur coast, Middle Andaman. Nevertheless, occurrence of the larger foraminifera in this sequence indicates a Palaeocene faunal event.

Discussion

The Palaeogene sequence of this area is classified into two groups viz. Mithakhari Group and Andaman Flysch Group (Table 2). Cretaceous to Upper Eocene age was assigned to the Mithakhari Group (Karunakaran et al. 1968; Ray 1982) though there remained arbitrariness in assigning the upper and lower boundaries due to limited fossil control. While discussing the foraminiferal biostratigraphy of India, Pakistan and Burma, Nagappa (1959) suspected non-development of Palaeocene and Lower Eocene sequence in Andaman and Nicobar Islands. His reasoning was based on the analogy that the oldest Tertiary rocks in Ramree Island towards north of the Andaman and Nicobar Islands is Middle Eocene. Towards South of Andaman and Nicobar Islands, in the Far East region, Middle Eocene sequences directly overly the Cretaceous rocks.

Limestone boulders within the melange zone of Betapur coast, Middle Andaman yielded Palaeocene larger foraminiferal assemblage of *Ranikothalia, Miscellanea, Daviesina* and *Discocyclina*. Fossil algae *Jania, Corallina* and *Distichoplax* are associated with the larger foraminifera indicating shallow marine high energy condition.

Fig. 4 Photomicrograph of foraminiferal limestone showing assemblage of *Ranikothalia, Miscellanea, Discocyclina, Daviesina* and fossil algae

Mithakhari Group in Middle and North Andaman can be classified broadly into two facies (Fig. 1b) viz. lower argillaceous facies with the limestone towards top and upper arenaceous facies. These two facies together constitute the Tertiary sequence of Middle Andaman; the argillaceous facies and the arenaceous facies broadly correspond to the Lipa Black Shale and Namunagarh Grit formations respectively. Foraminifera of Palaeocene affinity has been identified from basal and upper part of Lipa Black Shale. *Globigerina daubjergensis, Globorotalia compressa, Globigerina* cf. *fringa* and *Chiloguembelina* sp. cf. C. *wilcoxiensis* assemblage from Chandbagh Nala and *Ranikothalia–Miscellanea–Daviesina–Discocyclina* assemblage from the limestone exposed in Burmadera. The Planktonic foraminiferal assemblages of *Globigerina daubjergensis, Globorotalia compressa, Globigerina* cf. *fringa* and *Chiloguembelina* sp.cf. *C. wilcoxiensis* indicates Early Palaeocene age. As towards top of the sequence the limestone exposed near Tugapur has yielded Late Palaeoocene larger foraminifera, Palaeocene age for the argillaceous sequence can be assigned. This corroborates with the view of Allen et al. (2007) that the Mithakhari Group was deposited during Palaeocene–Eocene with a maximum age of ca. 60 Ma (Zr-radiometric dates derived from sandstones).

Slivers of dismembered ophiolite have been recorded at different stratigraphic levels within the argillaceous facies, represented by grey to dark grey shale and thin sandstone alternations. The report of Maastrichtian planktonic foraminiferal

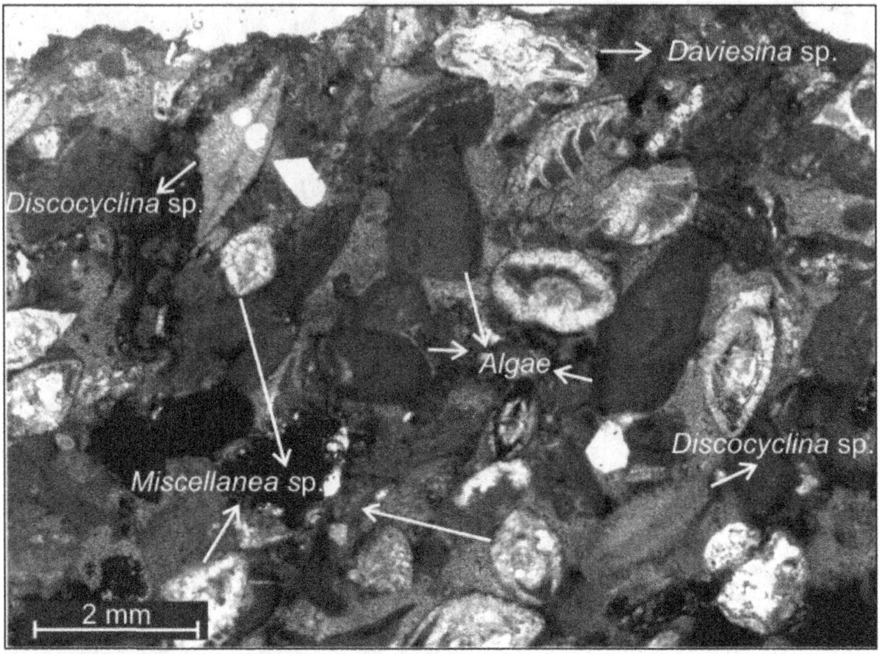

Fig. 5 Photomicrograph of foraminiferal limestone showing assemblage of *Ranikothalia, Miscellanea, Discocyclina, Daviesina* and fossil algae

assemblage by Pandey and Rao (1976) from Neali and Chandbagh Nala is probably from such pelagic sediments as present within the Mithakhari Group. In Middle Andaman, the Tertiary sequence represented by clastic sediments starts from Lower Palaeocene and the age of the pelagic sediments associated with the ophiolite can be extended up to Maastrichtian. It is mention worthy that Gee (1926) considered the cream colour porcellanitic limestone exposed near Tugapur and Burmadera as post-Eocene and correlated the limestone with the Yaw Limestone of Burma. The report of *Miscellanea* from similar type of limestone during the present work confirms the presence of the genus in the islands. *Miscellanea* has restricted stratigraphic range within Upper Palaeocene in the Indian Subcontinent of Indian, Pakistan and Burma (Nagappa 1951).

As already mentioned, limestone containing Upper Palaeocene larger benthic foraminifera are recorded from the arenaceous sediments of younger stratigraphic level and from the melange zone of Betapur coast, Middle Andaman. The larger foraminifers identified are from such limestone occurring with in the melange zone. Melange zones are important in reconstructing the Ocean Plate Stratigraphy (Wakita and Metcalfe 2005). Presence of larger benthic foraminifers indicates a Palaeocene event in Andaman. Hence, two prominent Palaeocene faunal events, viz. proliferation of planktonic foraminifers forming a remarkable foraminiferal assemblage comprising *Globigerina daubjergensis, Globorotalia fringa* and *Chiloguembelina*

Table 2 Generalized stratigraphy of Andaman and Nicobar Island (after Ray 1982)

Group	Formation	Litho units	Age
Archipelago Group		Limestone-Chalk-Claystone-sandstone—Conglomerate	Lower Miocene
Andaman Flysch Group		Sandstone—shale rhythmites with Bouma sequences	Oligocene-Eocene
Mithakhari Group	Namunagarh Grit Formation	Pebbly and gritty greywacke, green sandstone, siltstone	Middle to late Eocene
	Hope Town Conglomerate Formation	Polymictic matrix supported conglomerate alternated with pebbly-gritty sandstone and greenish claystone	
	Lipa Black Shale Formation	Pyritiferous black shale and radiolarite earth	
Ophiolite Group		Basic, ultrabasic and intermediate igneous rocks and pelagic sediments	Cretaceous to Palaeocene

sp. cf. C. *wilcoxiensis* in the basal part and development of *Ranikothalia, Miscellanea, Daviesina* and *Discocyclina* association in the upper part, can be readily be recognized in Middle Andaman. As a matter of support to the explanation of existence of Palaeocene sequence it may be noted that all these genera are reported from the Palaeocene sequence of Khasi and Jaintia hill, Garo hills and Pondicherry in India. In Pondichery Formation *Ranikothalia* is represented by *Ranikothalia nuttalli* Davies where as in Khasi Jaintia Hills this genus is represented by both *R. nuttalli* and *R. sidensis* (Samanta 1980). In Middle Andaman *Ranikothalia* is represented by species (Table 1) similar to those of Eastern India (Koley and Wanjarwadkar 2013). *Daviesina* is always reported to be associated with *Ranikothalia* and *Miscellanea* in these sequences. Hence, as of now, *Ranikothalia* and *Daviesina* assemblage in India is restricted only in Pondichery, Khasi-Jaintia and Middle Andaman; Middle Andaman being its easternmost boundary. Apart from India, the morphologically distinct genus *Ranikothalia* is recorded from Caribbean region, Oman, Venezuela, Salt Range. *Ranikothalia* (*Nummulites thalicus*) was also reported from Sarawak (Adams 1965) and Borneo (Van der Vlerk 1929), but proper taxonomic description and illustrations are lacking. Later workers (Nagappa 1959; Samanta 1980) did not agree with the presence of *Ranikothalia* in Far East region. Therefore, the occurrence of this genus in Middle Andaman indicates a possible limit of Ranikot sequence.

The argillaceous facies (=Lipa Black Shale Formation) is not developed in South Andaman (Fig. 2); only the arenaceous facies of the Mithakhari Group is developed. Adequate palaeontological evidence of the development of Palaeocene sequence in South Andaman is still lacking. Probably the Tertiary sequence in South Andaman starts with Middle Eocene which resembles the set up of far East where the lowermost Tertiary unit too is Middle Eocene. During Palaeocene, towards south of Middle Andaman sedimentation is negligible probably because of lack of sediment supply beyond Middle Andaman.

Pal et al. (2003) considered that accretionary prism formed during Lower to Late Eocene and Bandopadhyay (2005) suggested accretion broadly during Palaeogene. Flysch type trench fill sediments cover the pelagic sediments when it arrives in the trench (Wakita and Metcalfe 2005). During Cretaceous, pelagic sediments represented by chert, limestone and claystone were deposited on the subducting oceanic crust of Indian Plate. Presence of *Globotruncana* and Radiolaria in the Cretaceous sediments of Andaman Islands indicates deposition in deep water and beyond the influence of terrestrial sediment source. On its arrival in the trench during Palaeocene, shale dominated sediments started accumulating as trench fill deposits as evidenced in Middle and North Andaman. Accumulation of the clastic sediments marks the lower boundary of the argillacoues facies which is equivalent to Lipa Black Shale (Fig. 1b) well developed in Middle and North Andaman. The thick Palaeocene sequence developed in Middle Andaman and North Andaman indicates that the rate of sedimentation outweighed the rate of subduction. The change in the environment of deposition from deep water to high energy shallow marine condition during Palaeocene implies shallowing which in turn indicates uplift during the process of accretion. The process of accretion of the trench sediments started during the later part of the Palaeocene. Absence of the Palaeocene sediments suggests the South Andaman and the area further south were either geomorphologically positive or there was geomorphologic barrier which restricted the sediment supply to the locale during Palaeocene and even Lower Eocene.

Conclusions

1. Planktonic and larger foraminiferal assemblage recorded from Middle Andaman indicates that the age of the argillacous facies (=Lipa Black Shale Formation) is Palaeocene.
2. There was no deposition or negligible deposition of the clastic sediments in areas beyond Middle Andaman and further South during Palaeocene.
3. Depositional environment changes from deep marine during Early Palaeocene to shallow marine high energy condition during Late Palaeocene. Accretion of the trench sediments started after the Late Palaeocene.

Acknowledgements Authors are grateful to the Director General, GSI and the Deputy Director General & HOD, GSI, Eastern Region for providing opportunity to work in this project. Thanks are also due to Dr. S. K. Bharti, Senior Geologist, GSI for extending help for SEM imaging of the specimens.

References

Adams CG (1965) The foraminifera and stratigraphy of the Melinau Limestone, Sarawak, and its importance in Tertiary correlation. Q J Geol Soc Lond 121:283–338

Adams CG (1970) A reconsideration of the East Indian letter classification of the Tertiary. Bull Br Museum (Nat Hist) Geol 19(3):85–137

Allen R, Carter A, Najman Y, Bandopadhyay PC, Chapman HJ, Bickle MJ, Garzanti E, Vezzoli G, Andò S, Foster GL, Gerring C (2007) New constraints on the sedimentation and uplift history of the Andaman-Nicobar accretionary prism, South Andaman Island. Geol Soc Am 436:1–33 (special paper)

Bandopadhyay PC (2005) Discovery of abundant pyroclasts in the Namunagarh Grit, South Andaman: evidence of arc volcanism and active subduction during Palaeogene in the Andaman area. J Asian Earth Sci 25:95–107

Bandopadhyay PC (2011) Re-interpretation of the age and environment of deposition of Paleogene turbidites in the Andaman and Nicobar Islands, Western Sunda Arc. J Asian Earth Sci 45(2):126–137

Blow WH, Berggren WA (1969) Rates of evolution in Cenozoic planktonic foraminiferal correlation. Micropalaeontology 15(3):351–365

Bolli HM (1957) The genera *Globigerina* and *Globorotalia* in the Palaeocene Lower Eocene Lizard Spring Formation of Trinidad, B. W. I. Bull U S Nat Museum 215:61–81

Chatterjee AK (1964) The Tertiary fauna of Andamans. In: Sunduram RK (ed) International Geological Congress Report, 22nd Session, New Delhi, pp 303–318

Curray JR (2005) Tectonics and history of Andaman Sea region. J Asian Earth Sci 25:187–232

Gee ER (1926) The geology of the Andaman and Nicobar Islands with special reference to Middle Andaman Island. Rec Geol Surv India 59:208–232

Glaessner MF (1959) Tertiary stratigraphic correlation in the Indo-Pacific region and Australia. J Geol Soc India 1:53–67

Karunakaran C, Ray KK, Saha SS (1968) Tertiary sedimentation in the Andaman-Nicobar geosyncline. J Geol Soc India 9:32–39

Karunakaran C, Pawde MB, Raina VK, Saha SS (1964) Geology of South Andaman Island. In: Proceedings of international geological congress, 22nd session, India, vol 11, pp 79–100

Koley T, Anju CS (2013) Study of foraminifera and algae from Tertiary sediments of South Andaman Islands, Andaman, Unpublished GSI Report, FS 2012-13

Koley T, Wanjarwadkar KM (2013) First report of *Ranikothalia* Caudri from Middle Andaman Island, India and its significance. J Geol Soc India 81:549–555

Kumar P, Soodan KS (1976) Early Palaeocene planktonic foraminifera from Baratang Formation, Middle Andaman Island. In: Proceedings of Sixth Indian Colloquium on Micropalaeontology and Stratigraphy, pp 145–150

Kundal P, Wanjarwadkar KM (2002) On stratigraphy, age and depositional environment of algal limestone of Middle Andaman Island, Andaman, India. Gondwana Geol Mag 17(2):103–108

Mishra PK, Kumar P (1988) Fossil algae from the Cretaceous of Varagur, Tiruchirapalli district, Tamil Nadu. Palaeobotanist 37(1):36–51

Nagappa Y (1959) Foraminiferal biostratigraphy of the Cretaceous–Eocene succession in the India–Pakistan–Burma region. Micropalaeontology 5:145–192

Nagappa Y (1951) The stratigraphic value of *Miscellanea* and *Pellatispira* in India, Pakistan and Burma. In: Proceedings of Indian Academy of Science, Section B, vol 33, no 1, pp 41–44
Oldham RD (1885) Notes on the geology of the Andaman Islands. Rec of Geol Surv India 18(3):135–145
Pandey J, Rao KV (1976) Late Cretaceous foraminifera from the Middle Andaman Island. In: Proceeding of VI Indian Colloquium on Micropalaeontology and Stratigraphy, Varanasi, pp 182–205
Pal T, Chakraborty PP, Dutta Gupta T, Singh CD (2003) Geodynamic evolution of the outer arc—-forearc belt in the Andaman Islands, the central part of the Burma-Java Subduction Complex. Geol Mag 140:289–307
Rink PH (1870) Die Nicobar Inseln. Eine Geographische Skizze, mit speciieller Beruksichtigung der Geognosie, Kopenhagen. Translated selections. Rec Gov India 77:105–153
Ray KK (1982) A review of the geology of Andaman and Nicobar Islands. Mem Geol Surv India 41(2):110–125
Roy DK, Acharyya SK, Roy KK, Lahiri TC, Sen MK (1988) Nature of occurrence, age and depositional environment of the oceanic pelagic sediments associated with the ophiolite assemblage, South Andaman Island, India. Indian Miner 42(1):31–56
Samanta B (1974) The limits and subdivision of the Palaeocene with remarks on the marine occurrences recorded in the India-Pakistan Region. Q J Geol Min Metall Soc India 46:183–205
Samanta B (1980) Foraminiferal Genus *Ranikothalia* Caudri from the Pondicherry Formation, Pondicherry, South India. Q J Geol Min Metall Soc India 52(3&4):121–134
Tipper GH (1911) Geology of Andaman Islands, with reference to Nicobars. Mem Geol Sur India 35(3):195–213
Van Der Vlerk IM (1929) Groote foraminiferen van N.O. Borneo: wet. Meded. Dienst. Mijnb. Ned.—Oost- Indie 9:1–44
Wakita K, Metcalfe I (2005) Ocean plate stratigraphy in East and South East Asia. J Asian Earth Sci 24:679–702
Weeks A, Harbinson RN, Peter G (1967) Island arc system in Andaman Sea. Am Assoc Pet Geol Bull 51(8):1803–1815

Middle Eocene—Lower Oligocene Climatic Transition and Planktonic Foraminiferal Biostratigraphy at DSDP Sites 219 and 237, Arabian Sea and Western Tropical Indian Ocean

Prabha Kalia and L. R. Sahu

Abstract The study of late Middle Eocene-Early Oligocene planktonic *foraminifera* at the Arabian Sea DSDP Site 219 and the Site 237, western tropical Indian Ocean has led to significant advancement over the sporadic and sketchy Eocene-Oligocene biostratigraphy described earlier. *Orbulinoides beckmanni* Zone (E12) that serves as the biostratigraphic proxy for climatic warming phase, the middle Eocene climatic optimum (MECO), is recorded at the Site 237. Sudden warming resulted in the extinction of *Acarinina bullbrooki* and *A. topilensis* at the begining and decline of large acarininids at the top of the Zone E12 with the extinction of muricate surface dweller species *A. praetopilensis* and *Morozovelloides coronatus* and *M. lehneri.* Contrary to the conventional attribution to cooling for which there is no geochemical evidence, eutrophication of species in the warm interval approximating the biozone E12 best explains the decline and extinction of muricate large acarininids and morozovelloidids close to the MECO warming event. Although the onset of MECO is marked by the appearance of *Orbulinoides beckmanni* its culmination is coincident with the extinction of, *Globigerinatheka kugleri*, *G. euganea*, *Morozovelloides coronatus*, *M. lehneri* and *O. beckmanni.* Thus, there is a distinctive episode of extinction of five species close to the end of MECO with the drop in temperature. The change from the greenhouse to the icehouse world occurred during the latest Eocene–Oligocene Transition (EOT) ~34 Ma is expressed by the composition and pattern of change in the planktonic foraminiferal assemblages of the zones E13 to O2 at the two Sites. Almost complete Eocene-Oligocene stratigraphic record from zones E12 to O2 at the Deep Sea Sites studied in tropical Indian Ocean is consistent with other global reference sections. The late Eocene cooling and continental glaciation on Antarctica effected successive closely spaced events of extinction of cerroazulensis group of species and hantkeninids; only the first pulse of extinction of cerroazulensis group of species is recorded at the two Sites as the hantkeninid extinction at the E/O boundary is not documented in the present work due to the nonrecovery or non-availability

P. Kalia (✉) · L. R. Sahu
Department of Geology, University of Delhi, Delhi 110007, India
e-mail: prabha.kalia@gmail.com

P. Kalia
111, Shakti Apartments, Ashok Vihar Phase III, Delhi 110052, India

S. Bajpai et al. (eds.), *The Indian Paleogene*, Society of Earth Scientists Series,
https://doi.org/10.1007/978-3-319-77443-5_11

of core samples. The significant bioevent documented here at the two Sites is the early origin of *Globoquadrina tapuriensis*, *G. selli* and *G. venezuelana* in the late Eocene. Similar early appearance of these globoquadrines recorded from Tanzania, Spain and Italy extend down their range to Late Eocene.

Keywords Middle Eocene-Early Oligocene · Planktonic foraminifera middle Eocene climatic optimum · Eocene–Oligocene transition

Introduction

Subsequent to the globally recorded warmest marine temperatures in the late Paleocene to early Eocene, a long term cooling through the middle-late Eocene was initiated at the end of the Early Eocene Climate Optimum (EECO). This long term cooling trend through the middle-late Eocene was interrupted by several episodes of transient warming (Bohaty and Zachos 2003; Edgar et al. 2010; Katz et al. 2008) events. The longest and most intense of these warming events the middle Eocene climatic optimum (MECO) at about 40 Ma, initially recorded in southern Ocean cores (Bohaty and Zachos 2003) was subsequently identified in the Atlantic Ocean, central western Tethys and equatorial Indian Ocean (Bohaty et al. 2009; Edgar et al. 2010; Luciani et al. 2010; Spofforth et al. 2010; Gallazo et al. 2014, 2015; Fioroni et al. 2015). The change from the greenhouse to the icehouse world occurred during the latest Eocene–Oligocene Transition (EOT) ~34 Ma, the point at which the Earth's climate registered a transition to the late Paleogene—Neogene icehouse (Berggren and Prothero 1992). Increasing benthic foraminifer $\delta^{18}O$ values across EOT are regarded to reflect the onset of cooling impacting the extinctions of foraminifer species followed by the radiation of typical Oligocene fauna in the thermocline with the spread of continental glaciation on Antarctica (Pearson et al. 2008; Wade and Pearson 2008; Houben et al. 2012; Wade et al. 2012).

Detailed isotopic analysis has recorded distinct shifts in increasing $\delta^{18}O$ during EOT for which the nomenclature 'precursor shift (EOT-1)', subsequent transient shift (EOT-2) and the Oligocene peak values event O1 or O1-i (Miller et al. 1991, 2008; Zachos et al. 1996; Katz et al. 2008; Lear et al. 2008) have been used. These isotopic shifts were also referred as the 'interval of EOT $\delta^{18}O$ shifts' and the interval of Early Oligocene glacial maximum (EOGM) by Coxall and Pearson (2007). EOT precursor shift and the O1-i are 300 kyr apart as they fall within the ~400 kyr eccentricity cycles in the astronomic scale. Significant extinctions are recorded in *Turborotalia cerroazulensis* group at EOT-1 in central Pacific (Coxal et al. 2005), western Tethys (Houben et al. 2012) and Tanzania (Pearson et al. 2008). In the present study, late Middle Eocene to early Oligocene planktonic foraminifers have been studied across MECO and EOT at the Arabian Sea DSDP Site 219 and western tropical Indian Ocean Site 237 (Fig. 1) and described in the light of emended taxonomy, stratigraphic ranges following the biostratigraphic alphanumeric scheme of Berggren and Pearson (2006). Extinctions and faunal turnovers reflect the significant changes from warm to cold

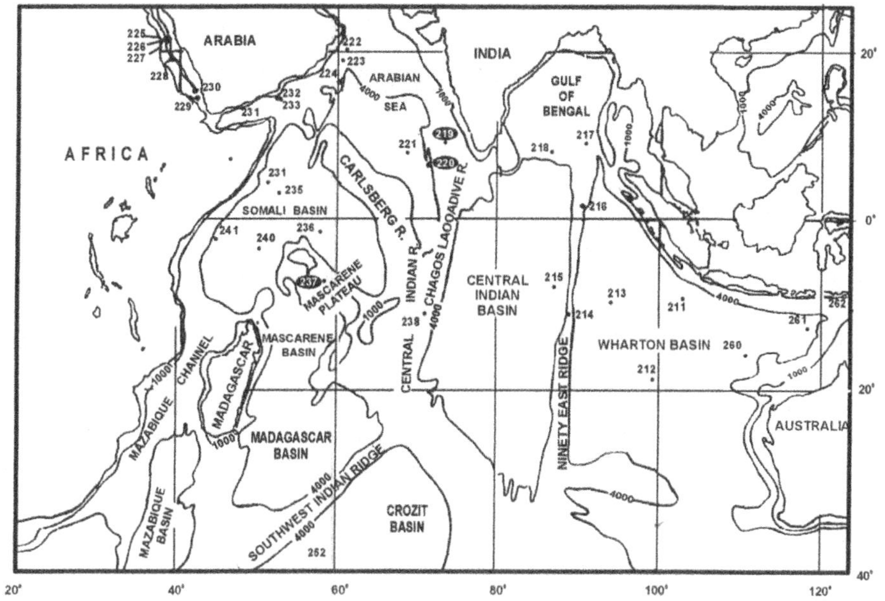

Fig. 1 Site Location map

climate and are recorded here across the MECO to EOT interval by their planktonic foraminiferal proxies at the two DSDP Sites in Indian Ocean. MECO is reflected by the record of *Orbulinoides beckmanni* Zone E12 at the Site 237 but represented by a hiatus at the Site 219. EOT is documented by the record of Zones E16 to O2 at the Site 219, while it is incompletely documented at the Site 237 by the partial record of the upper part of Zone E16 and lower Zone O1 due to non-recovery of core samples.

Initial report (Fleisher 1974) at the Site 219 recognised Zones P14–P19 and hiatuses were recognized by the absence of the Zone P16 and Zones P11–P13 in the middle to late Eocene core intervals. Later work considered the P17/P18 (=E16/O1) boundary to be marked by the dissolution event spanning the cores from 16-2 and 16-1, above which the hantkeninids were absent (Keller 1982–1983, 1983; Keller et al. 1992; Molina et al. 1993). A hiatus between Zone P12 and P14 is marked at the Site by the absence of the Zone P13 in the Middle Eocene stratigraphic record of Site 219. In the present study, hantkeninds are recorded in the core sample 16-1 above the dissolution level but these fragile index species were not found in the core 15-6 marked by the HO of cerroazulensis group of turboroatalid species and subbotinid (*S. linaperta*, *S. yaguaensis*) species. Although the sedimentation was continuous across the Eocene-Oligocene, the upper boundary of the Latest Eocene Zone E16 is difficult to mark as the HO of hantkeninids could not be ascertained due to the problem related to the preservation of fragile tests of hantkeninids and non-availability of samples from the cores 15-5 and 15-4 in the E-O boundary interval. The HO of the three turborotalid species of cerroazulensis group as recorded in the in Tanzania, Italy and Atlantic Ocean localities (Pearson et al. 2008; Houben et al. 2012; Wade et al. 2012) occurs about 0.1 Ma below the extinction level of *Hantkenina alabamensis* at the top of Zone E16 (Berggren and Pearson 2006). Eocene/Oligocene (Zone E16/O1) boundary is tentatively placed by us above the Highest Occurrence (HO) of *Turborotalia cerroazulensis, T. cocoaensis, T. pomeroli and T. cunialensis* in the core sample 15-6 and the study of the core samples between 15-5 and 15-4 at close interval may help defining the E/O boundary with precision if these samples are found to be devoid of *cerroazulensis* group of species and record hantkeninids and their highest occurrences at the Deep sea Site 219.

At the Site 237 previously identified hiatus (Hiemen et al. 1974) incorporating late Middle Eocene to Oligocene (i.e. P15–P18 Zones) is absent and the Zones E8–E16 are continuously recorded, with the exception of Zone E16/O1 boundary interval that could not physically be traced due to the non-recovery of Core sections 22-05 and 22-06. Zone E16 is marked in the Core 23-01 (007–008 cm) but the level of extinction of *Turborotalia cerroazulensis, T. cocoaensis* and *T. cunialensis* could not be marked at this Site due to non-recovery of sediments. Nevertheless, the sediment accumulation rate of 1.4 m/my (Hiemen et al. 1974) suggests the presence of Zones E16 (upper) and O1 (lower) within the unrecovered portion of the Core sections. Occurrence of the Middle Eocene to early Oligocene planktonic foraminifera (Fig. 3) is uninterrupted from core 25-4 to 22-1. *Orbulinoides beckmanni* Zone E12 that coincides with MECO, not previously recognized is documented and described in the present study in the samples 25-06 (002–003 cm) to 24–06 (148–149 cm). The total range of the marker species *Orbulinoides beckmanni* along with *Globigerinatheka eugaena, G. mexicana, G. barri, G. kugleri, G. index, Morozovelloides lehneri, M. crassatus, Hantkenina liebusi, H. dumblei, Acarinina praetopilensis, A. mcgowrani, A. rohri A. medizzai, Subbotina senni* and *Turborotalia pomeroli* characterise the assemblage of the Zone E12.

Location and Setting

Deep Sea section at Site 219 located (Fig. 1) in the crest of Laccadive Chagos Ridge (9° 01′ 75″N: 72° 52′ 67″E), extending southward from the eastern margin of Arabian sea at 1764 m depth, consists of sedimentary record from Paleocene to Pleistocene. During early Paleogene the location of this Site was at about 3°S (Sclater et al. 1977). The Middle Eocene to early Oligocene sediments mainly includes white biogenic ooze rich in foraminifers and calcareous nannoplaktons and radiolarians, chalk and chert (Fig. 2). The Site 237 is located on the Mascarene plateau (Fig. 1) between Seychelles and Saya de Malha bank (7° 04′ 99″N: 58° 07′ 48″E) at 1623 m depth. The 518 m thick Paleogene sequence consists of thick Paleocene-middle Eocene and a condensed late Eocene and Oligocene sections mainly composed of foram rich nannofossil ooze (Fig. 3).

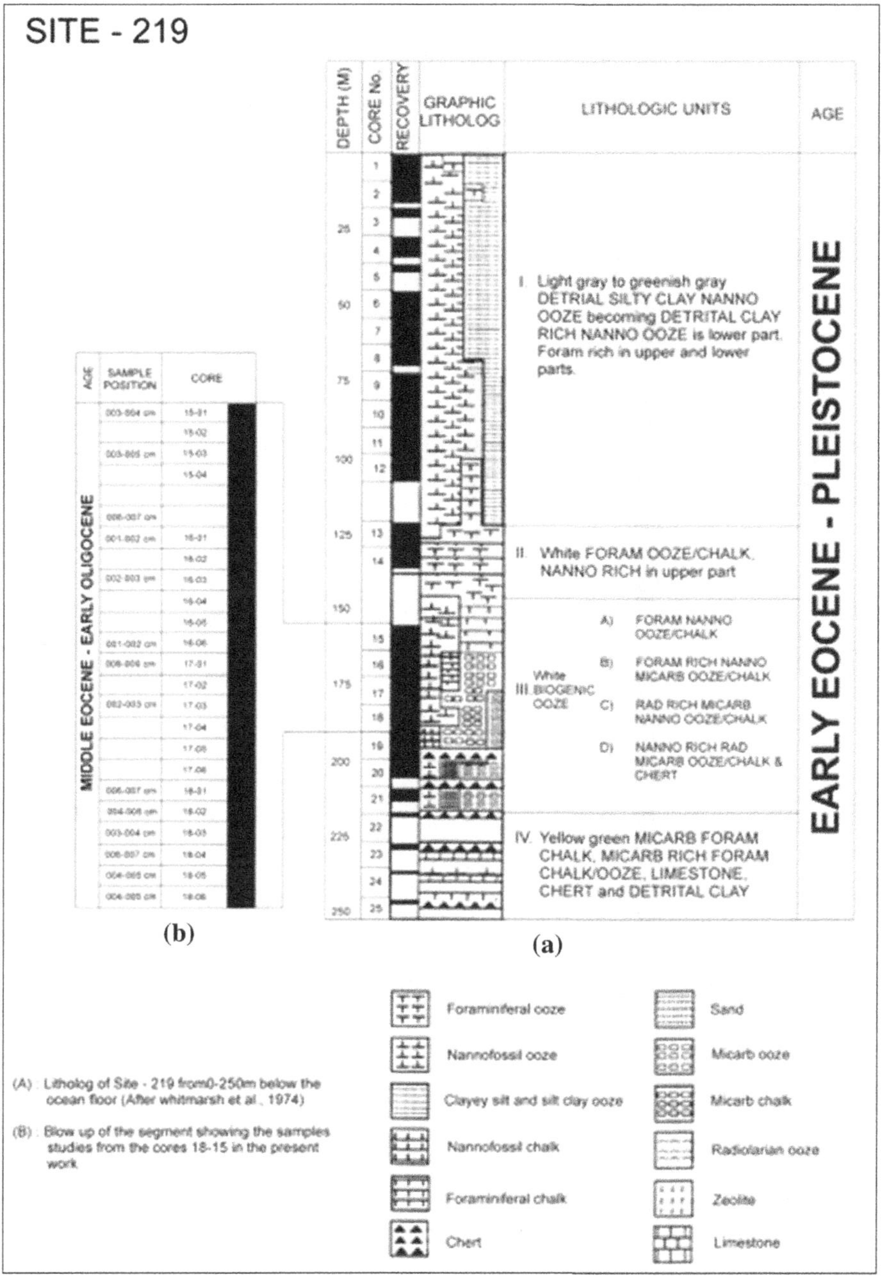

Fig. 2 Litholog DSDP Site 219

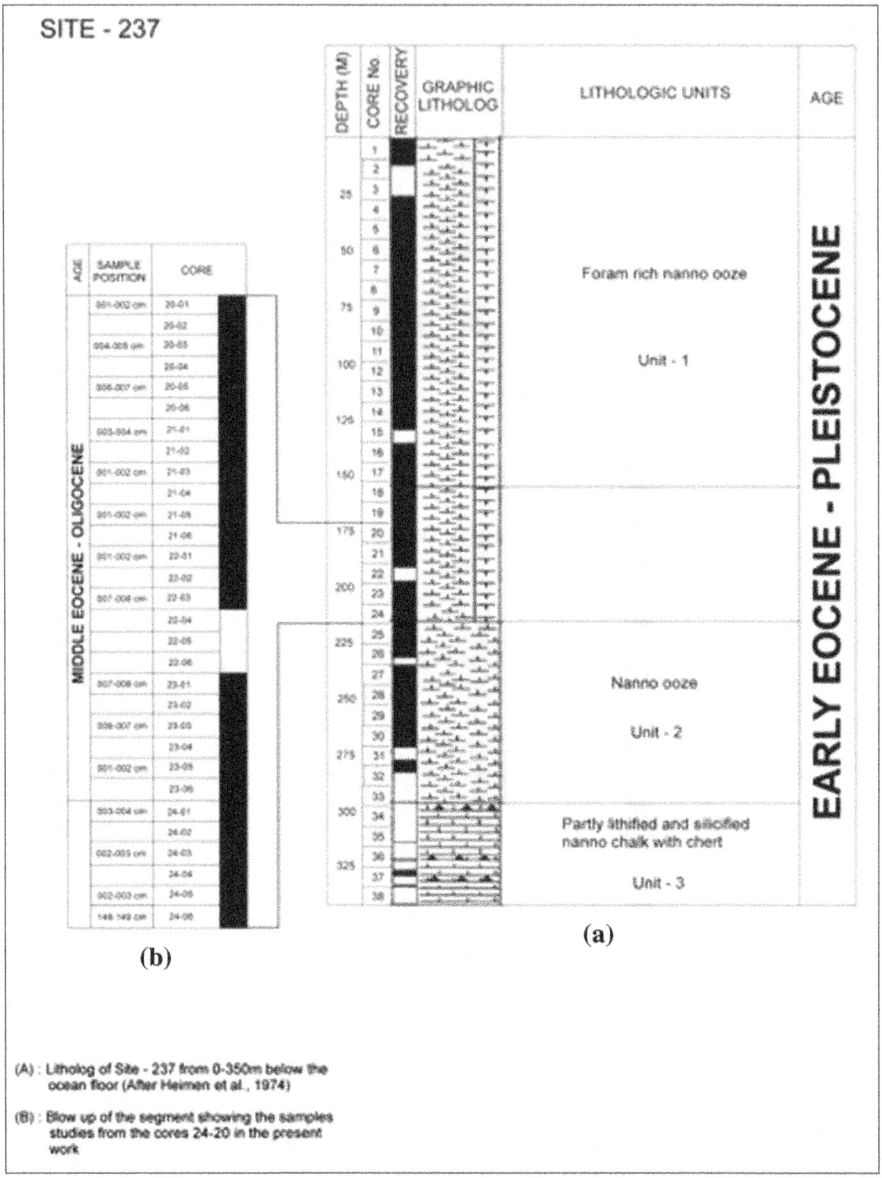

Fig. 3 Litholog DSDP Site 237

Fig. 4 Range chart, Site 219

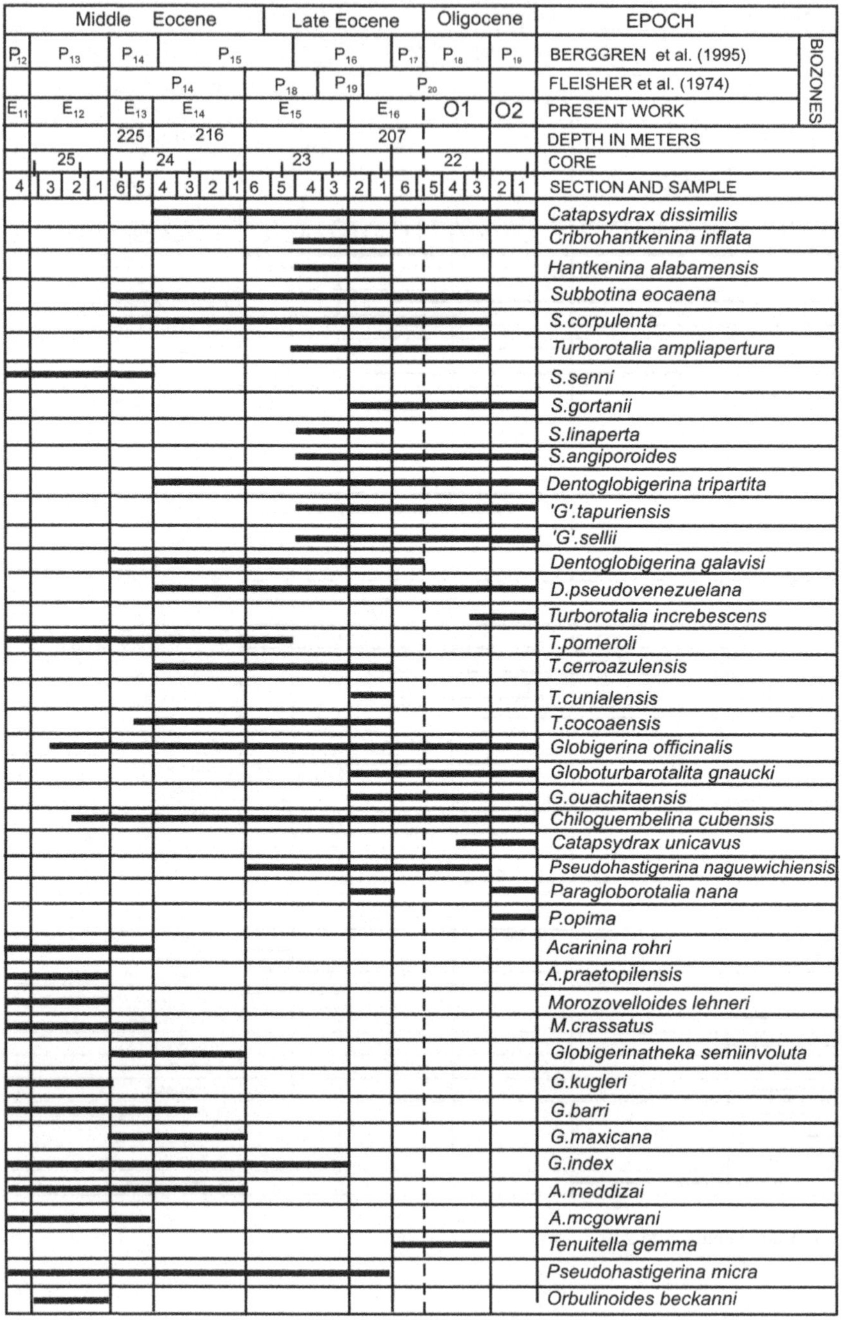

Fig. 5 Range chart, Site 237

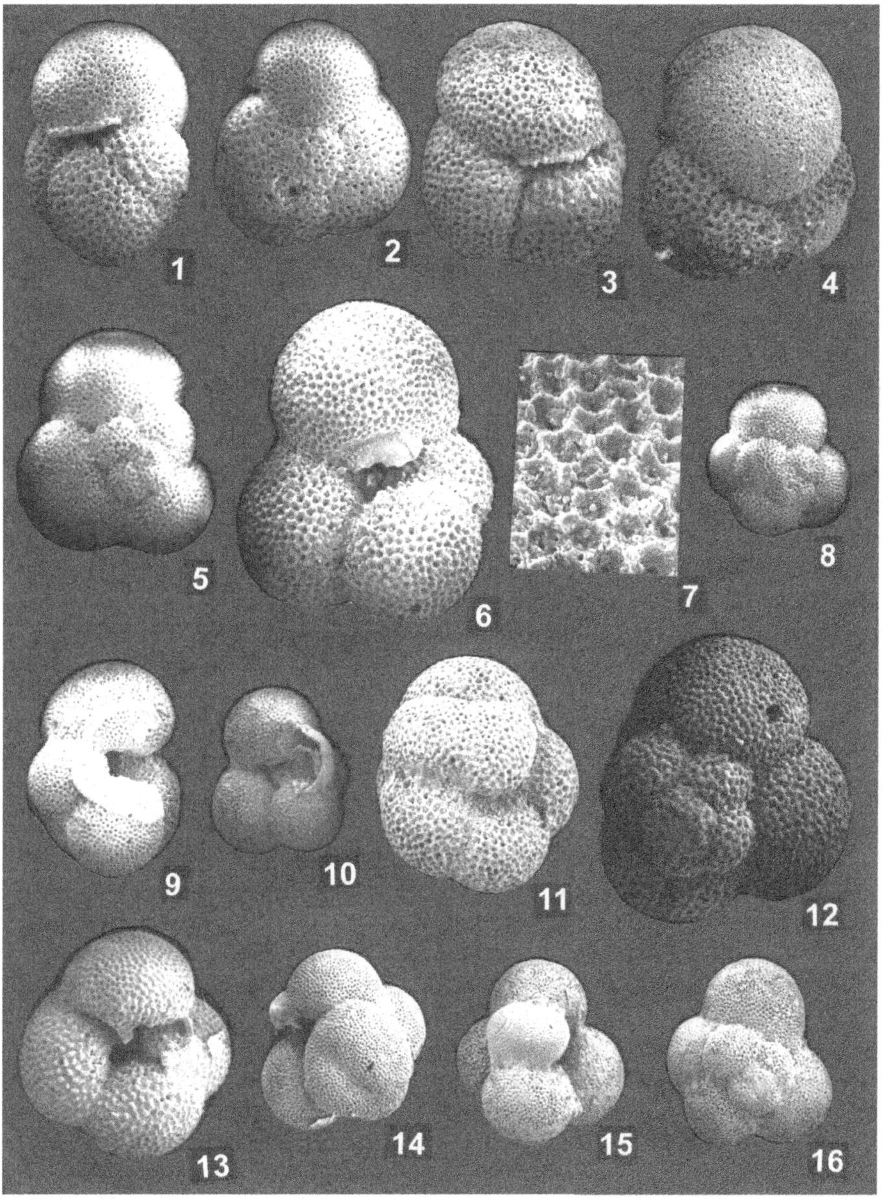

Plate 1 Figures 1, 2. *Subbotina linaperta*; Fig. 1 × 60; 2 × 130. Figures 3, 4. *Subbotina angiporoides*; ×130. Figures 5–7. *Subbotina eocaena*; ×140. Figures 8–11. *Subbotina corpulenta;* Figs. 8–10 × 65; Fig. 11 ×135. Figures 12, 13. *Subbotina hagni*; ×130. Figures 14–16 *Subbotina gortanii*; ×70

Materials and Methods

Middle Eocene to early Oligocene planktonic foraminifers were studied in the Core samples 18-06 to 15-01 from Site 219 (Fig. 2) and 24-06 to 22-03 from Site 237 (Fig. 3). Fairly to moderately preserved planktonic foraminifer were separated from the core samples by usual methods. At the Site 237 was selected to study the foraminiferal assemblages and the biozones are described across the MECO to EOT interval whereas at the Site 219 zones E13-O2 are described as a hiatus occurs below the Zone E13.

Planktonic Foraminiferal Assemblages

Planktonic foraminiferal assemblages and patterns of change through the Eocene to early Oligocene are similar in Tethyan, Atlantic, and Pacific regions (Wade and Pearson 2008) This interval was earlier considered to be marked by the successive disappearance of surface dweller acarininid and globigerinathekid taxa and increasing abundance of intermediate dwellers in response to global cooling (Keller 1982–1983; Molina et al. 1993). On the contrary, multispecies stable isotope data from Tanzania indicates no clear relationship between the extinctions and water-depth habitat (Wade and Pearson 2008). The present record of the extinction of three species of *Acarinina*, namely, *A. rohri*, *A. topilensis and A. mcgowrani* at the Site 237 coincides with the initiation of sudden warming (MECO) marked by the LO of *Orbulinoides beckmanni* which is linked significantly to the warming event rather than cooling. As documented in the present study and other low latitude ODP Sites 1051 and 1260 in Atlantic Ocean (Edgar et al. 2010) and Tethyan Alano section in NE Italy (Luciani et al. 2010), the extinction of *O. beckmanni*, *Globigerinatheka kugleri*, *G. euganea*, *Morozovelloides lehneri* and *M. coronatus* at the top of the biozone E12, are linked to the culmination of middle Eocene warming event (MECO). The late Middle Eocene to early Oligocene planktonic foraminiferal species recorded at the DSDP Site 219 and 237 are illustrated in plates 1 to 8 and their ranges are plotted in Figs. 4 and 5. At the Indian Ocean DSDP Site 237, the Middle Eocene to early Oligocene planktonic foraminifera are recorded in the Core samples 25-4 to 22-1 (Fig. 3). The Zone E12 that coincides with MECO is recorded in the samples 25-06 (002–003 cm) to 24-06 (148–149 cm). In addition to *Orbulininoides beckmanni* that is present throughout the zone, the species *Globigerinatheka eugaena*, *G. mexicana*, *G. barri*, *G. kugleri*, *G. index*, *Morozovelloides lehneri*, *M. crassatus*, *Hantkenina liebusi*, *H. dumblei*), *Acarinina praetopilensis*, *A. Mcgowrani*, *A. rohri and A. medizzai*, *Subbotina senni* and *Turborotalia pomeroli* are the characteristic members of the assemblage of the Zone E12. Following the extinction of *O. beckmani* and *A. praetopilensis*, *Globigerinatheka kugleri*, *G. euganea* and *Morozovelloides lehneri*, the succeeding assemblage is marked by the occurrence of *Turborotalia cocoaensis* and *Dentoglobigerina galavisi*. As shown in the Fig. 5, the HO of *A. rohri* marks the top of the Zone E13 in the core sample 24-04 and the HO of *Globigerinatheka semiinvoluta* in the Core sample 24-01 marks the E14 top. Hantkeninid record is sporadic at the Site 237; *Cribrohantkenina inflata* and *H. alabamensis* occur in the Core samples 23-05, 23-03 and 23-01 in the Zones E15 (upper part) and E16 (Fig. 5). The HO of *Turborotalia pomeroli*, *T. cerroazulensis*, *T. cocoaensis* and *T. cunialensis* is recorded in the core sample 23-1 along with hantkeninids and therefore, do not signify the end—Eocene time as the extinction of hankeninids post-dated the extinction of the cerroazulensis group of species (Molina et al. 2006; Wade and Pearson 2008). Higher up, from Core 23-03 onwards the species that joined the earlier association include *Subbotina angiporoides*, *Globoturborotalita ouachitaens*is, *G. gnaucki* and *Tenuitella gemma* is recorded from 23-01 to 22-03. *Pseudohastigerina naguewichiensis* characterises the Zones E16 and O1. *Tenuitella gemma*, *Dentoglobigerina tripartita*, *D. pseudovenezuelana*, *D.* galavisi, *Globoquadrina sellii*, *G. tapuriensis*, *Turborotalia ampliapertura*, *Globoturborotalita ouachitaensis* and *G. gnaucki* common in the Zone E15 and E16 become abundant in the early Oligocene Zones O1 and O2. HO of *Catapsydrax* is close to the upper boundary of the Zone O2. The assemblages and pattern of change in planktonic foraminifera through the Middle Eocene to Oligocene is in conformity with the records in Atlantic, Tethyan and Indian Ocean regions.

At the Site 219 planktonic foraminifer assemblages reported by the earlier workers (Keller 1982–1983; Keller et al. 1992; Molina et al. 1993) are recorded in the present work with some differences in the occurrence and stratigraphic range through the section (Fig. 4). *Cribrohantkenina inflata* and *Hantkenina alabamensis* are recorded from Core 17-3 to 16-3 and 18-6 to 16-1 respectively but not up to the top of the Zone E16. Molina et al. 1993 regarded core 16-2 to 15-6 mainly due to the occurrence of *Tenuitella gemma* and *G. tapuriensis* and *G. sellii* as the Oligocene Zone P18. The presence of these species does not indicate Oligocene Zone O1 because *T. gemma* ranges from upper E16 to O7 terminating at 22.96 Ma (Wade et al. 2011) and *G. tapuriensis/sellii* have a record from latest Eocene in Tanzania, Spain and Massignano GSSP, Italy (Coccioni et al. 1988; Molina et al. 1986; Pearson et al. 2008; Wade and Pearson 2008) and from Arabian Sea and western tropical Indian Ocean (present study). The extinction of the cerroazulensis group of species occurs a little before the extinction of hantkeninids (Wade and Pearson 2008; Pearson et al. 2008) but the extinction of hantkeninids have not been marked in the present study due to their sparse and sporadic occurrence and non-availability of samples above the Cores 15-5 to 15-4 at the Site 219 and non-recovery of Cores 22-5 t o 22-4 at the Site 237. *T. gemma* is recorded here at the Indian Ocean DSDP Sites within biozone E16 where it occurs with *Turborotalia cerroazulensis*, *T. cocoaensis* and *T. cunialensis* in Core 23-1 at the Site 237 and Core 15-6 at the Site 219. Previous workers placed the E/O boundary at the extinction level of the cerroazulensis group of species at the Site 219 as the hantkeninids were sporadic and rare in occurrence (Keller 1982–1983; Keller et al. 1992; Molina et al. 1993). *G. sellii* and *G. tapuriensis* were earlier regarded to appear only in the Oligocene. Previously, LO of *G. tapuriensis* was used to define the P18/P19 boundary (Blow 1969) and LO of *G. sellii* was regarded by Bolli and Saunders (1985) within the Zone P18 (=O1). These forms do appear within

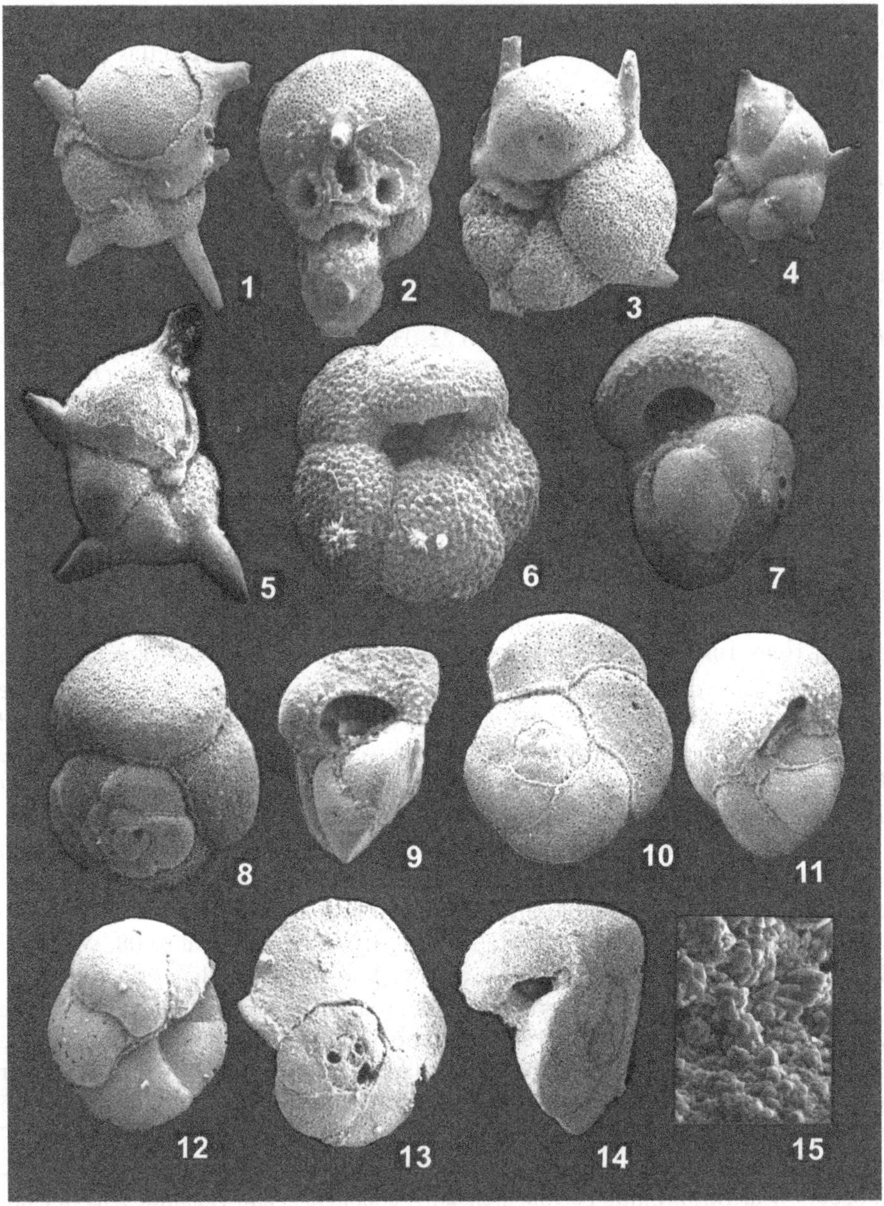

Plate 2 Figures 1–3. *Cribrohantkenina inflate*; ×130. Figures 4, 5. *Hantkenina alabamensis*; ×140. Figures 6, 8, 15. *Turborotalia pomeroli*; Figs. 6–8 ×130; 15 × 1050, wall shows diagenetic overgrowth. Figure 9. *Turborotalia cerroazulensis*; ×130. Figures 10, 11. *Turborotalia cocoaensis*; ×130. Figures 12–14. *Turborotalia cunialensis*; Fig. 12 × 60; 13, 14 × 130

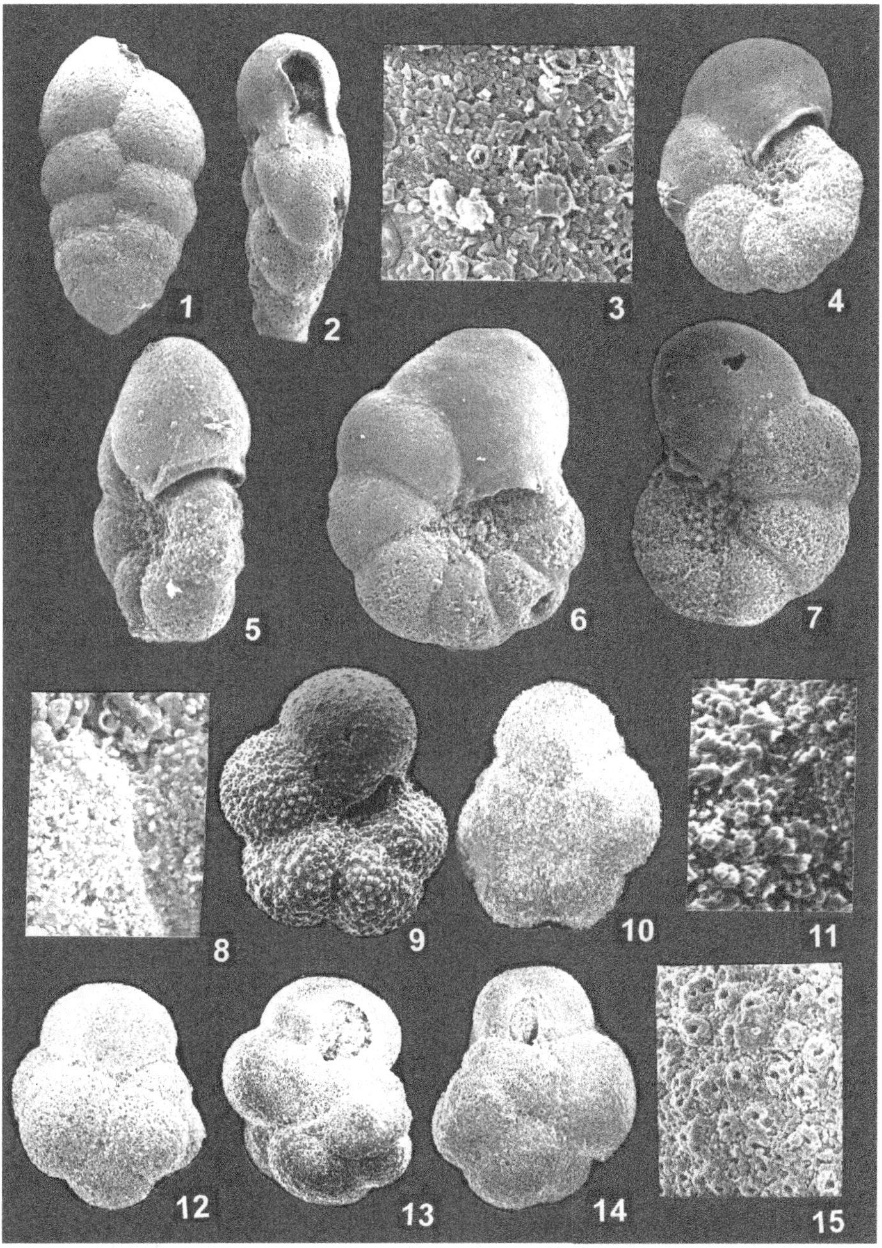

Plate 3 Figures 1–3. *Chiloguembelina cubensis*; ×130. Figures 4–8. *Pseudohastigerina naguewichiensis*. Figures 4–7 × 270; Fig. 8 × 2000 showing diagenetic overgrowth on the test surface. Figures 9–11. *Tenuitella gemma*; Figs. 9, 10 × 270; Fig. 11 × 1000 showing diagenetic overgrowth on the test surface. Figures 12–15. *Cassigerinella chipolensis*; Figs. 12, 13 × 200; 14 × 250; 15 × 1200 surface view showing the pore mounds on the microperforate test wall

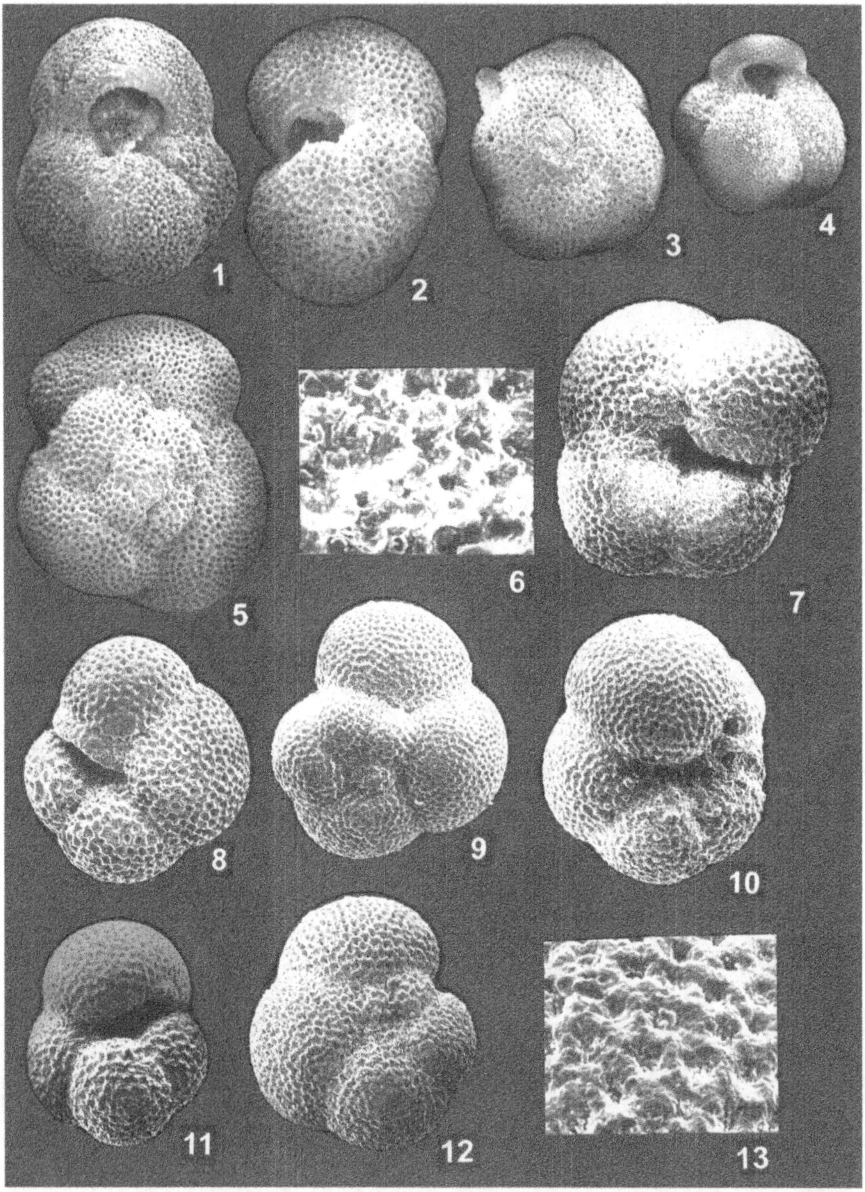

Plate 4 Figures 1–5. *Turborotalia ampliapertura*; × 130. Figures 6, 7. *Paragloborotalia opima*; Fig. 6 × 900 showing partial overgrowth on the noncancellate spinose wall; 7 × 140. Figures 8–10. *Paragloborotalia nana*; × 190. Figures 11–13. *Paragloborotalia increbescens*; Figs. 11, 12 × 150; 13 × 600 showing wall surface

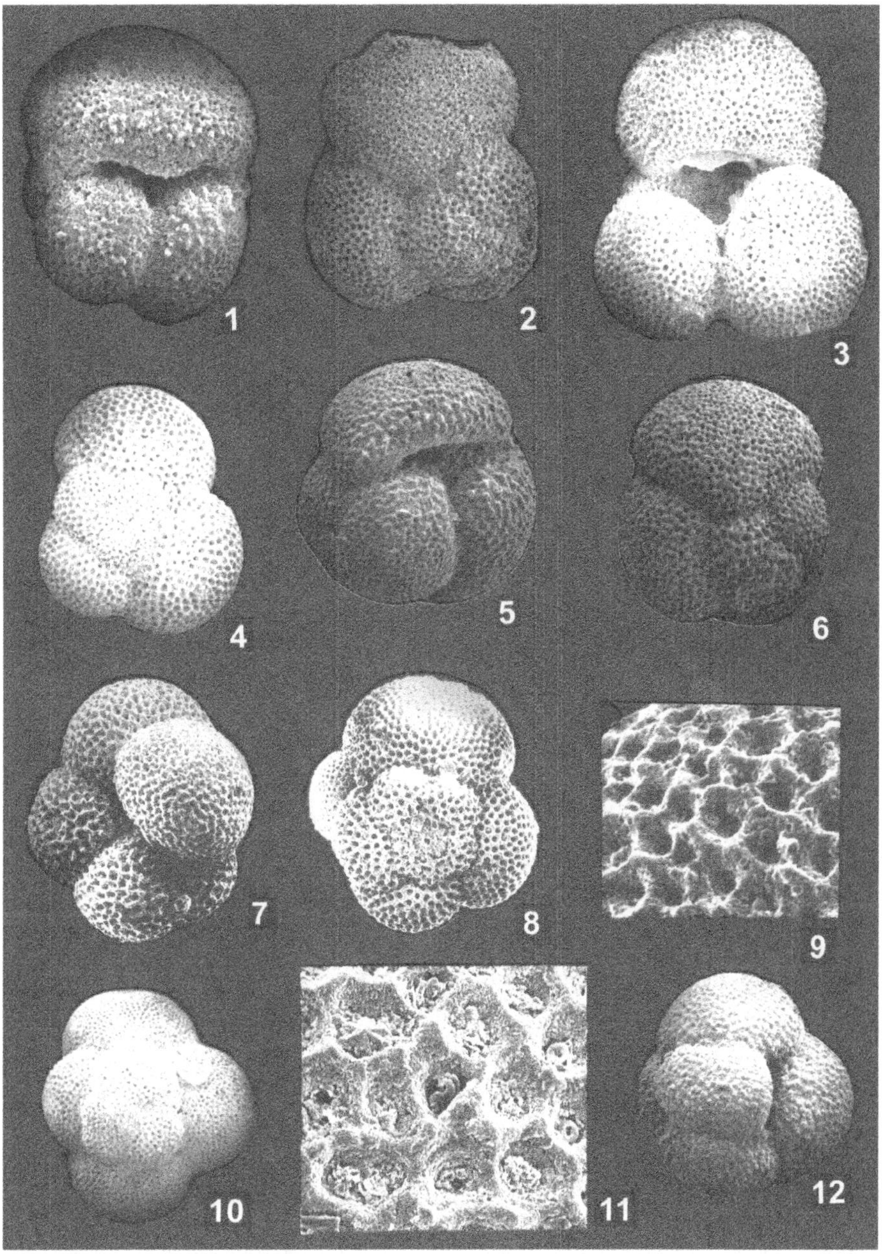

Plate 5 Figures 1, 2. *Dentoglobigerina pseudovenezuelana*; ×130. Figures 3, 4. *Dentoglobigerina galavisi*; ×130. Figures 5, 6. *Dentoglobigerina venezuelana*; ×130. Figures 7–9. *Catapsydrax unicavatus*; Figs. 7, 8 × 95; 9 × 650, showing cancellate wall with spine holes. Figures 10–12 *Catapsydrax dissimilis*; Figs. 10, 12 × 65; Fig. 11 x 880

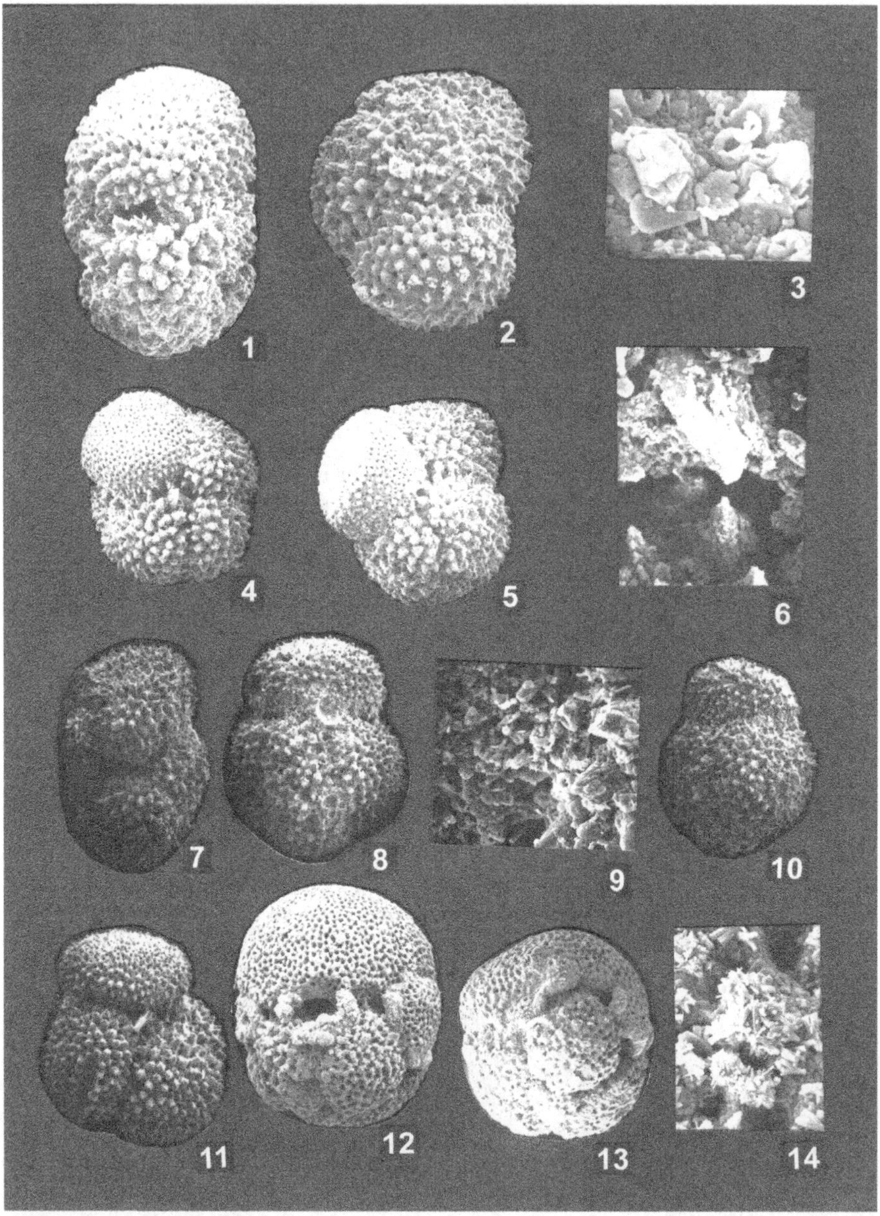

Plate 6 Figures 1–3. *Acarinina meddizzai*; Figs. 12 × 270; 3 × 1000 showing spinose surface. Figures 4–6. *Acarinina mcgowrani*; ×135. Figures 7–9. *Acarinina rohri*; Figs. 7, 8 × 135; Fig. 9 × 1000. Figures 10, 11. *Acarinina collectea*; ×135. Figures 12–14. *Orbulinoides beckmanni*; Figs. 12, 13 × 135; Fig. 14 × 1000 showing parly recrystallised wall surface

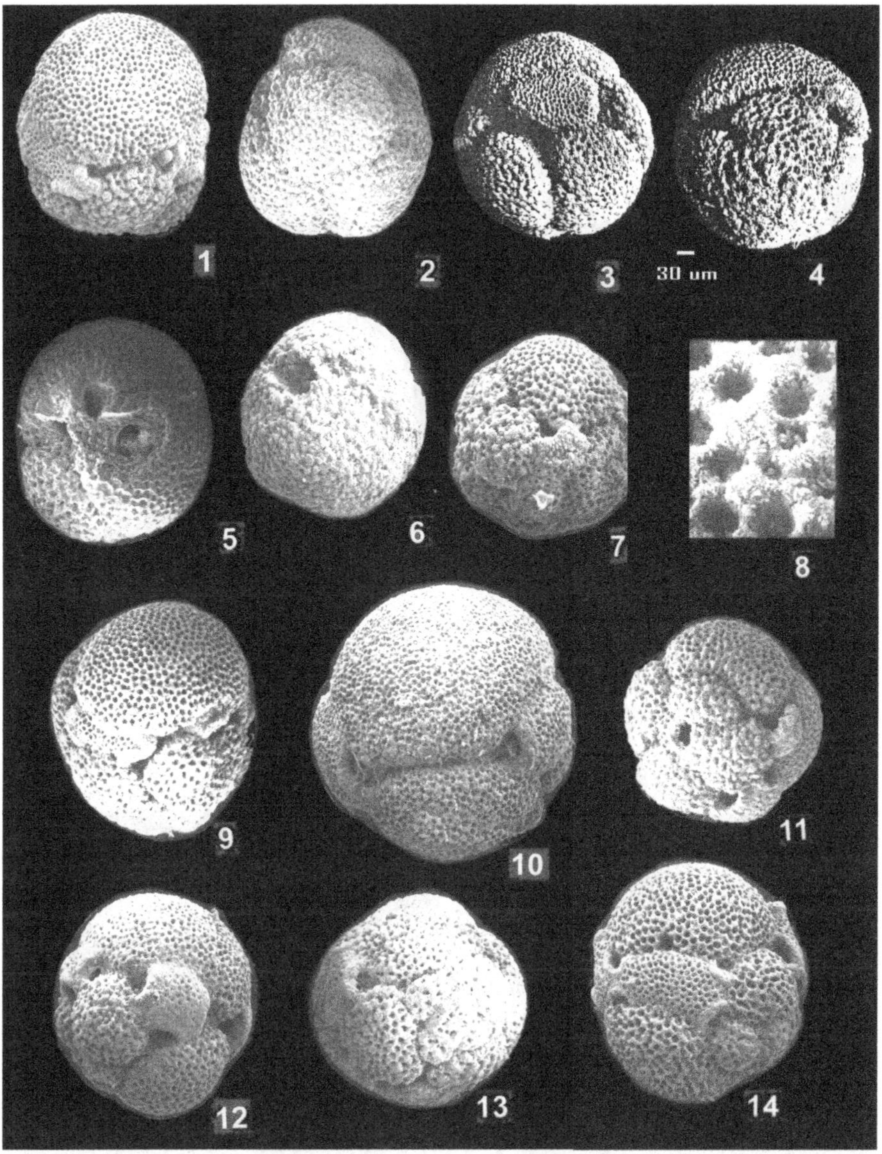

Plate 7 Figures 1–2. *Globigerinatheka index*; ×135. Figures 3–4. *Globigerinatheka subconglobata*; Fig. 3 × 120, Fig. 4 × 130. Figures 5–6. *Globigerinatheka semiinvoluta*; Fig. 4 × 1050 showing surface details; Fig. 5, 6 × 135. Figures 7–9. *Globigerinatheka mexicana*; Figs. 7, 9 × 135; 8 × 1000. Figures 10–12. *Globigerinatheka barri*; ×135. Figures 13–14 *Globigerinatheka euganea*; ×135

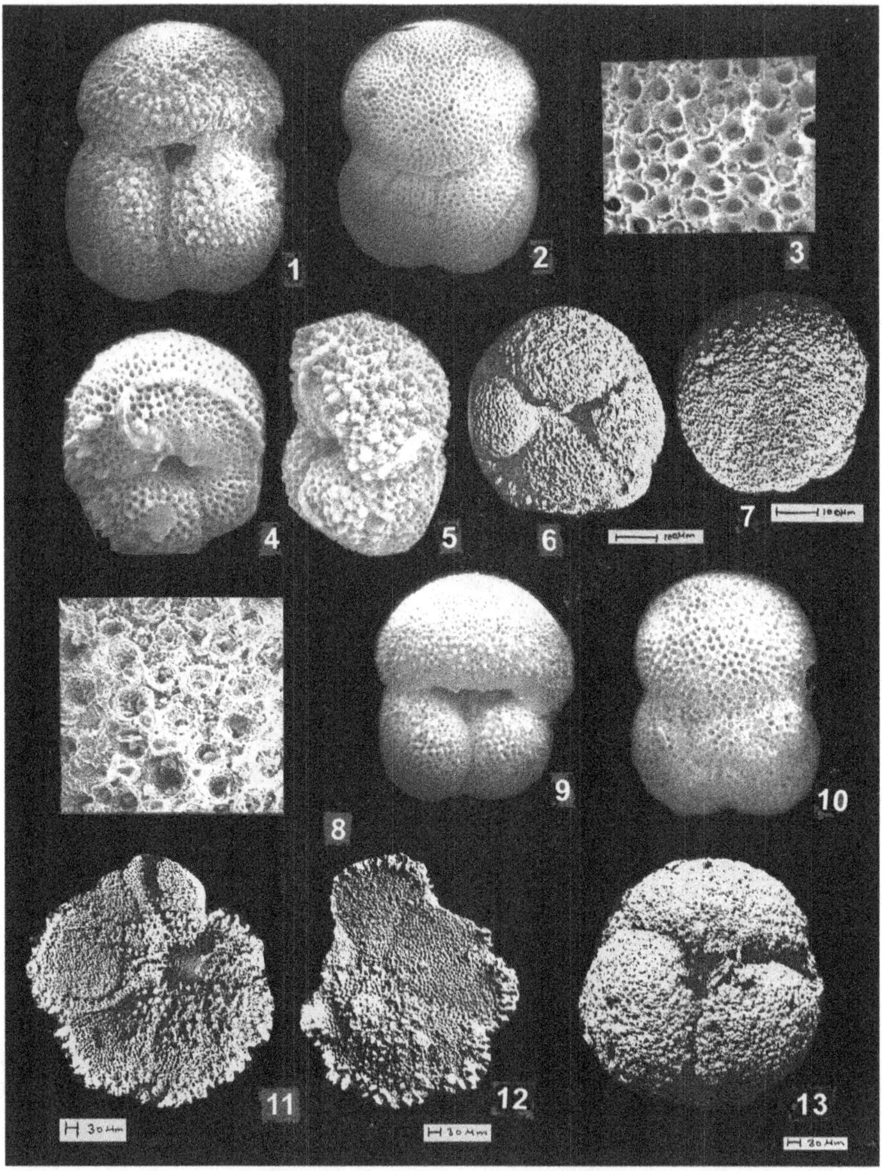

Plate 8 Figures 1, 2. *Dentoglobigerina galavisi* ×135. Figures 3–5. *Globoquadrina sellii* Fig. 3 × 60; Fig. 4, 5 × 135. Figures 6, 7. *Orbulinoides beckmanni* ×100. Figures 8–10. *Globoquadrina tapuriensis* Fig. 8 × 800 showing the wall texture; 9 × 65; 10 × 135. Figures 11–12. *Morozovelloides lehneri* ×100. Figure 13. *Globigerinatheka kugleri* ×135

the Eocene Zone E16 as reported from Italy, Spain and Tanzania and the present record of these species from Site 219 and 237. It is a significant bioevent recorded at the Site 237 and 219, as the species *Globoquadrina sellii*, and *G. tapuiensis* and *G. venezuelana* were earlier considered to be exclusively Oligocene in age. These species are here recorded from Zone E16 onwards at the Site 237 (core 23-05) and at the Site 219 in core samples 16-6 (*tapuriensis*) and 16-3 (*sellii*) and supports the similar, early stratigraphic LO of these species in Zone E16 from Tanzania (Pearson et al. 2008; Wade and Pearson 2008) and GSSP at Messignano, Italy (Coccioni et al. 1988).

Middle Eocene—Lower Oligocene Biostratigraphy

Transition from middle Eocene climatic optimum MECO to late Eocene climatic cooling triggered the extinctions among surface dweller planktonic foraminifera that culminated at the Eocene/Oligocene boundary. The stratigraphic interval of the biozone E12 defined by the total range of *Orbulinoides beckmanni* (Berggren and Pearson 2006) has been approximately calibrated to the MECO event (Bohaty et al. 2009). Based on the present study of the planktonic foraminifers at the low latitude Indian Ocean DSDP Sites 219 and 237, the biostratigraphy has been advanced in the context of emended taxonomy and stratigraphic ranges, following the alphanumeric scheme of Berggren and Pearson (2006). Biozones E13 to O2 at the Site 219 as the MECO proxy biozone E12 is absent due to a hiatus from P11–P13 and the zones E12 and E16 (base) and O1 (upper)—O2 at the Site 237 are recognized and described below. Earlier, Molina et al. 1993 recognized the biozones P15–P18 (Berggren and Miller 1988) in the core interval 18-1 to 15-6. Our study of the core samples from 18-6 to 15-1 has revised the biozonation (Fig. 4) following the E-O zonation scheme at the Site 219. At the Site 237, Hiemen et al. (1974), recognized Zone P14 in the cores 24-05 to 22-03 and a late Eocene to early Oligocene hiatus stretching across P15–P18 zones between cores 24-03 to 24-02; Zone P19 was marked by the presence of *Globoquqdrina selli*, *Catapsydrax unicavatus* and youngest occurrence of *Pseudohastigerina* spp. in the Core sample 24-2 to 23-02. Present study has recognized the Zones E12-O2 (Fig. 5) in the Core samples 25-4 to 22-1 at the Site 237 and significantly improved the tentative biozonation by Heimen et al. (1974) The revised and advanced biostratigraphy at the Site 219 and 237 following the scheme of Berggren and Pearson (2006) is described below.

Zone E12 *Orbulinoides beckmanni* Taxon Range Zone

Remarks: Relatively poor recovery of Zone E12 and almost global hiatus at ~40 Ma near C 18r/C18n boundary in the Geomagnetic polarity time scale (GPTS) in truncates the top of Zone E12 (Erbacher et al. 2004) the deep sea sequences. Present work at the Site 219 the Zone Zone E12 (=P13) is absent due to a hiatus between the Zones P12 and P14 as reported in the previous work (Fleisher 1974; Hiemen et al. 1974). At the Site 237, a complete succession of Eocene zones E-7 to E16

base is recognized in the present study. Total range of *Orbulinoides beckman*ni at the Site 237 defines this Zone E12 and the assemblage includes, *Globigerinatheka subconglobata, G. mexicana, G. euganea, G. kugleri, G. barri, G. index, Orbulinoides beckmanni, Morozovelloides lehneri, M. coronatus, M. crassatus, Acarinina rohri, A. praetopilensis, A. mcgowrani A. meddizai, Hantkenina dumblei, H. liebusi, Subbotina senni, S. eocaena, Turborotalia pomeroli.*). The base of this Zone is a thresh-hold marked by the extinction of large acarininids that disappeared at the initiation of the transient episode of Middle Eocene warming (MECO) when temperatures rose by ~4 °C (Bohaty et al. 2009). Stable isotopic record of MECO overlaps with the LO and HO of warm water species *O. beckmanni*. At the Site 219 there is a hiatus between Zone P12 and P14 and the Zone P13 is missing. Zone P12 is marked below the core 19-05 as in the previous work (Keller 1983; Keller et al. 1992; Molina et al. 1993). In the preliminary report this hiatus was interpreted as encompassing the Zones P11–P14 (Whitmarsh et al. 1974).
Occurrence: At the Site 237 Zone E12 is about 6 m thick (between Core samples 25-4 (002003 cm) to 24-6 (148–149 cm). Not present At the Site 219 as a hiatus is marked below Core 19-5.

Zone E13 *Morozovelloides crassatus* Highest Occurrence Zone

Remarks: *Morozovelloides crassatus* Zone is present at the Site 237 between the HO of *O. beckmanni* and HO of *Morozovelloides crassatus.* At the Site 219, Zone E13 is disconformable over the Zone E11 due to the absence of *O. beckmanni, Morozovelloides. lehneri, Globigerintheka kugleri* and *Acarinina praetopilensis* in the Core sample 19-05 The HO of *M. crassatus, Subbotina senni, Acarinina mcgowrani A. rohri* within the core 18-2 defines its upper limit. *Globigerinatheka index* disappears early in the zone E13 at the Site 219 as recorded previously (Keller 1982; Molina et al. 1993). The extinction of the genus *Morozovelloides* is marked by the HO of *M. cra*ssatus at the top of the Zone E13.
Occurrence: Core interval: At Site 237 from Core 24-6 (148–149) to 24-5(002–003 m); At 219 Core 18-3 to 19-4.

Zone E14 *Globigerina semiinvoluta* Highest Occurrence Zone

Remarks: Biostratigraphic interval defined by the HO of *Morozovelloides crassatus* and the HO of *G. semiinvoluta* is well documented at the Site 219 and 237. Three species of *Globigenatheka*, namely, *G. mexicana, G. barri* and *G. semiinoluta* became extinct within this Zone. *G. index* has an early disappearance at the Site 219 while at the Site 237, it occurs up to the top of Zone E15.
Occurrence: Site 237- Core interval 24-5 (002–003 cm) to 24-1(003–004 cm); Site 219-Core 17-2 to 18-2.

Zone E15 *Globigerinatheka index* Highest Occurrence Zone

Remarks: Zone E15 is recorded at the Site 237 between the Cores 24-6, 24-3 and 24-1 by the HO *G. semiinvoluta* and HO of *G. ind*ex. HO of *A. meddizzai* also marks the base of this biozone at the Site 219 where *G. index* has an early disappearance.

Cribrohantkenina inflata disappears close to the E15/E16 boundary at this Site and is not recorded beyond, possibly, due to the preservation problems associated with this fragile species.
Occurrence: Site 237- Cores 24-1 to 23-3; Site 219- Cores 17-2-16-3.

Zone E16 *Hantkenina alabamensis* Highest Occurrence Zone

Remarks: At the Site 237 *Globigerinatherka index* is present in the Cores 24-06 to 24-01 and its HO marks the base of this zone. The HO of *H. alabamannsis* could not be ascertained at both the Sites due to non-recovery of samples at the Site 237 and non- availability of samples between 15-05 an 16- 04 at the Site 219 for study. At the Site 219 the highest record of *Hantkenina alabamensis* is from core sample 16-1 and *Cribrohantkenina inflata* from sample 16-3. Previously, *Cribrohantkenina inflata* Zone P16 was considered to be absent denoting a hiatus between the zones P15 and P17 (Whitmarsh et al. 1974), However, *C. inflata* and *H. alabamensis* were recorded by Molina et al. (1993) up to the lower part Zone P16 in the core sample 17-5 and in the core 16-2 respectively. Core 16-2 was regarded to be close to Eocene top as the Zone P17/P18 boundary was supposed to be affected by dissolution (Molina et al. 1993). At the Site 237, both these hantkeninid species are recorded here in the core sample from 23-05 to 23-01 but their HO could not be ascertained as the core samples from above (from 22-04 to 22-03) were not recovered. In earlier biozonation schemes, the HO of *T. cerroazul*ensis was regarded as a definitive criterion to demarcate the P17/P18 (E/O) boundary (Berggren and Miller 1988; Berggren et al. 1995; Molina et al. 1993). Recent biostratigraphic studies (Pearson et al. 2008; Wade and Pearson 2008; Houben et al. 2012) have clearly demonstrated that the HO of cerroazuelesis group of species is close to the Eocene top and is estimated to be about 0.1 Ma earlier to the extinction (HO) of *Hantkenina alabamensis* that defines the top of zone E16 (Berggren and Pearson 2006). HO of *H. alabamensis* could not be ascertained at the Site 219, due to the non-availability of samples from core 15-05 to 15-04 and non-recovery of core samples at the Site 237 from core 22-04 and 22-03. Highest common occurrence (HCO) of *Pseudohastigerina micra* in upper Eocene sediments at both the Sites is observed in the present study. Decrease in the size of *P. micra* has earlier been recorded at the Site 219 (Keller 1982–1983; Molina et al. 1993). The significant size decrease of *P. micra* that parallels the extinction of *Hantkenina* appears to be coeval between the Indian Ocean and the Gulf of Mexico (Wade and Pearson 2008; Miller et al. 2008). *Tenuitella gemma* has an early appearance at both the Sites studied within the Zone E16.
Occurrence: Site 237- Core 23-3 to 22-5, upper boundary not placed due to non-recovery of core samples from Core 22-6 to 22-4 m, the interval which probably corresponds to upper part of the Zone E16 and lower part of O1. At Site 219, the lower boundary lies in Core 16-3 and the upper boundary may fall between Core 15-5 and 15-4 as the core 15-3 bears some Oligocene species like *Cassigerinella chipolensis*, also reported in previous work (Molina et al. 1993).

Zone O1 *Pseudohastigerina naguewichiensis* Highest occurrence Zone

Remarks: At the Site 237, the Zone is marked by the presence of *P. naguewichiensis* in the samples devoid of *Turborotalia cerroazulensis*, *T. cocoaensis*, *T. cunialensis* and hantkeninids and the upper boundary is marked by the HO of *P. naguewichiensis*. At the Site 219 the base of the Zone may lies below the Core sample 15-4 and the top of Zone lies in the Core sample 15-3 marked by the HO of *P. naguewichiensis*
Occurrence: Site 237- Core samples 22-3 (007–008 cm), base not defined due to non-recovery of core above 22-6 to 22-4. Site 219 core samples 15-4 to 15-3.

Zone O2 *Turborotalia ampliapertura* Highest occurrence Zone

Remarks: At the Site 237, the biozone is marked by the HO *P. naguewichiensis* in Core 22-2 and HO of *T. ampliapertura* in Core 22-3. At the Site 219, HO of *P. naguewich*iensis is recorded in the core 15-3 and HO of *T. ampliapertura* occurs in Core15-1 above which the samples were not studied. This zone is characterized by the occurrence of typically Oligocene species *Paragloborotalia opima* at both the Sites.
Occurrence: Site 237- Core sample 22-3 (001–002 cm)-22-1 (007–008 cm); Site 219 core samples 15-3 to 15-1.

Discussion and Conclusions

The Middle Eocene Climatic Warming Event (MECO) that interrupted the long term Eocene cooling was initially documented at high latitudes (Bohaty and Zachos 2003). The Zone E12, which almost coincides with MECO, was poorly known from low and mid–latitudes. The record of MECO got extended to mid-latitudes by the study of land based marine sections in central Italy, Atlantic Ocean (Javane et al. 2007) and ODP Site 1263 southeast Atlantic Ocean (Galazzo et al. 2014). These records from multiple deep-sea sites revealed that a large negative $\delta^{18}O$ excursion, a prominent feature of all the continuous stratigraphic sections ~40 Ma, underlines the fact that MECO event (between 40–40.5 Ma; C18r/18n boundary) with a duration of 500 ka, was globally ubiquitous (Bohaty et al. 2009). This warming event was poorly known in the low latitudes till the isotopic and palaeomagneic and biostratigraphic study of ODP sections at the Sites 1051 and 1260 (Edgar et al. 2010) and the Tethyan marginal section at Alano, NE Italy (Luciani et al. 2010) was done. The isotopic data and palaeomagnetic record at the low latitude, Atlantic Ocean ODP Site 1051 and the equatorial Site 1260 demonstrated that the stratigraphic range of *O. beckmanni*, although approximating the duration of the MECO, is diachronous (Edgar et al. 2010) as the LO of this index species of Zone E12 is ~500 Ka before (40.5 Ma) in the equatorial Atlantic than in the subtropics (40.0 Ma). The HO of this species in low latitudes at the Site 1051 (Edgar et al. 2010) is later in the magnetochron C18n than its previous calibration in C18n.2n to 40.0 Ma (Wade 2004) and is younger by 600 ka. (Luciani et al. 2010). *O. beckmanni* has been regarded as an approximate proxy but not the ‘excursion taxon’ for MECO as the HO of the species at the Site 1051 post-dates the MECO by 600 Ka (Edgar et al. 2010).

Biostratigraphy of MECO

Stratigraphic range of *Orbulinoides beckmanni* defining the Zone E12 is diachronous and LO and HO of the species are not synchronous in the equatorial and the subtropical regions. *O. beckmanni* Zone E12 is present at the DSDP Site 237 but this biostratigraphic proxy for MECO is not recorded at the Site 219. A hiatus was indicated between the zones P12 and P14 by the presumed absence of *O. beckmanii* (Fleisher 1974). The present study records the index species, *O. beckmanni* at the Site 237 from the core samples 25-02, 25-04 and 24-06 defining an expanded (~6 m) Zone E12. Although there exists no magnetostratigraphic and stable isotopic control over the DSDP section 237, the present record of the Zone E12 serves as the proxy for the MECO, the sudden warming at the beginning of which resulted in the extinction of *Acarinina bullbrooki* and *A. topilensis* (Fig. 4) and decline of large acarininids was set-in at the top if the Zone E12 with the extinction of muricate surface dweller species *A. praetopilensis Morozovelloides coronatus* and *M. lehneri.* Though the base of the Zone E12 is marked by the appearance of *Orbulinoides beckmanni* its top is coincidental with the extinction of *Globigerinatheka kugler*i, *G. euganea*, *Morozovelloides coronatus*, *M. lehneri* and *O. beckmanni.* Thus the decline and culmination of MECO coincided with the distinctive event of extinction of six morozovlloidid and globigernathekid species at the end or close to the end of MECO. During the late Middle Eocene, *M. crassatus* marks the ultimate extinction of the genus *Morozovelloides* at the E13 top at the E13/E14 boundary (Site 237, Site 219). Luciani et al. (2010), regarded the decline and evolutionary extinction of muricate large acarininids and morozovelloidids close to the MECO warming event near the biozonee E12, as connected with the eutrophication of species in the warm interval, contrary to the conventional attribution to cooling (Berggren 1969; Keller 1982–1983; Boersma and Premoli Silva 1986; Keller et al. 1992; Molina et al. 1993; Pearson 1996) for which there exists no geochemical evidence. Post-MECO cooling is indicated by the ultimate decline in the abundance of muricate taxa subsequent to the $\delta^{18}O$ negative peak in MECO, above which the cool water, deep dwelling *Catapsydrax* and temperate groups (Pseudohastigerinids, globigerinathekids and *Turborotalita*) register an increase in abundance (Luciani et al. 2010). Similar pattern of change in the planktonic foraminiferal assemblages has been observed in both the deep sea sections studied in the present work. Globogerinathekids survived beyond MECO and were quite common in Zones E13 and E14. *G. semiinvoluta* is common in zone E13 to E14 and its HO marks the E14/15 boundary in core 24-01 at the Site 237 and in core 17-03 at the Site 219. HO of *G. index* in Core 23-03 at Site 237 marks the E16 base (E15/16 boundary) along with the LO of *Turborotalia cunialensis*. *G. index* disappears much earlier than its total stratigraphic range at approximately the same level as *G. semiinvoluta* at the Site 219 and *G. tropicalis* also has an early disappearance at both the Sites much before the Eocene/Oligocene boundary. *G. index* that is not found to occur at the Site 219 beyond core sample 17-04, occurs till the top of the zone E15 and defines the E15/E16 boundary at the Site 237.

Biostratigraphy of Eocene-Oligocene Transition

Due to non-recovery and non-availability of core samples record HO of hantkeninid the end Eocene hantkeninid extinction event could not be determined at both the Sites. At the Site 237, there is a break in record due to the non-recovery of samples covering the upper part of E16 and lower part of O1 zones. The highest occurrence of cerroazulensis group of species is close to the E-O boundary and represents the bioevent keyed to EOT-1 also known as precursor EOT in Tanzania, western Tethys and Atlantic Ocean Sites (Pearson et al. 2008; Wade and Pearson 2008; Houben et al. 2012; Wade et al. 2012) marks the Zone E16 (upper part) but does not define the E/O boundary. There is a distinctive size decrease recorded in *Pseudhastigerina micra* (Keller 1982–1983) in the Zone P17 (=to E16). The extinction of hantkeninis at the E/O boundary could not be documented at the Site 237 and 219 due to problems with preservation of fragile tests of hntkeninids and sampling constraints. HO of *Pseudohastigerina neguewichiensis* demarcates the O1/O2 Zonal boundary, and the HO of *T. ampliapertura* marks the top of the biozone O2 in the DSDP section at the Site 219 and O2/ O3 boundary in the section at the Site 237.

The genera *Dentoglobigerina* and *Globoquadrina* having nonspinose cancellate, honeycomb wall texture were included by Olsson et al. (2006) in the family Globoquadrinidae. The taxa earlier regarded to be exclusively Oligocene in age such as *G. tapuriensis* and *G. venezuelana* are recorded in the latest Eocene at the Sites 219 and 237. The origination of *G. tapuriensis* and *G. venezuelana* in latest Eocene were earlier reported from Tanzania, and their appearance in the latest Eocene and abundance in the earliest Oligocene was considered to have occurred within the thermocline and not in surface waters (Wade and Pearson 2008). *Dentoglobigerina galavisi* and *D. pseudovenezuelana* occur consistently through upper Eocene to early Oligocene interval in the sections studied at the two Sites in Indian Ocean. Above the E/O boundary *G. tapuriensis* increases in abundance and dentoglobigerinids record a radiation with the advent of additional species e.g. *G. venezuelana* and *G. praedehiscens* in Oligocene as reported from Tanzania. *Chiloguembelina cubensis* is well preserved and occurs abundantly at 219 from Zone P15 to early Oligocene zones but occurrence is rare at the Site 237. At the Indian Ocean Sites studied here, the record of *Globoquadrina tapuriensis, G. sellii* and *G. venezuelana* in the latest Eocene (From Zone E16) and increase in their abundance in the early Oligocene following the E/O boundary is consistent with previous studies from Spain, Messignano Type section, Italy and Tanzania (Molina et al. 1986, Coccioni et al. 1988 and Wade and Pearson 2008). With the radiation of *Globoquarina* group of species following the E/O boundary extinction events, the planktonic foraminiferal diversity returned to pre-extinction levels (Wade and Pearson 2008).

Although the extent of temperature change and timings remained controversial, pronounced climatic change during the Eocene-Oligocene transition at ca. 33–34 Ma is provided by the positive $\delta^{18}O$ isotopic values in the deep sea benthic foraminifera (Kennet and Shekleton 1976; Miller et al. 1991; Zachos et al. 1996; Coxal et al. 2005) and planktonic foraminifera (Pearson et al. 2008; Wade and Pearson 2008; Wade et al.

2012 and Houben et al. 2012). The largest $\delta^{18}O$ increase referred to as Oligocene isotope-1 (Oi-1; Miller et al. 1991) represents the instance of Antarctic continental glaciation indicated by a significant drop in global sea level (Wade et al. 2012; Houben et al. 2012). A pronounced (3–4 °C) reduction in sea surface temperature occurred in the latest Eocene associated with 'precursor' $\delta^{18}O$ increase at the EOT-1 (Wade et al. 2012). Across the EOT, the estimates for cooling vary at different mid- and low-latitude locations ranging from 8 °C cooling in central North America (Zenazzi et al. 2007), to 2.5 °C cooling at EOT-1 in the western Indian Ocean (Wade and Pearson 2008) but no cooling at Oi-1 (Lear et al. 2008). In the Gulf of Mexico region, extensive 3–4 °C cooling through EOT-1 in surface and bottom waters is followed by a further decrease in temperature by 2 °C in the early Oligocene coincident with the isotopic peak Oi-1 has been estimated (Wade et al. 2012). The decrease in temperatures across EOT affected the extinctions and change in the composition of foraminiferal assemblages (Pearson et al. 2008; Wade and Pearson 2008; Wade et al. 2012; Houben et al. 2012). Sea level fell in the latest Eocene at the first isotopic shift EOT-1 by ~20 m, whereas,~50–60 m fall at the Oligocene isotopic event Oi-1 reflecting the expansion of the Antarctic ice cover (Houben et al. 2012) explains the scarcity and absence of on-land sedimentary record of the Eocene–Oligocene transition. The Indian subcontinent particularly has sketchy Late-Eocene-Early Oligocene marine stratigraphic record and the present study of deep sea sections in the Indian Ocean provides the planktonic foraminiferal bioevents consistent with those from the GSSP, Messignano, Italy and other Atlantic and Indian Ocean Sites and marginal marine onland sections in Tanzania and Alano NE, Italy.

Taxonomic Appendix

Acarinina mcgowrani (Wade and Pearson 2006)
Acarinina meddizzai (Toumarkine and Bolli 1975)
Acarinina rohri (Bronnimann and Bermudez 1953)
Catapsydrax dissimilis (Cushman and Bermudez 1937)
Cassigerinalla chipolensis (Cushman and Ponton 1932, emended Li 1986)
Catapsydrax unicavatus (Bolli, Loeblich and Tappan 1957)
Chiloguembelina cubensis (Palmer 1934)
Cribrohantkenina inflata (Howe 1928)
Dentoglobigerina galavisi (Bermudez 1961)
Dentoglobigerina pseudovenezuelana (Blow and Banner 1962)
Dentoglobigerina tripartita (Koch 1926)
Globigerina officinalis (Subbotina 1953)
Globigerinatheka barri (Bronnimann 1952)
Globigerinatheka index (Finlay 1939)
Globigerinatheka kugleri (Bolli, Loeblich and Tappan 1957)
Globigerinatheka maxicana (Cushman 1925)
Globigerinatheka semiinvoluta (Keijer 1945)

Orbulinoides beckmanni (Saito 1962)
Globoquadrina sellii (Borsetti 1959)
Globoquadrina tapuriensis (Blow and Banner 1962)
Globoquadrina venezuelana (Hedberg 1937)
Globoturborotalita gnauki (Blow and Banner 1962)
Globoturborotalita ouachitaensis (Howe and Wallace 1932)
Hantkenina alabamensis (Cushman 1924)
Paragloborotalia nana (Bolli 1957)
Paragloborotalia opima (Bolli, 1957)
Pseudohastigerina micra (Cole 1927)
Pseudohastigerina naguewichiensis (Myatliuk 1950)
Subbotina angiporoides (Hornibrook 1965)
Subbotina corpulenta (Subbotina 1953)
Subbotina eocaena (Guembel 1886)
Subbotina gortanii (Borsetti 1959)
Subbotina hagni (Gohrbandt 1967)
Subbotina linaperta (Finlay 1939)
Morozovelloides crassatus (Cushman 1925)
Morozovelloides lehneri (Cushman and Jarvis 1929)
Turborotalia ampliapertura (Bolli 1957)
Turborotalia cerroazulensis (Cole 1928)
Turborotalia cocoaensis (Cushman 1928)
Turborotalia cunialensis (Toumarkine and Bolli 1970)
Turborotalia increbescens (Bandy 1949)
Turborotalia pomeroli (Toumarkine and Bolli 1970)
Turborotalia pomeroli (Toumarkine and Bolli 1970).

Acknowledgements Authors are thankful to Directorate of Science and Technology, and Council of Scientific and Industrial Research, Government of India for funding. SEM images were acquired at the Science Instrumentation Centre, Delhi University by Dr. N. C. Mehra and Mr. Rajesh at the Sophisticated Instrumentation Facility for Electron Microscopy (DST) at All India Institute of Medical Sciences. Special thanks are expressed to Sri Rooprai for preparing the coral Draw figures.

References

Berggren WA (1969) Rates of evolution in some Cenozoic planktonic foraminifera. Micropaleontology 15:351–365

Berggren, WA, Prothero DR (1992) Eocene-Oligocene climatic and biotic evolution. In: Prothero DR, Berggren WA (eds) Eocene-Oligocene climatic and biotic evolution Princeton University Press, pp 1–28

Berggren, WA, Person PN (2006) Tropical and subtropical planktonic foraminiferal zonation of the Eocene and Oligocene. In: Pearson PN, Olsson RK, Huber BT, Hemleben C, Berggren WA (eds) Atlas of eocene planktonic foraminifera, no 41. Special publication—Cushman Foundation for Foraminiferal Research, pp. 29–40

Berggren WA, Miller KG (1988) Paleogene tropical planktonic foraminiferal biostratigraphy and magnetobiochronology. Micropaleontology 34(4):362–380

Berggren WA, Kent, DV, Swisher III CC, Aubrey MP (1995) A revised Cenozoic geochronology and chronostratigraphy. In: Berggren WA, Kent DV, Aubrey MP, Hardenbol J (eds) Geochronology, time scales and global stratigraphic correlation: a unified temporal framework for an historical geology, vol 54. Special publication—Society of Economic Paleontologists and Mineralogists, pp 129–221

Blow, WH (1969) Late Middle Eocene to recent planktonic foraminiferal biostratigrahy. In: Bronnimann P, Renz HH (eds) Proccedings of the first international conference of planktonic microfossils vol 1. EJ Brill, Leiden, Geneva, 1967, pp 199–422

Boersma A, Premoli Silva I (1986) Terminal Eocene event: planktonic foraminifera and isotopic evidence. In: Pomerol C, Premoli Silva I (eds) Terminal Eocene events, Elsevier, pp 233–224

Bohaty SM, Zachos JC (2003) A significant Southern Ocean warming event in the late middle eocene. Geology 31:1017–1020

Bohaty SM, Zachos, C, Florindo, F, Delaney ML. (2009) Coupled greenhouse warming and deep sea acidification in the Middle Eocene. Paleoceanography, v.24 PA 2207, doi: 10. 209/2008PA 001676

Bolli HM, Saunders JB (1985) Oligocene to Holocene low latitude planktonic foraminifera. In: Bolli HM, Saunders JB, Perth Nielson K (eds) Plankton stratigraphy. Cambridge University Press, Cambridge, pp 15–262

Coccioni R, Monaco P, Monechi S, Nocchi M, Parisi G (1988) Biostratigraphy of the Eocene-Oligocene boundary at Messignano (Ancona, Italy) In: Premoli Silva IR, Coccioni A, Montanari (eds) The Eocene-Oligocene boundary in the Marche-Umbria basin (Italy). International Subcommission on the Paleogene stratigraphy, Ancona, Special Publication, pp 59–80

Coxal HK, Pearson PN (2007) The Eocene-Oligocene transition In: Williams M, Hayward AM, Gregory FJ, Schimidt DN (eds) Deep-time perspective on climate change: marrying the signals from computer models and biological proxies, no 2. Special Publications Micropalaeontlogical Society, pp. 297–351

Coxal HK, Wilson PA, Palike H, Lear CH, Backman J (2005) Rapid stepwise onset of Atlantic glaciation and deeper calcite compensation in Pacific Ocean. Nature 433:53–57

Edgar KM, Wilson PA, Sexton PF, Gibbs SJ, Roberts AP, Norris RD (2010) New biostratigraphic, magnetostratigraphic and isotopic insights into the Middle Eocene climatic optimum in low latitudes. Palaeogeogr Palaeoclimatol Palaeoecol 297: 670–682

Erbacher J, Mosher DC, Malone MJ et al (2004) In: Proceedings of the Ocean drilling Program Initial Reports 207, College station, Texas (ODP)

Fioroni C, Villa G, Persico D, Jovane L (2015) Middle Eocene-Lower Oligocene calcareous nannofossil biostratigraphy and palaeoceanographic implications from Site 711 (Equatorial Indian Ocean). Mar Micropaleontol 118:50–62

Fleisher RL (1974) Cenozoic planktonic foraminifera and biostratigraphy, Arabian Sea, Deep Sea Drilling Project, leg 23-A. In: Whitmarsh et al Initial reports, Deep Sea Drilling Project, vol 23. Government Printing Office, Washington, D.C, pp 1001–1072

Gallazo FB, Thomas E, Pagani M, Warren C, Luciani V, Giusberti L (2014) The Middle Eocene Climatic Optimum (MECO): A multiproxy record of Paeeoceanographic changes in the southeast Atlantic (ODP Site 1263, Walvis Ridge). Paeoceanography 29:1143–1151

Gallazo FB, Thomas E, Giusberti L (2015) Benthic foraminiferal response to the Middle Eocene Climatic Optimum (MECO) in the south-eastern Atlantic (ODP Site 1263). Palaeogeogr Palaeoclimatol Palaeoecol 417:432–444

Hiemen M, Frerichs WE, Vincent E, (1974) Paleogene planktonic foraminifera from the western tropical Indian Ocean, Site 237, Deep Sea Drilling Project, Leg, 24. In: Fleicher et al (eds) Initial Report of the DSDP XXIV Project vol 24. U. S. Government Printing office, Washington D.C., pp 851–858

Houben AJP, Mourick AV, Montanari A, Coccini R, Brinkhuis H (2012) The Eocene-Oligocene transition: changes in sea level, temperature or both? Palaeogeogr Palaeoclimatol Palaeoecol 335–336:75–83

Javane L, Florindo F, Cocccioni R, Dinares-Turell J, Marsili A, Monechi S, Roberts AP, Spovieri M (2007) The Middle Eocene climate optimum event in the Contessa highway section, Umbrian Appenines, Italy. Geol Soc Am Bull 119:413–427

Katz ME, Miller KG, Wright JD, Wade BS, Browning JV, Cramer BS, Rosenthal Y (2008) Stepwise transition from the greenhouse to the Oligocene icehouse. Nat Geosci 1:329–334

Keller, G. (1982–1983) Biochronology and paleoclimatic implications of Middle Eocene to Oligocene foraminiferal faunas. Mar micropaleontol 7:463–486

Keller G (1983) Paleoclimatic analysis of Middle Eocene through Oligocene planktonic foraminiferal faunas. Palaeogeogr Palaeoclimatol Palaeoecol 43:73–94

Keller G, Macleod N, Barrera E (1992) Eocene-Oligocene faunal turnover in Planktic foraminifra and Antarctic glaciations. In: Prothero DR, Berggren WA (eds) Eocene-Oligocene climatic evolution. Princeton University Press, New Jersey, pp 1–28

Kennet JP, Shackleton NJ (1976) Oxygen isotopic evidence for the development of the psychrosphere 38 Mys. ago. Nature 260:513–515

Lear C, Bailey TR, Pearson PN, Coxall HK, Rosanthal Y (2008) Cooling and ice growth across the Eocene-Oligocene transition. Geology 36:252–254

Luciani V, Giusberti L, Agnini C, Fornaciari E, Rio D (2010) Ecological and evolutionary response of Tethyan planktonic foraminifera to the middle Eocene climatic optimum (MECO) from the Alano section (NE Italy). Palaeogeogr Palaeoclimatol Palaeoecol 292:82–95

Miller KG, Wright JD, Fairbanks RG (1991) Unlocking the icehouse Oligocene-Miocene Oxygen isotopes, eustasy and margin erosion. J Geophys Res 96:6829–6848

Miller KG, Browning JB, Aubrey MP, Wade BS, Katz ME, Kulpecz AA, Wright JD (2008) Eocene-Oligocene global climate and sea level changes: ST Stephens quarry, Alabama. Geol Soc Am Bull 121(1):34–53

Molina E, Monaco, Nochhi M, Parisi G (1986) Biostratigraphic correlation between the central Subbetic (Spain) and Umbro-Marchean (Italy) pelagic sequences at the Eocene/Oligocene boundary using foraminifera. In: Pomerol C, Premoli Silva I (eds) Terminal Eocene events: developments in Palaeontology and Stratigraphy vol 9. Elsevier Science Publishers BV, Amsterdam, pp 75–85

Molina E, Gonzalvo C, Keller G (1993) Eocene-Oligocene planktonic foraminiferal transition: extinctions, impacts and hiatuses. Geol Mag 130:483–499

Molina E, Gonzalvo C, Ortiz S, Cruz LE (2006) Foraminiferal turnover across the Eocene-Oligocene transition at Fuente Caldera, Southern Spain: no cause-effect relationship between meteorite impacts and extinctions. Mar Micropaleontol 58:270–286

Olsson RK, Hemlebem C, Pearson PN (2006) Taxonomy, biostratigraphy and phylogeny of Eocene *Dentoglobigerina*. In: Pearson PN, Olsson RK, Huber BT, Hemleben C, Berggren WA (eds) Atlas of planktonic foraminifera, no 41. Special Publication—Cushman Foundation, pp 401–408

Pearson PN (1996) Cladogenetic, extinctions and survivorship patterns from lineage phylogeny: the Paleogene planktonic foraminifera. Micropaleontology 42:179–188

Pearson PN, Mcmillan IN, Wade BS, Jones D, Coxall HK, Brown PR, Lear CH (2008) Extinction and environmental change across the Eocene-Oligocene boundary in Tanzania. Geology 36(2):179–182

Sclater JG, Anderson R, Thiede J (1977) Paleobathymetry and sediments of the Indian Ocean. In: Heirtzler JR, Bolli HM, Davies TA, Saunders JB (eds) Indian Ocean Geology and Biostratigraphy. American Geophysical Union, Washington, D.C., pp 25–39

Spofforth DGA, Agnini C, Palike H, Rio D, Fornaciari E, Giusberti L, Luciani V, Lanci L, Muttoni G (2010) Organic carbon burial following the Middle Eocene Climatic Optimum (MECO) in the central western Tethys. Paleocenography. https://doi.org/10.1029/2009PA001738

Wade BS (2004) Plaktonic foraminiferal biostratigraphy and mechanisms in the extinction of *Morozovella* in the late middle Eocene. Mar Micropaleontol 51:23–38

Wade BS, Pearson PN (2008) Planktonic foraminiferal turnover diversity fluctuations and geochemical signal across the Eocene-Oligocene boundary in Tanzania. Mar Micropaleontol 68:244–255
Wade BS, Pearson PN, Berggren WA, Palike H (2011) Review and revision of Cenozoic tropical planktonic foraminiferal biostratigraphy and calibration to the geomagnetic polarity and astronomical scale. Earth-Sci Rev 104:111–142
Wade BS, Houben AJP, Willamijn Quaijtaai W, Schouten S, Rosenthai Y, Miller KG, Katz ME, Wright JD, Brinkhuis H (2012) Multiproxy record of abrupt sea-surface cooling across the Eocene-Oligocene transition in the gulf of Mexico. Geology 40(2):159–162
Whitmarsh RB, Weser OE, Ali Syed, Boudreaux JE, Fleisher RL, Kidd RB, Jipa, Dan, Malik TK, Matter A, Nigrini C, Siddiquie HN, Stoffers P (1974) Initial Reports of the Deep Sea Drilling Project, vol 23. Government Printing Office, Washington D.C., p 1180
Zachos JC, Quinn RM, Salamy K (1996) High resolution (104 yr) deep sea foraminiferal isotope records at the Eocene Oligocene climate transition. Paleoceanography 11:251–266
Zanazzi A, Kohn MJ, Macfadden BJ, Terry DO (2007) Large temperature drop across Eocene-Oligocene transition in central North America. Nature 445:639–642

The Oligocene Corals Had Circumtropical Distribution

Patrali Sinha and Kalyan Halder

Abstract An analysis of the palaeobiogeographic distribution of the Oligocene corals of the world reveals strong specific endemism whereas large generic pandemism. Four palaeobiogeographic provinces are identified here based on this distribution: the Western Indian Province (WIP) represented by Kutch, the Mediterranean-Iranian Province (MIP) consisting of Greece, Italy and Iran, the Caribbean-Northern South American Province (CNSAP) composed of Antigua, Puerto Rico and Venezuela, and the Northwestern American Province (NAP) represented by the state of Washington, USA. Different basins within a province show some specific similarity whereas specific provincialism is nearly absolute. Genera show wide distribution—the WIP shows 82% similarity with the MIP and 36% with the CNSAP; the CNSAP has 69% similarity with the MIP. However, the NAP shows significant generic endemism. The tropical affinity, and wide and rapid dispersability of the Oligocene corals are evident from this distribution pattern. Dispersal beyond tropics was apparently limited. This explains the relatively higher endemism of the NAP. Generic exchange between the WIP and the MIP, both belonging to the Tethys Realm, has been known for gastropods. Affinity of these provinces with the CNSAP is worth noticing. This similarity reflects the presence of a trans-Atlantic current during the Oligocene. The circumtropical distribution of the coral genera also evinces protracted planktotrophic larval ontogeny. However, endemism in the species level indicates rapid evolution. This distribution pattern having tropical yet wide longitudinal extent of genera and provincialism of species resembles the present day distribution of corals. It reflects that the scleractinian coral biology had already attained modern aspects in the Oligocene.

Keywords Coral · Oligocene · Kutch · Palaeobiogeography · Tethys · Tropic

---

P. Sinha · K. Halder (✉)
Department of Geology, Presidency University, 86/1, College Street, 700073 Kolkata, India
e-mail: kalyan.geol@presiuniv.ac.in

P. Sinha
e-mail: patralisinha.geol@gmail.com

S. Bajpai et al. (eds.), *The Indian Paleogene*, Society of Earth Scientists Series,
https://doi.org/10.1007/978-3-319-77443-5_12

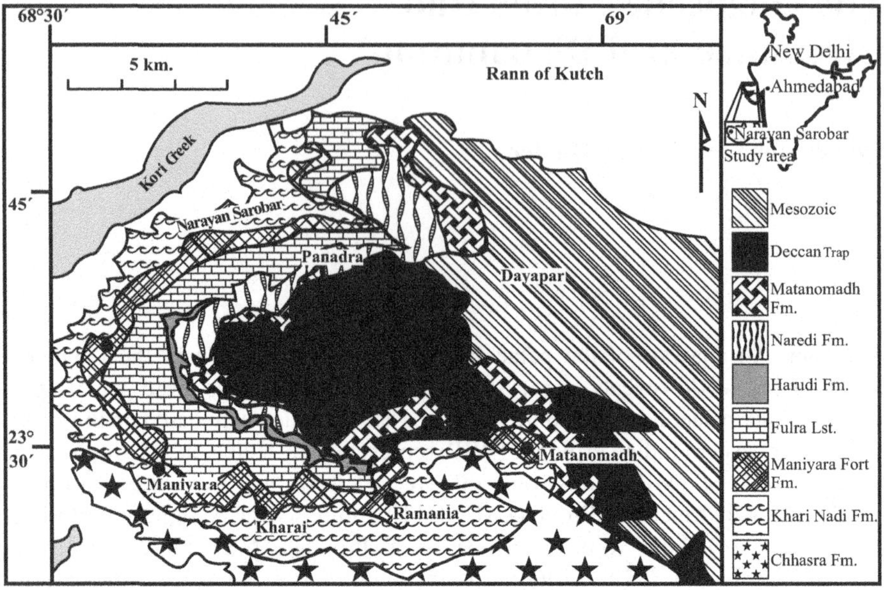

Fig. 1 Geological map of the Cenozoic of Kutch with coral collection sites marked by solid black circles

Introduction

The Oligocene marine sedimentary deposits of Kutch, Gujarat are known to house a diverse coral fauna (Biswas 1992). However, the composition of this fauna largely remains unknown to the palaeontological fraternity till date. We intend to make a rigorous systematic study of this coral fauna. Towards this goal we have collected corals from five localities that are dispersed along the entire transect of a narrow arcuate exposure of the Oligocene rocks at Kutch (Fig. 1).

The systematic study of our collection from one of the localities, which is exposed near village Ramania, has so far been completed. It has been observed during this study that almost all the species from Kutch are new whereas nearly all the genera to which they belong were reported from other parts of the world. The genera are known mainly from the Middle-East, southern European countries and the Caribbean archipelago. This strong specific endemism and large generic pandemism of the Kutch fauna prompted us to analyse global distribution patterns of the Oligocene reef corals. We would like to see through this analysis:

(1) whether this pattern of specific endemism and generic pandemism is true for other contemporaneous coral faunas from different areas of the world,
(2) if formal palaeobiogeographic units can be erected based on coral faunas, and
(3) if the units can be compared to the existing palaeobiogeographic units, mainly based on molluscs.

Table 1 Geographic areas from where taxonomic data have been gathered along with number of species and genus known from each

Name of the Area	No. of species	No. of genera
Kutch	24	11
Iran	61	38
Greece	38	27
Italy	20	18
Antigua	45	26
Venezuela	32	18
Puerto Rico	25	13
Washington	46	35
California	14	7

Further, we intend to investigate the factors that were responsible for the development of paleobiogeographic distribution patterns of the Oligocene corals and possible routes and mechanism of their dispersal.

Materials and Methods

The Oligocene deposits of Kutch are included in the Maniyara Fort Formation (Biswas 1992). It is composed of claystone, foraminiferal limestone and thick coral biostrome. This formation is underlain by a thick foraminiferal limestone formation (Fulra Limestone) and overlain by the Khari Nadi Formation, which is composed of shale, siltstone and sandstone. The rocks of the Maniyara Fort Formation outcrop in a nearly continuous arc, roughly paralleling the present day coastline (Fig. 1).

Our systematic study revealed that the coral fauna from Ramania (Fig. 1) is constituted of 24 species belonging to 11 genera. We have gathered systematic and geographic information of the Oligocene corals from some major publications pertaining to different areas of the world. The sources of information used in this study are: Vaughan (1900), Durham (1942), Geister and Ungaro (1977), Schuster and Wielandt (1999), Budd (2000), Schuster (2002), Kolodziej and Marcopoulou-Diacantoni (2003), Johnson (2007), Johnson et al. (2009), Champagne (2010) and López-Pérez (2012).

The areas for which taxonomic data could be gathered are Iran, Italy, Greece, USA, Puerto Rico, Antigua and Venezuela (Table 1). All the data were analysed in specific and generic levels to see the degree of compositional similarities between different geographic areas (Tables 2, 3 and 4). The data were plotted in palaeogeographic maps of Smith et al. (1994) (Figs. 2 and 3). Formal palaeobiogeographic units are erected based on the global distribution of corals.

Table 2 Similarity matrix showing specific similarities among different geographic areas

	Iran	Greece	Italy	Antigua	Venezuela	Puerto Rico	Kutch	Washington	California
Iran	61	11	4	1	0	3	2	1	0
Greece		38	7	0	0	5	1	1	0
Italy			20	0	0	3	0	0	1
Antigua				45	17	5	0	0	0
Venezuela					32	6	0	0	0
Puerto Rico						25	0	3	0
Kutch							24	0	0
Washington								46	1
California									14

Table 3 Similarity matrix showing generic similarities among different geographic areas

	Iran	Greece	Italy	Antigua	Venezuela	Puerto Rico	Kutch	Washington	California
Iran	38	19	12	19	13	12	7	10	1
Greece		27	11	14	9	9	3	5	1
Italy			18	11	9	9	8	4	1
Antigua				26	16	12	4	9	2
Venezuela					18	10	3	8	2
Puerto Rico						13	2	6	2
Kutch							11	1	1
Washington								35	6
California									7

Table 4 Similarity matrix showing specific/generic similarities among different provinces

	Western Indian Province (WIP)	Mediterranean-Iranian Province (MIP)	Caribbean Northern South American Province (CNSAP)	Northwestern American Province (NAP)	California
WIP	24/11	2/9	0/4	0/1	0/1
MIP		98/49	8/20	1/11	1/2
CNSAP			76/29	3/9	0/3
NAP				46/35	1/6
California					14/7

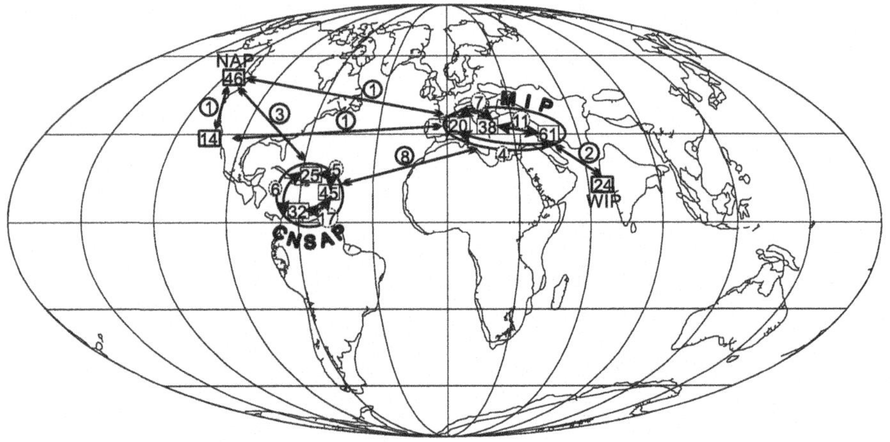

Fig. 2 Global distribution of coral species. Intra- and inter-provincial similarity in number of species are shown

Results and Discussion

Species Level Analysis

1. Twenty Four species have been identified from the locality of Ramania, Kutch, of which only 2 are already known—*Astrocoenia nana* Reuss, 1868 from Greece and Iran, and *Astrocoenia* cf. *bistellata* (Catullo, 1856) from Iran. The rest are new species (Table 2).
2. Data on the global distribution of coral species reveal high degree of similarity among faunas from Italy, Iran and Greece (Table 2; Fig. 2).
3. Substantial similarity also characterises the distribution of corals among Antigua, Venezuela and Puerto Rico, the three localities bordering the Caribbean Sea for which we could gather data (Table 2; Fig. 2). Puerto Rico shares some similarity

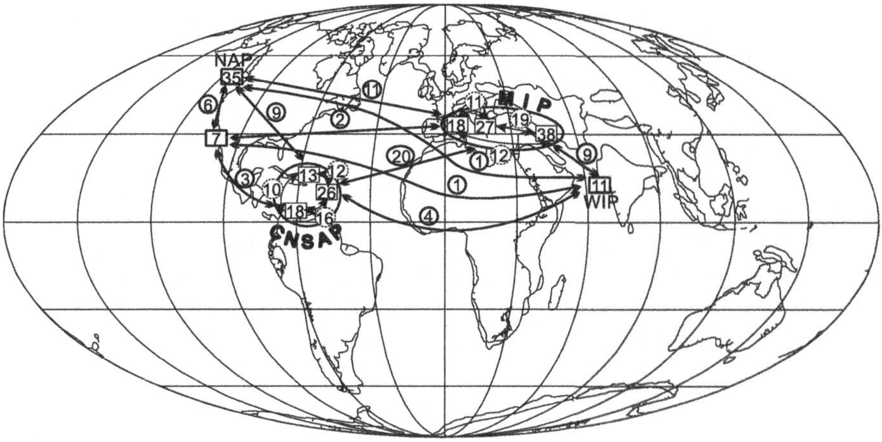

Fig. 3 Global distribution of coral genera. Intra- and inter-provincial similarity in number of genera are shown

with the areas of Europe and Middle East to which Antigua and Venezuela share almost nothing.

4. The state of Washington, USA is the only other area from where substantial number of Oligocene corals are known, almost all of which are endemic (Table 2; Fig. 2).
5. The Oligocene corals are also known from a more tropical part of USA, i.e. California. Only 14 species are known, which also show strong endemism (Table 2; Fig. 2).

Strong specific endemism of the Oligocene corals from Kutch necessitates establishment of a palaeobiogeographic province. The Western Indian Province (WIP) was erected by Harzhauser (2007) based mainly on the Oligocene gastropods from Pakistan. Harzhauser et al. (2009) included the Oligocene gastropods from Kutch in this province based on faunal similarity. Contemporaneous mollusc faunas from Kutch and other adjoining areas of western India have long been known to bear strong similarity with the Pakistan fauna (Sowerby 1840; D'Archiac and Haime 1854; Vredenburg 1925, 1928) and were always considered together with the latter. Corals are not known from the Oligocene deposits of Pakistan. Here we designate the Oligocene corals from Kutch to the same palaeobiogeographic unit, i.e. the WIP in order to avoid multiplication of similar unit names.

Large sharing of species between Iran, Greece and Italy evinces close biogeographic relationship. Italy shares 35% of its coral species with Greece and Greece shares about 29% with Iran. This pattern of specific similarity of the coral faunas correspond to that of the Oligocene gastropods. Harzhauser et al. (2002) erected the Mediterranean Iranian Province (MIP) to include basins of northern Italy, Greece and central Iran based on the specific similarity of Oligocene gastropods. Here, we upheld the same province name for contemporaneous coral faunas.

Significant specific similarity exists among Antigua, Venezuela and Puerto Rico. Puerto Rico shares 20% and 24% of its species with Antigua and Venezuela respectively. About 53% of Venezuelan coral species are also known from Antigua. We erect here the Caribbean-Northern South American Province (CNSAP) to include these three areas based on similarity of the Oligocene coral faunas.

Very strong endemism of the coral fauna from Washington necessitates erection of a new faunal province. We christen this province as the Northwestern American Province (NAP). Strong endemism of the Californian fauna also suggests the presence of a separate biogeographic province. However, the information from this area is relatively poor and only 14 species are known. Hence, we refrain from establishing a formal biogeographic unit based on such poor data and refer to the Californian fauna informally.

Genus Level Analysis

1. The 24 species that we have so far identified from Kutch, western India belong to 11 genera. Most of the genera are known from Italy, Iran and Greece. The Kutch assemblage also shows some generic similarity to the faunas from Antigua, Venezuela and Puerto Rico (Table 3).
2. Intraprovincial generic similarity among the basins is expectedly very high (about 70–90%) (Table 3). Interestingly, interprovincial generic similarity is also quite high. The CNSAP and the WIP share about 69% and 82% of their genera with the MIP respectively. The WIP bears about 36% generic similarity with the CNSAP (Table 4).
3. The NAP shows descent similarity with the MIP (about 31%) and the CNSAP (about 26%). It has only one common genus with the WIP (Table 4). Generic similarities among the CNSAP, the MIP and the WIP are, however, much higher.
4. Out of 7 genera that were recorded from California, 6 are also known from the NAP. California also shares about 43% of its genera with the CNSAP and about 29% with the MIP. One genus is common between California and Kutch.

The strong generic similarity between the WIP and the MIP parallels the trend shown by contemporaneous gastropod molluscs (Harzhauser et al. 2009). This trend reflects Tethyan heritage and indicates that considerable connectivity still existed between these provinces during the Oligocene. These provinces were included in the Western Tethys Region of the Tethys Realm based on benthic mollusc fauna (Harzhauser et al. 2002 and references therein).

The strong Tethyan affinity of the Oligocene coral fauna from the CNSAP is worth noticing. The relationship between the CNSAP and the Tethyan provinces, especially the MIP lying on the other shore of Atlantic, has never been adequately explored (see Harzhauser et al. 2002). Harzhauser et al. (2002) showed significant commonness in the gastropod composition of the Caribbean and adjacent tropical areas with that of the MIP. Hence, the CNSAP should be considered as the western fringe of the Tethys Realm.

The Washington coral fauna appears to be unique in composition. It shares relatively less with any of the other provinces. The California fauna, which shares majority of its genera with the NAP, has little in common with the latter when specific composition is considered. The uniqueness of the NAP calls for its inclusion in a different region or even realm. The question of affinity of the California fauna remains unresolved because of the relatively poor nature of information. Unfortunately, we could not access information on the faunas from Indo-Pacific basins like Java and Borneo (Gerth 1921, 1923, 1925). This information can be crucial in delineating palaeobiogeographic relationship of the NAP and deciphering possible routes of dispersal of its constituent corals.

Remarks

The Oligocene coral species are seldom found beyond province boundaries. Most of them are also often endemic to a single basin. In contrast, majority of the genera appear to migrate great distances beyond province territories, often even crossing significant geographic barriers. This strong specific endemism and large generic pandemism indicate wide migration followed by rapid allopatric speciation.

The global distribution of the Oligocene corals clearly shows wide longitudinal vis-a-vis relatively restricted latitudinal occurrence mainly within palaeo-tropics and subtropics. Present day reef corals are known to be strongly sensitive to climatic factors and restricted to the tropical zone. Apparently, this attribute was already acquired by the Oligocene corals. The diverse coral fauna from Washington, USA, which flourished in higher latitudes, however is an exception. Presence of corals there perhaps indicates that the prevailing climate of the Oligocene was warmer and more equable than the present day. Palaeo-temperature data suggests the same (Hansen et al. 2013). However, the unique nature of the composition of the Washington fauna with large specific as well as generic endemism indicates the presence of a strong climatic barrier to the dispersal of the tropical corals to higher latitudes. Only more cold-tolerant genera could permeate into the higher latitudes and survive.

The presence of a seaway connection and regular migration between the WIP and the MIP during the Paleogene has been known for a long time in relation to the molluscs (Eames 1951, 1952; Iqbal 1969a, b, 1972; Halder 2012; Halder and Sinha 2014; Halder and Bano 2015; Harzhauser et al. 2002, 2009). The migration took place through the relict Tethys seaway. It may be pointed out that many early Paleogene molluscs of western India appear in younger strata of the Mediterranean Europe. However, Iqbal (1969a, b) declined to consider a westward migration of the fauna in the absence of more definitive evidences. Harzhauser et al. (2002) demonstrated the presence of an easterly surface current in the Tethys that bathed the western shores of the Indian Peninsula during the Paleogene (see also Popov 1993). The information analysed in the present study is not adequate for the reconstruction of possible direction of migration of the coral fauna between the WIP and the MIP.

The presence of a coral fauna in the Western Atlantic CNSAP, similar to that of the MIP and the WIP, indicates an extension of the Tethyan faunal influence across the Atlantic Ocean and hence, the presence of a trans-Atlantic current system. The direction of migration, again, cannot be inferred from the present dataset. An analysis of the information on the contemporaneous gastropods also could not conclusively reveal the migration direction. However, a westward current from the MIP to the CNSAP appears to be the stronger possibility in terms of the gastropod distribution (Harzhauser et al. 2002). The relatively lower similarity between the WIP and the CNSAP may be attributed to the large geographic distance between these provinces and the relatively weak present state of the knowledge on the western Indian fauna.

It is more difficult to understand the relationship of the Pacific coastal basins of the USA with other provinces in the absence of data from other Pacific basins. California, however, having about 43% generic similarity with the CNSAP perhaps got its coral fauna from the latter. Presumably, this migration had taken place through the connection that existed at that time between the Caribbean Sea and the Pacific Ocean. The Isthmus of Panama severed this tie later. Washington, in spite of its unique faunal composition, might have got early settlers from California. Relatively cold-tolerant forms from the latter area perhaps migrated to Washington and flourished there in a relatively unusual setting as opportunist taxa. Data from the western Pacific archipelago is required to assess if and to what extent faunal similarity existed between the two shores of the Pacific. However, a migration route across the huge Pacific Ocean is difficult to perceive in the absence of hopping sites.

Conclusion

The global distribution of the Oligocene coral faunas indicates an almost circum-tropical occurrence in nearly all coastal basins including the western Pacific. Significant absence of corals was from the tropical African countries. Whether it was real or reflects lack of study is not known. The distribution shows very strong Tethyan affinity of most of the provinces—including that on the western shore of the Atlantic and even the tropical eastern Pacific—elucidating a trans-Atlantic surface current and possible passage through the sea between the two American continents. However, strong endemism of the coral species to the provinces and even to the basins reflects rapid speciation was operative in allopatry.

It is interesting to note that several of the Oligocene coral genera are still extant. The Oligocene corals had a climate controlled distribution mainly in the tropical and the subtropical areas. They were generally associated with fairly shallow water shelf sediments (Bosellini 1998). They also had longitudinally wide geographic distribution. All these features correspond to the distribution of the present day reef corals (Veron 2016). Shallow water occurrence of the present day corals in the tropical zone is known to be controlled by the symbiotic association with photosynthetic organisms. Protracted planktotrophic larval ontogeny of majority of the reef corals is responsible for their wide geographic distribution (Veron 2013). Apparently, these modern aspects of the coral biology had been attained by the reef corals already during the Oligocene.

Acknowledgements Tahthagata Roy Choudhury helped in drawing. KH got financial support from the Department of Science and Technology, India (Project No. SR/S4/ES-653/2012).

Appendix 1

Global distribution of Coral species reported from different basins/provinces of the Oligocene

Global distribution of Coral species reported from different basins/provinces of the Oligocene

Sl. no.	Species name	MIP			CNSAP			WIP	California	NAP
		Greece	Italy	Iran	Antigua	Venezuela	Puerto Rico	Kutch		Washington
1	*Astrocoenia nana*	√		√				√		
2	*Astrocoenia* cf. *bistellata*			√				√		
3	*Astrocoenia* cf. *zitteli*			√						
4	*Astrocoenia portoricensis*				√	√				
5	*Astrocoenia guantanamensis*				√	√				
6	*Astrocoenia sp. B*				√					
7	*Agathiphyllia sp. 1*			√						
8	*Agathiphyllia sp.2*			√						
9	*Agathiphyllia gregaria*	√	√				√			
10	*Agathiphyllia tenuis*				√	√	√			
11	*Agathiphyllia hilli*				√		√			
12	*Agathiphyllia browni*				√					
13	*Agathiphyllia splendens*				√					
14	*Agathiphyllia antiguensis*					√	√			
15	*Agathiphyllia anguillensis*						√			
16	*Agathiphyllia bosniaca*						√			
17	*Agathiphyllia robusta*						√			
18	*Agathiphyllia sp. 3*				√					
19	*Antiguastrea cellulosa*				√	√	√			
20	*Antiguastrea elegans*						√			
21	*Antiguastrea alveolaris*						√			
22	*Antiguastrea sp.*			√						
23	*Antiguastrea lucasiana*	√	√							
24	*Actinacis rollei*		√	√						
25	*Actinacis sp. A*				√					
26	*Astreopora stellaris*			√						
27	*Astreopora meneghiniana*	√		√						
28	*Astreopora tecta*		√							
29	*Astreopora decaphyllia*	√								
30	*Astreopora antiguensis*				√					
31	*Astreopora goethalsi*					√				
32	*Alveopora sp.*			√			√			
33	*Alveopora? sp.*			√						
34	*Alveopora tampae*				√	√				
35	*Alveopora sp. A*				√					
36	*Alveopora rudis*		√							
37	*Astrangia persica*			√						
38	*Acanthastrea sp.*			√						
39	*Asterosmilia sp.*			√						
40	*Acropora saludensis*				√	√				

41	*Acropora sp. A*					√				
42	*Acropora sp.*	√	√	√						
43	*Colpophyllia eocaenica*			√						
44	*Colpophyllia longicollis*			√						
45	*Colpophyllia maeandrinoides*	√								
46	*Colpophyllia willoughbiensis*				√	√				
47	*Caulastrea pseudoflabellum*	√								
48	*Caulastrea sp.*	√					√			
49	*Caulastraea farsis*			√						
50	*Caulastrea fusinieri*		√							
51	*Caulastraea portoricensis*				√					
52	*Cyathoseris sp.*	√	√							
53	*Cyathoseris* cf. *appennina*	√								
54	*Cyathoseris hypocrateriformis*	√								
55	*Ceratotrochus (Conotrochus) sp.*			√						
56	*Cricocyathus annulatus*			√						
57	*Ceriosmilia iranica*			√						
58	*Cereiphyllia tenuis*		√							
59	*Diploastrea coronata*			√						
60	*Diploastrea costata*			√						
61	*Diploastrea crassolamellata*				√	√				
62	*Diploastrea magnifica*				√	√				
63	*Diploastrea nugenti*				√					
64	*Diploastrea sp. T*				√					
65	*Diploastrea sp.*					√				
66	*Diploria antiguensis*				√					
67	*Diploria sp.*	√								
68	*Euphyllia caliculata*	√								
69	*Favites insignis*			√						
70	*Favites cf.inaequiseptata*			√						
71	*Favites macrocalyx*	√								
72	*Favites oligocenia*	√								
73	*Favites polygonalis*				√					
74	*Favites ambigua*	√								
75	*Favites sp.*			√			√			
76	*Fungophyllia sp. A*				√					
77	*Favia sp. 2*	√								
78	*Favia sp. 1*	√								
79	*Gardinoseries? sp.*			√						
80	*Goniopora* cf. *nodulosa*	√		√						
81	*Goniopora sp.*	√					√			
82	*Goniopora sp. 2*	√		√						
83	*Goniopora sp. 3*			√						
84	*Goniopora sp. 4*			√						
85	*Goniopora rudis*		√							
86	*Goniopora imperatoris*				√	√				
87	*Galaxea sp.*			√						
88	*Goniastrea canalis*				√	√				
89	*Hydnophora solidior*			√						
90	*Hydnophora* cf. *inaequalis*			√						
91	*Hydnophora* cf. *rudis*			√						
92	*Hydnophora pulchara*	√								
93	*Hydnophora sp.*			√						
94	*Hydnophora sp. A*				√					
95	*Heliopora sp. A*				√					
96	*Ilariosmilia subcurvata*			√						
97	*Leptoseries irregularis*			√						
98	*Leptoseries* cf. *dinarica*									
99	*Leptoseris sp. C*				√					
100	*Leptoseris portoricensis*				√					
101	*Leptoseris alternas*			√						
102	*Leptoria* cf. *concentrica*			√						
103	*Leptoria bithecata*	√								
104	*Leptoria sp.*	√		√						
105	*Leptomussa variabilis*	√		√						

106	*Leptomussa sp.*		√	√						
107	*Montipora sp. A*			√	√					
108	*Montastrea sp.*	√		√						
109	*Montastraea canalis*				√	√	√			
110	*Montastraea endothecata*				√		√			
111	*Montastraea imperatoris*					√	√			
112	*Montastraea limbata*					√	√			
113	*Montastrea endothecata*						√			
114	*Montastraea sp.*					√				
115	*Montastrea guettardi*		√							
116	*Pironastraea antiguensis*				√					
117	*Pindosmilia brunni*	√								
118	*Porites regularis*				√					
119	*Porites macdonaldi*				√					
120	*Porites portoricensis*				√	√				
121	*Porites microscopica*				√					
122	*Porites reguiaris*				√					
123	*Porites baracoaensis*				√	√				
124	*Porites trinitatis*					√				
125	*Porites waylandi*					√				
126	*Pocillopora arnoldi*					√				
127	*Porites sp.*	√	√	√			√			√
128	*Plocophyllia bartai*		√							
129	*Pavona bronni*	√	√							
130	*Platycoenia iranica*	√		√						
131	*Plesiastrea sp.*	√		√						
132	*Placosmiliopsis fimbriatus*			√						
133	*Placosmiliopsis multisinuous*			√						
134	*Rhizangia sp.*	√								
135	*Siderastrea conferta*				√	√				
136	*Siderastrea (Siderofungia) sp.*			√						
137	*Solenastrea sp. B*					√				
138	*Stephanocyathus (Odontocyathus) sp.*			√						
139	*Stephanocoenia duncani*				√	√				
140	*Stylophora affinis*					√				
141	*Stylophora granulate*					√				
142	*Stylophora undata*					√				
143	*Stylophora imperatoris*				√					
144	*Stylophora minor*				√					
145	*Stylophora sp. C*				√					
146	*Stylophora sp.*	√	√				√			
147	*Stylophora thirsiformis*			√						
148	*Stylofora conferta*	√								
149	*Stylophora cf. sokkohensis*			√						
150	*Acanthastrea n. sp.1*							√		
151	*Acanthastrea n. sp. 2*							√		
152	*Stylaster antiquus*		√							
153	*Tethocyathus persicus*	√								
154	*Trachyphyllia sp.*					√				
155	*Turbinaria sp.*	√								
156	*Turbinaria* cf. *sitaensis*			√						
157	*Tarbellastrea carryensis*			√						
158	*Pavona n. sp. 1*							√		
159	*Pavona n. sp. 2*							√		
160	*Pavona n. sp. 3*							√		

References

Biswas SK (1992) Tertiary stratigraphy of Kutch. J Palaeont Soc India 37:1–29

Bosellini FR (1998) Diversity, composition and structure of Late Eocene shelf-edge coral associations (Nago Limestone, Northern Italy). Facies 39:203–226

Budd AF (2000) Diversity and extinction in the Cenozoic history of Caribbean reefs. Coral Reefs 19:23–35

161	*Tarbellastrea hamedanii*	√		√						
162	*Tarbellastrea* cf. *distans*			√						
163	*Balanophyllia elongate*								√	
164	*Stylophora ponderosa*									√
165	*Stephanocyathus holcombensis*									√
166	*Flabellum hertleini*									√
167	*Archohelia weaveri*									√
168	*Astrangia clarki*									√
169	*Balanophyllia fulleri*									√
170	*Caryophyllia woodmanensis*									√
171	*Favia n. sp. 1*							√		
172	*Favia n. sp. 2*							√		
173	*Favia n. sp. 3*							√		
174	*Favia n. sp. 4*							√		
175	*Favia n. sp. 5*							√		
176	*Favia n. sp. 6*							√		
177	*Favia n. sp. 7*							√		
178	*Favia n. sp. 8*							√		
179	*Favia n. sp.9*							√		
180	*Paracyathus sp.*									√
181	*Stylaster milleri*									√
182	*Trochocyathus townsendensis*									√
183	*Tubastrea nomlandi*									√
184	*Dendrophyllia hannibali*									√
185	*Caryophyllia blakeleyensis*									√
186	*Balanophyllia teglandae*									√
187	*Montastrea n. sp. 1*							√		
188	*Eusmilia bainbridgensis*									√
189	*Paracyathus sp.*									√
190	*Platycyathus? sp.*									√
191	*Siderastrea washingtonensis*									√
192	*Flabellum rhomboideuui*									√
193	*Plesiastrea n. sp. 1*							√		
194	*Turbinolia insiguitica*									√
195	*Steriphonotrochus pulcher*									√
196	*Oculiua aldrichi*									√
197	*Oculiua harrisi*									√
198	*Balanophyllia caulifera var. nuiltigrauosa*									√
199	*Trochocyathus californianus*								√	
200	*Stylophora minutissima*									√
201	*Astrocoenia pumpellyi*									√
202	*Flabellum remoudianum*								√	
203	*Flabellum califonicum*								√	
204	*Trochocyathus striatus*								√	
205	*Trochocyathus zitteli*								√	
206	*Trococyatbus stantoni*								√	
207	*Pocillopora damicornis*								√	
208	*Pocillopora meandrina*								√	
209	*Siderastrea californica*								√	
210	*Astreopora occidentalis*									√
211	*Astreopora sanjuanensis*									√
212	*Goniopora n. sp. 1*							√		
213	*Pocillopora verrucosa*								√	
214	*Siderastrea clarki*								√	
215	*Pocillopora sp. 1*									√
216	*Hydnophora n. sp. 1*							√		
217	*Hydnophora n. sp. 2*							√		
218	*Colpophyllia sp.*						√			√
219	*Montipora schencki*									√
220	*Pavona sp.*		√						√	

Champagne TAN (2010) Oligocene coral evolution in Puerto Rico and Antigua: morphometric analysis of Agathiphyllia, Antiguastrea, and Montastraea. Master of Science thesis, University of Iowa. http://ir.uiowa.edu/etd/1128

221	*Turbinolia sp.*								√	√
222	*Platycoenia sp.*									√
223	*Astrangia sp.*									√
224	*Platytrochus sp.*									√
225	*Discotrochus sp.*									√
226	*Madracis sp.*									√
227	*Dichocoenia sp.*									√
228	*Stephanocoenia sp.*									√
229	*Spenotrochus sp.*									√
230	*Leptoria n. sp. 1*							√		
231	*Endopachys sp.*									√
232	*Eupsammia sp.*									√
233	*Euphyllia sp.*		√							
234	*Galaxea sp.*		√							
235	*Montipora sp.*			√						
236	*Rhabdophyllia sp.*			√						
237	*Stylocoenia sp.*			√						
238	*Pocillopora sp.*				√	√				
239	*Siderastrea sp.*						√			
240	*Astreopora sp.*						√			
241	*Flabellum sp.*						√			√
242	*Astrohelia sp.*									√
243	*Eupsammid sp.*									√
244	*n. gen. 1 n. sp. 1*							√		
245	*n. gen. 2 n. sp. 1*							√		
	Total	**38**	**20**	**61**	**45**	**32**	**25**	**24**	**14**	**46**

D' Archiac V, Haime J (1854) Description des animaux fossiles du Groupe Nummulitique de l'Inde, Seconde Livraison: Mollusques, Gide et J. Baudry, Paris, France

Durham JW (1942) Eocene and Oligocene coral faunas of Washington. J Paleont 16:84–104

Eames FE (1951) A contribution to the study of the Eocene in western Pakistan and western India. B. The description of the Lamellibranchia from standard sections in the Rakhi Nala and Zinda Pir areas of the western Punjab and in the Kohat District. Phil. Trans Royal Soc London B Bio Sci 235:311–482

Eames FE (1952) A contribution to the study of the Eocene in western Pakistan and western India, C. The description of the Scaphopoda and Gastropoda from standard sections in the Rakhi Nala and Zinda Pir areas of the western Punjab and in the Kohat district. Phil. Trans Royal Soc London B Bio Sci 236:1–168

Geister J, Ungaro S (1977) The Oligocene coral formations of the ColliBerici (Vicenza, Northern Italy). Eclogae Geol Helv 70:811–823

Gerth H (1921) Die Fossilien von Java auf Grund einer Sammlung von Dr. R.D.M. Verbeek und von anderen bearbeitet durch Dr. K.Martin,Professor der Geologie an der Universitat zu Leiden. Anthozoen von Java und die Mollusken der Njalindungschichten, erster Teil. Samml. Geol. Reichs-Mus. Leiden, N.F.1, 2. Abt Heft 3:387–445

Gerth H (1923) Die Anthozoen des Jungtertiars von Borneo. Samml Geol Reichs-Mus Leiden. 10 (Ser. 1):37–136

Gerth H (1925) Jungtertiare Korallen von Nias, Java und Borneo, nebst einer Uebersicht uber die aus dem Kanozoikum des indischen Archipels bekannten Arten. Leidsche Geol Meded 1:22–81

Halder K (2012) Cenozoic fossil nautiloids (Cephalopoda) from Kutch, Western India. Palaeoworld 21:116–130

Halder K, Bano S (2015) Cenozoic Corbulidae (Bivalvia, Mollusca) from the Indian subcontinent—-palaeobiogeography and revision of three species from Kutch. India Arab J Geosci 8:2019–2034

Halder K, Sinha P (2014) Some Eocene cerithioids (Gastropoda, Mollusca) from Kutch, Western India, and their bearing on palaeobiogeography of the Indian subcontinent. Hindawi Publishing Corporation. Paleont J 2014:11 pages (Article ID 673469). http://dx.doi.org/10.1155/2014/673469

Hansen J, Sato M, Russell G, Kharecha P (2013) Climate sensitivity, sea level, and atmospheric carbon dioxide. Phil Trans R Soc A 371:20120294. https://doi.org/10.1098/rsta.2012.0294

Harzhauser M (2007) Oligocene and Aquitanian gastropod faunas from the Sultanate of Oman and their biogeographic implication for the early western Indo-Pacific. Palaeontographica 280:75–121

Harzhauser M, Piller WE, Steininger FF (2002) Circum-Mediterranean Oligo-Miocene biogeographic evolution—the gastropods' point of view. Palaeogeogr Palaeoclim Palaeoeco 183:103–133

Harzhauser M, Reuter M, Piller WE, Berning B, Kroh A, Mandic O (2009) Oligocene and Early Miocene gastropods from Kutch (NW India) document an early biogeographic switch from Western Tethys to Indo-Pacific. Paläont Z 83:333–372

Iqbal MWA (1969a) Bibliography of tertiary pelecypod and gastropod species of West Pakistan. Rec Geol Surv Pakistan 18:1–63

Iqbal MWA (1969b) The Tertiary pelycopod and gastropod fauna from Drug, Zindapir, Vidor (district D.G. Khan), Jhalar and Chharat (district Campbellpore), West Pakistan. Mem Geol Surv Pak Palaeontol Pak 6:1–77

Iqbal MWA (1972) Palaeocene bivalve and gastropod fauna from Jherruk-Lakhra-Bara Nai (Sind), Salt Range (Punjab) and Samana Range (N. W. F. P.). Pak Palaeontol Pak 9:1–105

Johnson KG (2007) Reef-coral diversity in the Late Oligocene Antigua Formation and temporal variation of local diversity on Caribbean Cenozoic Reefs. In: Hubmann B, Piller WE (eds) Fossil Corals and Sponges. Proceedings of the 9th international symposium on Fossil Cnidaria and Porifera. Osterr. Akad. Wiss., Schriftenr. Erdwiss. Komm., vol 17, pp 471–491

Johnson KG, Sanchez-Villagra MR, Aguilera OA (2009) The Oligocene-Miocene Transition on coral reefs in the Falcó N Basin (NW Venezuela). Palaios 24:59–69

Kolodziej B, Marcopoulou-Diacantoni A (2003) Late Eocene-Oligocene corals from Evros (Thrace Basin, NE Greece). Ber Inst Geol Paläont K.-F.-Univ Graz 7:45

López-Pérez RA (2012) Late Miocene to Pleistocene reef corals in the Gulf of California. Bull Am Paleont 383:1–78

Popov SV (1993) Zoogeography of the Late Eocene basins of Western Eurasia based on bivalve molluscs. Strat Geol Correl 2:103–118

Schuster F (2002) Taxonomy of Oligocene to Early Miocene scleractinian corals from Iran, Egypt, Turkey, and Greece, vol 239. Courier Forschungsinstitut Senckenberg, pp 1–161

Schuster F, Wielandt U (1999) Oligocene and Early Miocene coral faunas from Iran: palaeoecology and palaeobiogeography. Int J Earth Sci 88:571–581

Smith AG, Smith DG, Funnel DM (1994) Atlas of Mesozoic and Cenozoic Coastlines. Cambridge University Press, Cambridge

Sowerby JC (1840) Explanations of the plates and wood-cuts. Plates XX to XXVI, to illustrate Capt. Grant's Memoir on Cutch. Trans Geol Soc London 5:1–289

Vaughan TW (1900) The Eocene and lower Oligocene coral faunas of the united states with descriptions of a few doubtfully Cretaceous species. Monographs of the USGS 39:1–263

Veron J (2013) Overview of the taxonomy of zooxanthellate Scleractinia. Zool J Linnean Soc 169:485–508

Veron J (2016) Patterns of diversity. Corals of the World—about corals and reefs. Australian Institute of Marine Sciences. Viewed online March 2016. http://www.coral.aims.gov.au/aboutcorals

Vredenburg E (1925) Description of Mollusca from the Post-Eocene Tertiary formations of North-Western India. 1. Mem Geol Surv India 50:1–350

Vredenburg E (1928) Description of Mollusca from the Post-Eocene Tertiary formations of North-Western India. 2. Mem Geol Surv India 50:351–463

Molluscan Biostratigraphy and Palynological Assemblage of Paleogene Disang Formation, Manipur, India

Y. Raghumani Singh, Umarani Sijagurumayum and B. P. Singh

Abstract In the present paper we tried to highlight Paleogene events of the Manipur region on the basis of fossil evidences. The Paleogene succession of the Manipur region is represented by the Disang and Barail groups that are mainly exposed in the Imphal valley. In the present investigation, 108 species (80 bivlaves and 28 gastropods) belonging to 51 genera of bivalves and 26 genera of gastropods have been recovered from the Upper Disang Formation of Changamdabi area of Imphal valley. Based on the faunal assemblage, three biostratigraphic zones namely Zone-I *Castocorbula* (*Parmicorbula*) *regulbiensis–Flemingostreapharoanum* var-*abivulina* with age ranging from Late Paleocene-Early Eocene (Thanetian-Ypresian), Zone-II *Vulsellapakistanica-Corbula* (*Varicorbula*) *daltoni* with age ranging from Middle Eocene-Late Eocene (Lutetian-Bartonian) and Zone-III *Callista* (*Callista*) *yawensis-Lucina Yawensis* with Late Eocene age (Priabonian) were identified in ascending order of stratigraphy. Two ecological events in the Upper Disang Formation are identified i.e. shallowing of the sea-aerated water corresponding to zones I and open sea conditions with well aerated warm water condition corresponding to zones II and III. The occurrence of dinoflagellates (Eocene), flasher bedding and lenticular bedding in the shale, siltstone and sandstone containing succession of the Upper Disang of Gelmoul quarry (Churachandpur) suggests a shallow marine (tidal flat) depositional environment. The overlying plant fossils containing Barail sequences (Oligocene) of Kaina Hills, Imphal valley indicate coastal/terrestrial environmental condition under warm and humid tropical climate. Thus, there is a shallowing up in the depositional environment from Upper Disang Formation to Barail Group.

Keyword Paleogene · Zones · Bivalves · Gastropods · Imphal valley
Upper Disang Formation

Y. Raghumani Singh (✉) · U. Sijagurumayum
Department of Earth Sciences, Manipur University, Imphal 795003, India
e-mail: yengmani@gmail.com

B. P. Singh
CAS in Geology, Banaras Hindu University, Varanasi 221005, India

S. Bajpai et al. (eds.), *The Indian Paleogene*, Society of Earth Scientists Series,
https://doi.org/10.1007/978-3-319-77443-5_13

Introduction

The Paleogene period (65–24 ma) is characterized by the drastic change in climate and the transformation from a greenhouse to an icehouse state. This transformation was not gradual and characterized by a numerous extreme transient climatic events (Zachos et al. 2001, Singh et al. 2017). The Paleogene environmental reconstruction and biostratigraphy has been carried out by a number of workers based on dinoflagellate cysts (Pross and Brinkhuis 2005; Sluijs et al. 2005); larger foraminiferal (Serra-Kiel et al. 1998; Berggren and Pearson 2005); molluscan fauna (Dockery III 1998; Hartman and Roth 1998); and calcareous nannoplankton (Martini 1971; Bukry 1973, 1975).

In India, Paleogene successions are exposed at a number of places in Tethys and Lesser Himalaya (Subathu and Dagshai/Lower Murree formations), hills of Meghalaya (Jaintia and Barail groups; Tura, Siju and Rewak formations of Garo Hills), Indo-Myanmar range (Ophiolite suite, Jaintia, Disang and Barail groups), Arunachal Pradesh (Yinkiong Group of Sing valley), Andaman (Ophiolite, Mithakhari, Andaman Flysch groups), Rajasthan (Palana, Khuiala, Bandah and Jagira formations), Gujarat (Madha, Berwati and Lakhapat formations) and Pondicherry (Niniyur and Pondicherry formations). Manipur is a part of the Indo-Myanmar Range which share international border with Myanmar-Burma on the eastern and southern sides. The remaining half of state has border line with the states of Nagaland, Assam and Mizoram respectively on the northern, western and south-western sides where Cretaceous and Tertiary sedimentary rocks occur in Manipur State (Table 1). They are associated with minor igneous and metamorphic rocks, associated with pelagic sediments like chert, limestone, shale and sandstone. The northeast and eastern part of the state is occupied by the older group of the Metamorphic Complex and the Ophiolite Mélange Zone. The central and western parts are composed of Tertiary flysch sediments including Disang, Barail, Surma and Tipam groups (Fig. 1). The lower Paleogene rocks are well exposed in the Imphal valley and eastern part of Manipur and represented by the Disang and Barail groups. The Disang Group is subdivided into two formations namely; Lower Disang and Upper Disang formations. The Lower Disang Formation is comprised mainly of sandstone and shale, while the Upper Disang Formation is represented by a rhythmic series of sandstone, siltstone and shale possessing fossils and slices of ophiolitic rocks (G.S.I 2011).

Considering the above, we here aim to provide biostratigraphy and depositional environment based on Paleogene molluscan fossils and palynofossils.

Materials and Methods

For molluscan investigation, the samples were drawn from three measured stratigraphic sections of the Upper Disang Formation namely which Canal Section (US_1

Table 1 Generalised stratigraphic succession of Manipur (After Soibam 2000; G.S.I 2011; Singh et al. 2015)

Age	Group	Formation	Lithology
Quaternary-Pleistocene	Alluvium		Clay, silt, sand, gravel, pebble, boulder deposits
Late Miocene	Tipam group		Mottled clay, mottled sandy clay, sandy shale, clayey shale and sandstone. Greenish to blue, moderate to coarse ferruginous sandstone with sandy shale, clay. Molasse type of deposits
Miocene to Late Oligocene	Surma group	Bokabil formation (~1400 m) Bhuban formation (~1400 m)	Shale, sandy shale, siltstone, ferruginous sandstone, massive to bedded ferruginous sandstone Alternations of sandstone and shale with minor conglomerates. Transitional characters from Flysch to molasse sediments
Unconformity			
Oligocene to Late Eocene	Barail group	Renji formation (~800 m) Jenam formation (~1200 m) Laisong formation (~1200 m)	Massive to thickly bedded sandstone. Alternations of shale and sandstone with carbonaceous matters. Intercalation of bedded sandstone with shale. Flysch sediments
Late Paleocene to Late Eocene	Disang group	Upper Disang formation(~2000 m)	Splintery shale and intercalation of shale, siltstone and sandstone showing occasionally rhythmite characters with fossils. Flysch sediments
Late Cretaceous to ?Late Paleocene		Lower Disang formation (~2000 m)	Dark grey to black shale with minor sandstone bands. Flysch sediments

(continued)

Table 1 (continued)

Age	Group	Formation	Lithology
Unconformity			
Cretaceous to ?Early Eocene	Ophiolite Mélange zone		Ultramafics with minor mafic—felsic rocks and marine sediments comprising radiolarian chert and limestone along with podiform chromitites
Unconformity			
(Pre-Mesozoic or Older)	Metamor-phic complex		Low to medium grade metamorphic rocks of various composition-phyllitic schist, quartzite, micaceous quartzite, quartz-chlorite-mica-schist and marble
(?) *Unconformity*			
?Early Mesozoic rocks or Pre-Cambrian rocks	Basement complex		Unseen

Section; GPS N24° 41′ 54.4″: E94° 06′ 05.9″), Maning Ching (US_2 Section; GPS 24° 41′ 34″N: E94° 05′ 07″E) and Khongjai Chingkhong Section (US_3 Section; GPS 24° 41′ 42″N: E94° 0.5′ 57″E) located in and around Changamdabi, Manipur (Fig. 1). The specimens of molluscan fossils recorded here are mainly external casts and moulds from Changamdabi area only. All the litho-units were searched for molluscan fossils and after a thorough search few fossiliferous beds were delineated in these sections (Figs. 2b, 3). The molluscan fauna has been studied in detail after the samples were prepared and cleaned for their identification with the help of relevant literature on fossil fauna.

Sixteen samples were collected for this study from the upper part of the Upper Disang Formation exposed at Gelmoul quarry (Fig. 1, 2a; GPS N24° 20′ 40.4″: E93° 39′ 39.6″). The samples were chemically processed by using HCl, HF and HNO_3 acids in the different duration. Out of these, nine samples were found productive for palynomorphs. They were detailed examined under microscope. The plant fossils has been recorded from the Barail sediment of Kaina Hills, Manipur (Fig. 2d; GPS N24° 40′ 45.9″; E94° 01′ 12.9″).

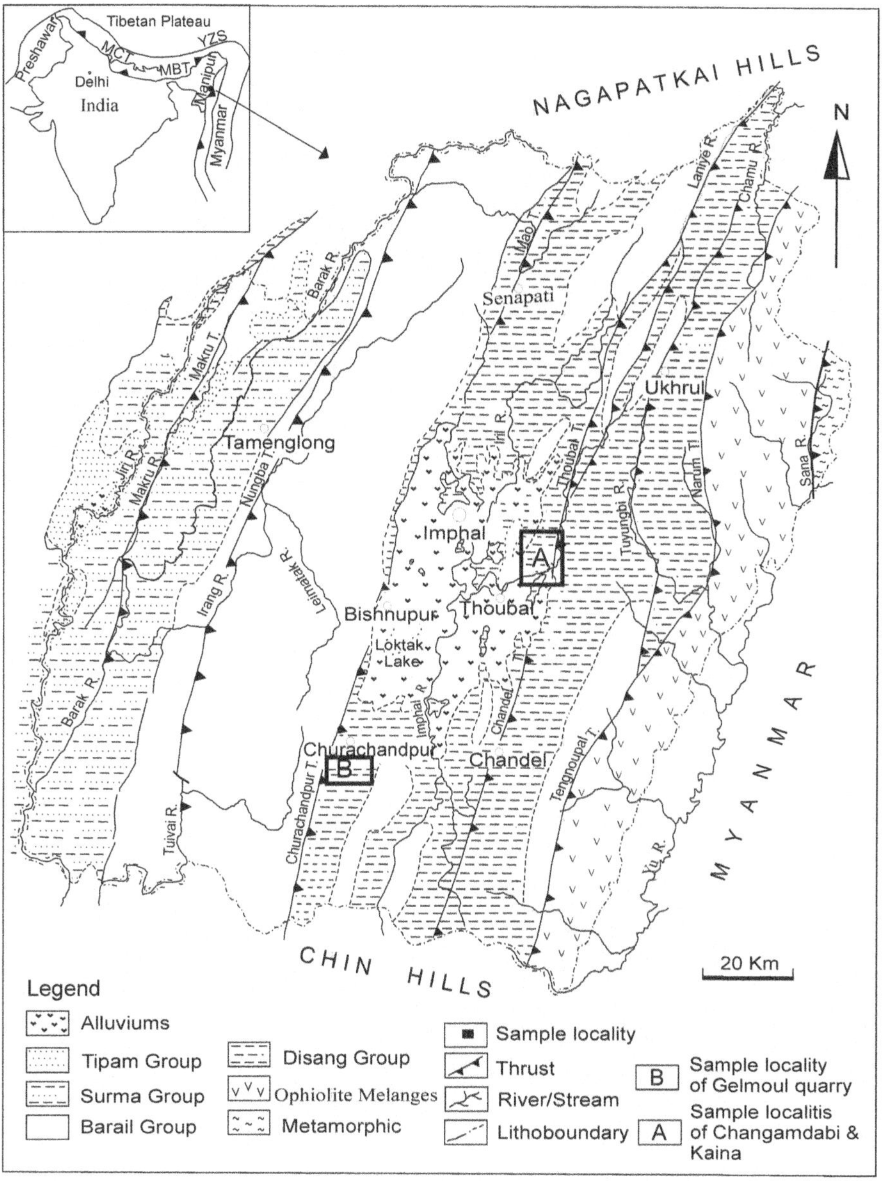

Fig. 1 Geological map of Manipur showing sample localities (After Singh et al. 2013)

Fig. 2 **a** Alteration of shale and sandstone/siltstone of Upper Disang Formation of at Gelmoul quarry (**b**). Molluscan fossils at Changamdabi (**c**). Ripple marks at Angtha (**d**) Plant fossils from Barail sediment at Kaina Hills (**e**, **f**). Trace fossils at Gelmoul quarry

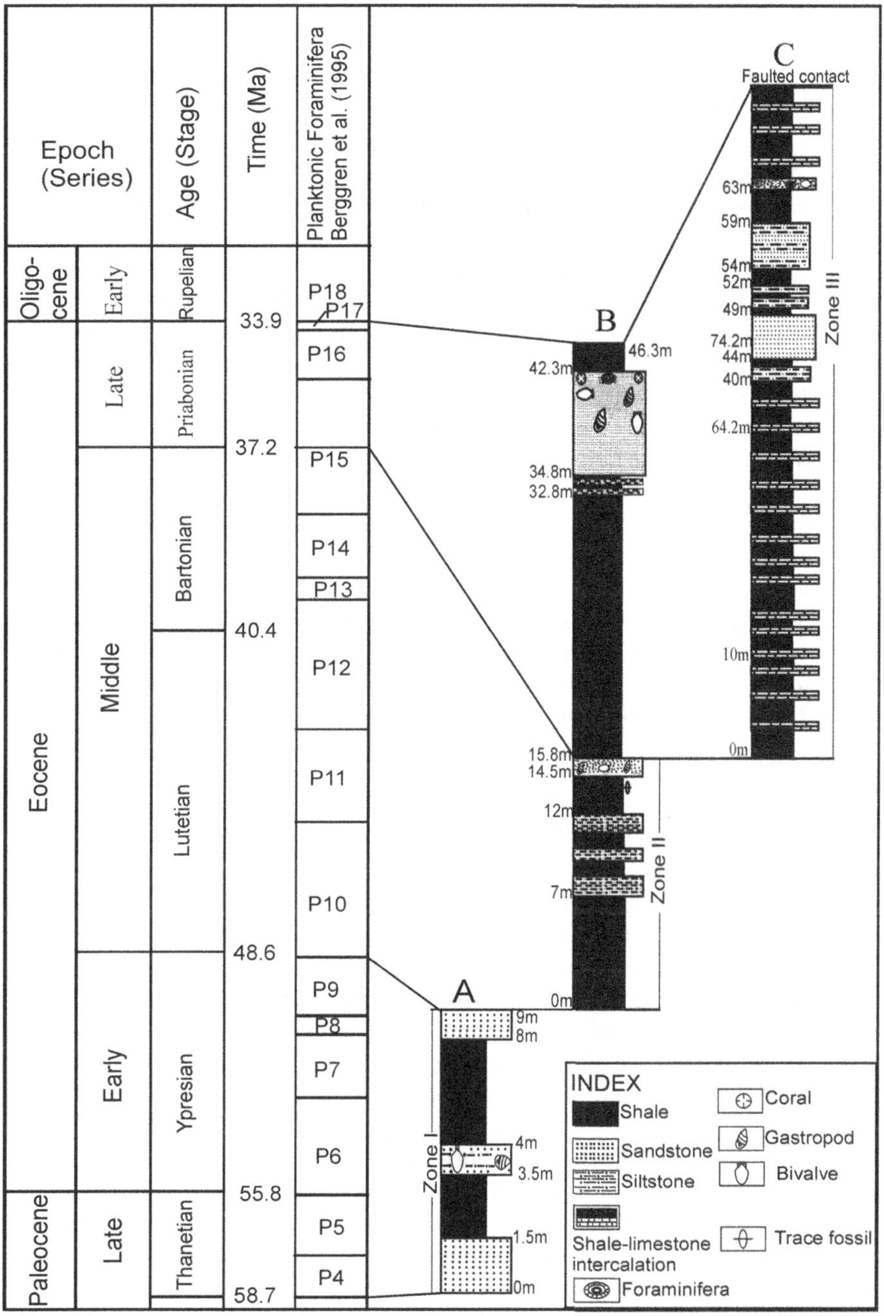

Fig. 3 Biostratigraphic zones of the Upper Disang Formation at the fossil localities (A, US_3, Khongjai Chingkhong Section; B, US_2, Maning Ching and C, US_1 Canal Section). Standard time-scale is given in the left

Fossil Assemblage

Molluscan Fossils

Fossil collection comprised more than a thousand individuals of which a sizable component was discarded due to very poor preservation or fragmentary nature. Systematically, the identified taxa of molluscan fossils contains 108 species (80 bivalves and 28 gastropods) belonging to 51 and 26 genera respectively. They have only been recorded from the Upper Disang Formation exposed at Changamdabi, Manipur associated with *Nummilites* in the form of mould. In addition to above, a variety of unidentifiable fauna remains are present in this area. Biostratigraphic scale for the area is to be erected solely on the basis of bivalves due to the less representation of the gastropods. Some of taxa of bivalves are represented by *Chlamys senatoria*, *Chlamys* sp. cf. *multistriata*, *Spondylus rououlti*, *Flemingostrea haydeni*, *Venericardia hollandi*, *Nemocardium thetregyinense*, *Sunetta yethama*, *Meretrix yawensis*, *Nucula* sp., *Barbatia* sp., *Cucullaea* sp., *Protonoetia manipurensis*, *Septifer* sp., *Aviculoperna changamdabiensis*, *Venericardia spondyliformis*, *Trachycardium yairipokensis*, *Tellina* sp., *Callista* sp., *Lentidium* sp., *Pholas* sp. and that some taxa of gastropods are characterised by *Globularia brevispira*, *Rimella fissurrella*, *Rimella pakistanica*, *Tetrastomella pseudohumilis*, *Tibia* sp., *Patella Changamdabiensis*, *Natica manipurensis*, etc. (Singh et al. 2010, 2017; Sijagumayum et al. 2011, 2014). The selected molluscan fossils are shown in Fig. 4.

Palynological Assemblage

Microforaminiferal linings are acid resistant organic remains of foraminifera preserved in geological sediments. The frequent occurrence of them in palynological preparations led many people to study them for their classification and other purposes such as stratigraphy and palaeoenvironment (Stancliffe 1989). However, up till now there is not a proper system to assign microforaminiferal linings a name at generic or specific level. Here we adopt Stancliffe's informal classification which classifies the microforaminiferal linings to various morphological forms or types. Gelmoul is marked by five main types viz., Planispiral Type III, Planispiral Type IV, Biserial Type II, Trochospiral Type I and Trochospiral Type II (Fig. 5a–h). In addition to a few dinoflagellate cysts and pollen and spores are counted in the assemblage (Fig. 5i–k). Among these fossils, the dinoflagellate cyst *Hystrichokolpoma* cf. *rigaudiae* (Fig. 5j) is the most useful in biostratigraphy which supports an Eocene age for the studied formation (Singh et al. 2013). Besides, the trace fossils have recorded from the upper Disang Formation exposed at Gelmoul quarry of Churachandpur (Fig. 2e, f) and Angtha areas (Fig. 2C; GPS N24° 41′ 30.8″: E94° 01′ 23.8″). These are *Skolithos*, *Thalassinoides*, *Ophimorpha*, *Phycodes*, *Helminthopsis*.

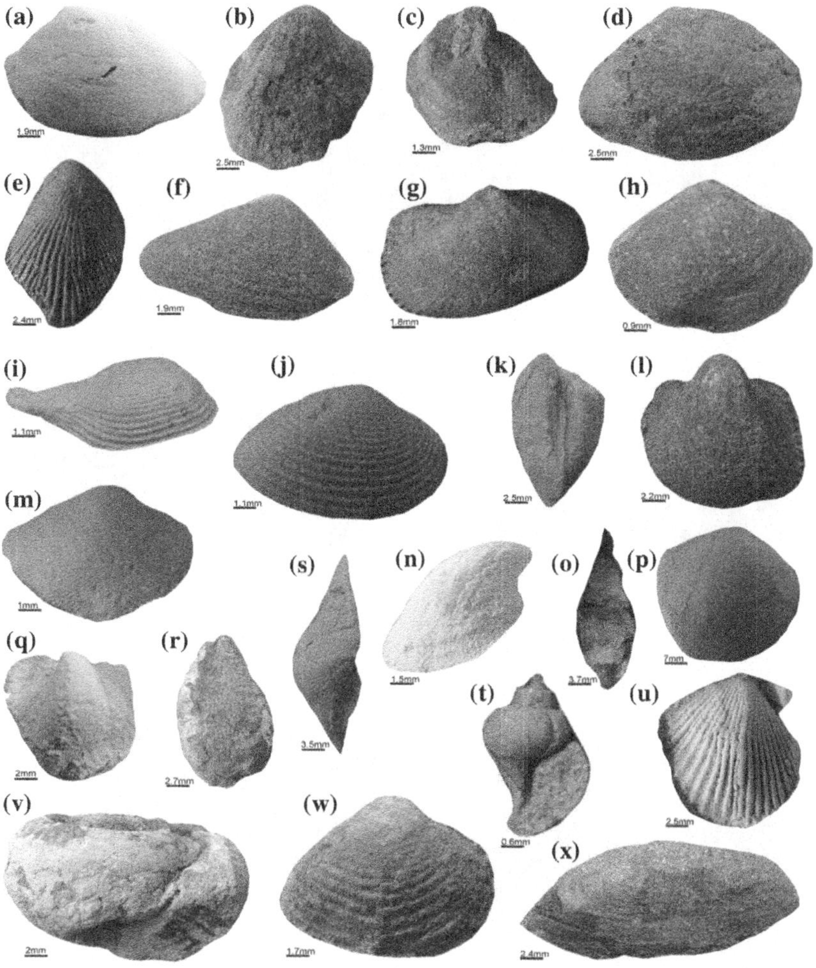

Fig. 4 **a** *Nucula* (*Leionucula*) *baboensis*, internal mould of right valve; **b**, *Venericardia* (*Venericardia*) *gilli*, external mould of left valve; **c**, *Flemingostrea pharaonum* Oppenheim var. *aviculina*, external mould of left valve; **d**, *Tellina* (*Tellina*) *pakistanica*, external mould of left valve; **e**, *Septifer* (*Septifer*) cf. *denticulatus*, external mould of left valve; **f**, *Corbula* (*Varicorbula*) *daltoni*, external mould of left valve; **g**, *Venericardia* (*Venericardia*) *hollandi*, external mould of left valve; **h**, *Venerella parva*, external mould of left valve; **i**, *Caestocorbula* (*Permicorbula*) *regulbiensis*, external mould of right valve; **j**, *Astarte* (*Astarte*) *filigera*, external mould of left valve; **k**, *Callista* (*Costacallista*) *punjabensis*, external mould of right valve; **l**, *Vepricardium (Hedecardium) sharpie*, external mould of left valve; **m**, *Boeuvia pulchella*, external mould of left valve; **n**, *Glycymeris* (*Glycymeris*) *brevirostris*, internal mould of left valve; **o**, *Terebellum fusiformopse*, Apertural view; **p**, *Lucina* (*Lucina*) *yawensis*, external mould of left valve; **q**, *Noetia (Noetia) harudiensis*, external mould of right valve; **r**, *Volvaria* cf. *birmanica*; **s**, *Rimella pakistanica*, apertural view; **t**, *Clavilithes songoensis*; **u**, *Chlamys* (*Chlamys*) *soriensis*, external mould of left valve; **v**. *Galeodea archiaci* apertural view; **w**, *Meretrix agrestris*, external mould of left valve; **x**, *Cyrtodaria rutupiensis*, external mould of right valve (After Singh et al. 2017). Scale bars represent 1.9 mm (**a**, **f**); 2.5 mm (**b**, **d**, **k**, **u**); 1.3 mm(**c**); 2.4 mm (**e**, **x**); 1.8 mm (**g**); 0.9 mm (**h**); 1.1.mm (**i**–**j**); 2.2 mm (**l**); 1 mm (**m**); 1.5 mm (**n**); 3.7 mm (**o**); 7 mm (**p**); 2 mm (**q**, **v**); 2.7 mm (**r**); 3.5 mm (s); 0.6 mm (**t**) and 1.7 mm (**w**). Colour online

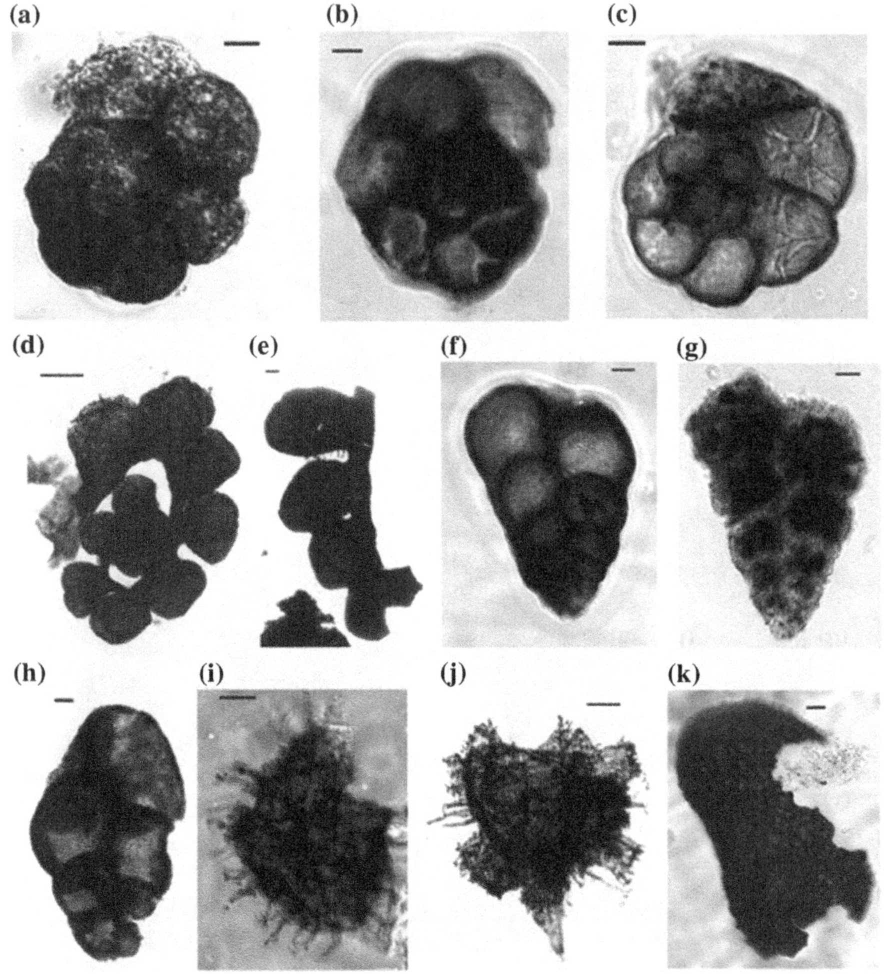

Fig. 5 **a** Trochospiral type II; **b**, **c** Trochospiral type I; **d** Planispiral type III; **e** Planispiral type III (broken); **f**, **g**, **h** Biserial type II; **i**.? *Cordosphaeridium* sp.; **j** *Hystrichokolpoma* cf. *rigaudiae*; **k** Spore (fragment) (After Singh et al. 2013 and bar represent 10 μm)

Plant Fossils

We recorded plant fossils from one horizon of Barail sequences. However, the identification of these fossils are under progress. The assemblage are mainly composed of dicotyledonous leaves, palm leave, fruiting shoot, bark (Guleria et al. 2005). In the same lithology and section of the present investigation, Singh et al. (2012) described gymnospermous (*Podocarpus oligocenicus*), Angiosperms (*Phoenicites* sp., *Prunus palaeoarmeniaca, Dilcherocarpon eocenicum*).

Discussion

Biostratigraphy

The Paleogene Period is subdivided into the Paleocene, Eocene and Oligocene epochs. This period is further assigned seven formal stages with GSSP. The present correlation of Paleogene stages is compared with megnetostratigraphy and standard marine zonation based on the calcareous nannoplankton zones of Martini (1971), planktonic foraminifera zones of Berggren et al. (1995) and dinoflagellate cyst Viborg zone of Heilmann-Clausen (1988) and Michelsen et al. (1998). Unfortunately in the study area planktonic foraminifera, larger foraminifera and calcareous nannoplankton are either rare or completely absent. Therefore, one can easily envisage that bivalves and gastropods can be used for the purpose of biostratigraphy. The age of Disang succession has been interpreted as Paleogene age (Pascoe 1912), Eocene (Mishra 1990), and Middle to Late Eocene (Mishra 1990; Kachhara et al. 2000).

Biostratigraphy has been attempted only to the upper Disang Formation exposed at Changamdabi area due to lack of fossil evidence to the other sections. Based on their geological and geographical distribution majority of the taxa are observed to be long ranging, in spite of that few are found to be suitable for biozonation because of their short geologic range. After a thorough study, three bivalves assemblage zone are recognised in the study area (Fig. 3). The details of zones are given below:

Zone I: Caestocorbula (Parmicorbula) regulbiensis-Flemingostrea pharoanum aviculina

Altogether 41 taxa of molluscan fauna are recorded from this zone of which:

Restricted to zone I: The *Caestocorbula* (*Parmicorbula*) *regulbiensis* (Morris) is common and is restricted to Paleocene and *Flemingostrea pharaonum aviculina* (Mayer-Eymer) is typical form of Late Paleocene to Early Eocene hence, name of the zone. The other restricted taxa are: *Nucula* (*Leionucula*) *rakhiensis* (Early Paleocene-Middle Eocene), *Nucula* (*Saccella*) *ukhrulensis* (Late Paleocene-Early Eocene), *Glycymeris* (*Gycymeris*) *brevirostris* (Early Eocene), *Lucina* (*Lucina*) *exquisite* (Early Eocene), *Boeuvia pulchella* (Middle Eocene), *Venericardia* (*Venricardia*) *mutabilis* (Eocene), *Astarte* (*Astarte*) *filigera* (Late Paleocene-Early Eocene), *Crassinella blandfordi* (Early Paleocene-Early Eocene), *Tellina* (*Lyratellina*) cf. *lyra* (Recent), *Macrosolen cyclopeus* (Early Eocene), *Venerella parva* (Middle Eocene), *Corbula* (*Bicorbula*) *subexarata* (Eocene), *Corbula* (*Varicorbula*) *harpa* (Early Paleocene-Early Eocene), *Kummelia Americana* (Eocene), *Turritella subathooensis* (Early-Middle Eocene), *Vermetus* cf. *hanguensis* (Eocene), *Batillaria coronate* (Early Paleocene-Early Eocene), *Seila stracheyi* (Early-Middle Eocene), *Fusinus buddhaicus* (Eocene), *Euthriofusus malcomsoni* (Early-Middle Eocene), *Volutocorbis daviesi* (Paleocene-Eocene), *Strepsidura tipperi* (Paleocene). In view of majority of the taxa leaving aside a few, all taxa have restricted range of Late Paleocene, Early Eocene, Paleocene-Early Eocene, Paleocene and Eocene. Hence, this zone is assigned to Late Paleocene-Early Eocene i.e. Thanetian-Ypresian age (Table 2).

Table 2 Biostratigraphic biozonation of the Upper Disang formation exposed at Changamdabi area

Epoch	Age	Formation	Indian stage	Biostratigraphic assemblage zone
Late Eocene	Priabonian	Upper Disang	Tarapurian	*Callista (Callista) yawensis-Lucina yawensis* Zone III
Middle Eocene–Late Eocene	Lutetian-Bartonian	Upper Disang	Kakdian–Babian	*Vulsella pakistanica-Corbula (Varicorbula) daltoni* Zone II
Late Paleocene–Early Eocene	Thanetian-Ypresian	Upper Disang	Khasian–Kakdian	*Caestocorbula (parmicorbula) regulbiensis-Flemingostrea pharoanum aviculina* Zone I

Zone II: Vusella pakistanica—Corbula (Varicorbula) daltoni

Both *Vulsella pakistanica* and *Corbula (Varicorbula) daltoni* are quite common and the former is typical Middle Eocene whereas latter with range of Middle-Late Eocene thus, the name of the Zone. The other restricted taxa to this zone are: *Noetia (Noetia) harudiensis* (Middle Eocene), *Noetia (Noetia) magnifica* (Middle-Late Eocene), *Noetia (Noetia) pondaungensis* (Middle-Late Eocene), *Protonoetia manipurensis (Middle Eocene)*, *Chlamys (Chlamys) pakistanica* (Late Eocene), *Chlamys (Chlamys) soriensis* (Middle Eocence), *Chlamys (Chlamys) wynnei* (Early-Middle Eocene), *Flemingostrea haydeni* (Early Paleocene-Middle Eocene), *Diplodonta (Diplodonta) pakistanica* (Middle Eocene), *Venericardia (Venericardia) hollandi* (Paleocene), *Venericardia (Glyptoactis) gilli* (Early Eocene), *Venericardia trachycardiiformis* (Late Eocene), *Venericardia (Venericardia) dufrenoyi* (Early-Middle Eocene), *Loxocardium kanleqnum* (Late Eocene), *Loxocardium thetregyinense* (Late Eocene), *Vepricardium (Hedecardium) sharpie* (Early Paleocene-Early Eocene), *Trachycardium (Trachycardium) yairipokensis (Late Paleocene-Early Eocene)*, *Tellina (Tellina) pakistanica* (Paleocene), *Trapezium daviesi* (Eocene), *Venus (Venus) pasokensis (Late Eocene)*, *Pitar (Calpitaria) pseudosubcyrenoides* (Middle Eocene), *Cyrtodaria rutupiensis* (Early Eocene), *Turritella (Stiracolpus) harraiensis* (Early-Middle Eocene), *Rimella fissurella* (Middle Eocene), *Terebellum fusiformopse* (Early Eocene), *Natica (Cochlis) manipurensis (Late Paleocene-Early Eocene)*, *Galeodea archiaci* (Early Eocene), *Tetrastomella pseudohumilis* (Early-Middle Eocene), *Volvaria* cf. *birmanica* (Early Oligocene), *Roxania pseudosemistriata* (Late Eocene), *Turbonilla soriensis* (Late Eocene).

In this zone only sixty one taxa has been recorded and most of taxa are restricted to Middle-Late Eocene. Therefore, this zone has been assigned as Late—Middle Eocene i.e. Lutetian—Bartonian in age (Table 2).

Zone III: Callista (Callista) yawensis-Lucina (Lucina) yawensis Zone
Both the taxa *Callista (Callista) yawensis* and *Lucina (Lucina) yawensis* are quite common and have restricted range of Late Eocene, hence the name of this zone. Restricted to zone III: The other restricted taxa are *Aviculoperna Changamdabiensis* (Late Eocene), *Ostrea (Ostrea) velata* (Middle Eocene-Early Oligocene), *Chama (Chama) brimonti* (Eocene), *Trachycardium (Trachycardium)* sp. juv. *cotteri* (Late Eocene-Recent), *Cardium subfragile* (Late Eocene), *Nemocardium (Discors) bunburyi* (Late Paleocene-Middle Eocene), *Meretrix agrestris* (Late Eocene), *Corbula (Corbula) paukensis* (Late Eocene), *Corbula (Bicorbula) subexarata* var. *lituus* (Late Eocene) and *Venus pasokensis* (Late Eocene).

From analysis of geological ranges of most of the species occurring in this zone, it falls in Late Eocene time. Out of 23 taxa reported in this zone, most of taxa are of Late Eocene. The occurrence of foraminifer taxon of *Nummulites fabiani* (Prever) is also supporting Late Eocene age. Hence, this zone has been assigned as Late Eocene i.e. Priabonian age (Table 2).

Depositional Environment

The depositional environment of Disang Group of rocks has variously been interpreted as deep-water geosynclinal (Mathur and Evans 1964) and shallow water depositional environment (Ranga Rao 1983; Mishra 1990; Pandey and Dave 1998; Kachhara et al. 2000; Singh et al. 2008). Based on ichno-assemblage, the transition of the Disang and Barail groups indicates a shallow-marine environment, with occasional high-energy conditions (Singh et al. 2008). Singh et al. (2010) on the basis of molluscan fossils suggested that the Upper Disang sequences of the Changamdabi area were deposited under shelf warm marine environmental condition under temperate to subtropical climates. Recently, Singh et al. (2013) suggested that the upper part of upper Disang formation of the Gelmoul quarry of Churachandpur, Manipur were deposited in a warm, shallow marine environment of deposition on the basis of foraminiferal linings, palynomorphs and associated flaser bedding and lenticular bedding.

In the present investigation, palaeoecological studies are based on the biotic communities and their relationship to the environments.

Zone-I

Amongst biotic community, habitat of silty substratum are dominating like *Nucula (Leionucula), Bathyarca, Astarte, Crassinella, Tellina, Macrosolen, Callista, Costacallista, Macrocallista, Venerella, Corbula, Bicorbula, Varicorbula, Fusinus, Euthriofusus, Volutacorbis*. At the same time, there are warm water genera e.g. *Nucula (Leionucula), Mytilaster, Tellina, Macrosolen, Corbula, Bicorbula, Varicorbula, Martesia, Turbo, Batillaria, Vermicularia, Fusinus, Euthriofusus, Clavilithes* and *Volutocorbis*. Depthwise few genera prefer only shallow water e.g. *Bath-*

yarca, *Lucina*, *Barbatia*, *Trigonodesma*, *Glycymeris* and *Crassinella*. The genus *Flemingostrea* thrives in reduced salinity and *Septifer* is a nestler habitat of lower subtidal zone with water depth of about 20 m. Likewise *Barbatia* flourishes in depth less than 36 m whereas genera like *Tellina*, *Turritella*, *Callista*, *Costacallista*, *Macrocallista* and *Venerella* prefer a depth less than 60 m (Perreau 1978; Mathur and Juyal 2000). The substratum was not uniformly sandy but at places soft supporting shallow burrowers (*Nucula*, *Leionucula*, *Lucina*, *Bathyarca*, *Astarte*, *Crassinella*, *Caestocorbula*, *Parmicorbula* and *Lentidium*) to deep burrowers (*Macrosolen* and *Tellina*), sometime epifaunal (*Barbatia*, *Glycymeris*, *Flemingostrea*, *Strepsidura*, *Callista*, *Costacallista*, *Macrocallista* and *Venerella*). The substratum has also supported the genus *Boeuvia* which burrow by means of surface sculpture as observed in fine sands, as well as the siphonal suspension feeder *Nuculana* (*Saccella*) on the other hand, *Barbatia* which prefers depression, *Septifer* a nestler, *Lucina* which have mucus tube at interaction of substratum with water and long siphoned *Solen*.

Through this ongoing discussion the above assemblage of organisms is quite diversified and overall characteristic of near shore, offshore shelf habitat development on silty and sandy substrate. The overall water temperature might have been around 20–25 °C and water depth during deposition of this succession in low tide to as much as 40 m. It seems that bottom currents were sufficient enough which could not allow formation of uniform laminae resulting into splintery shale. Further shallowing of the sea resulted into deposition of brownish medium-grained sandstone under very shallow well-aerated water.

Zone-II

46 taxa of bivalves belonging to 37 genera (subgenera) and 16 forms of gastropods representing same number of genera (subgenera) are collected from this fossiliferous horizon. Single piece of *Crab chella* is also collected. Crabs are epifaunal animals move freely on sandy bottom and used to live at depth between 5 to 15 m.

Amongst bivalve taxa of *Nucula* (*Leionucula*), *Glycymeris*, *Septifer*, *Flemingostrea*, *Venericardia*, *Solen*, *Tellina*, *Callista*, *Costacallista*, *Corbula*, *Varicorbula*, *Cucullaea*, *Noetia*, *Protonoetia*, *Lithophaga*, *Vulsella*, *Chlamys*, *Cubitostrea*, *Diplodonta*, *Cardium*, *Acanthocardia* (*Schedocardia*), *Loxocardium*, *Vepricardium* (*Hedecardium*), *Trachycardium*, *Trapezium*, *Venus*, *Meretrix*, *Pitar* (*Calpitaria*), *Cyrtodaria*, *Pholas*, *Pholadomya*, *Laternula* and *Picchiolia* are representative of zone II. *Cucullaea* is a thick shell genus with vestigeal byssus and of subquadrate outline. It is a byssate epifaunal nestler thriving in shallow water not more than 40 m depths. It prefers under water rock crevices. The genus *Noetia* is an infaunal bivalves lives at depth of about 30 m whereas the genus *Protonoetia* is widely spread in shallow seas and young ones are attached by byssus and become free on attaining adulthood. The genus *Vulsella* is an epifaunal bivalve lying freely on substrate habitat of shallow water where sufficient current prevails. The genus *Chlamys* is also free lying organism on the substrate preferring sandy bottom in marine condition with water depth not less than 25 m in case of any danger it can swim by clapping both the valves and by ejecting out water near the hinge (Perreau 1978). *Lithophaga* is a good borer. Its cylindrical elongate shape helps in boring through especially compacted mud and is restricted to warm waters. *Cubitostrea* is also habitat of warm water, shallow seas with depth less than 60 m preferring rocky bottom. *Diplodonta* is a semi permanent borrower connected to interface by inhalant tube, living on sandy bottom under shallow marine condition with water column less than 60 m in height (Perreau 1978; Mathur and Juyal 2000). *Cardium* can tolerate normal to reduce salinity thus habitat of brackish to marine water with sandy to muddy substratum. The taxa of *Acanthocardia*, *Schedocardia*, *Loxocardium*, *Hedecardium* and *Vepricardium* are habitat of sandy substrate. *Trachycardium* prefers outer neritic shelf with water depth of about 72 metres (Prashad 1932; Mathur and Juyal 2000). *Trapezium* is a nestler and prefers crevices. The taxa of *Venus*, *Meretrix*, *Pitar* and *Calpitaria* live preferably in tropical water of normal salinity and burrows deeply in sandy or muddy bottom. *Cyrtodaria* is of burrowing and nestling habit. *Pholas* is one of the most dangerous bivalves as it can bore into relatively hard substrata such as peat, wood, nuts, woody plant stems, stiff clays, soft or friable rocks, corals and shells. Due to this it is difficult to maintain vessel in harbour areas. This genus occurs mostly in marine condition but selective taxa can tolerate brackish and fresh water condition. *Pholadomya* is a deep burrowing animal under shallow marine condition with a depth less than 60 m (Perreau 1978; Murray 1985; Mathur and Juyal 2000). *Laternula* is habitat of warm water condition and *Pecchiolia* with spirally coiled umbo with convex and curved valves surface must be lying freely over the substratum like *Eryphae*.

Amongst gastropoda community *Turritella* (*Stiracolpus*) and *Clavilithes* are present in this zone. The occurrence of taxa of *Tibia*, *Natica* (*Cochlis*), *Polinices*, *Globularia*, *Cassis*, *Galeodea* and *Latirus* are indicative of warm water condition. The genus *Globularia* is a predator and preyed upon shallow burrowing bivalves. *Rimella* live on muddy bottom in warm tropical water generally under shallow marine condition in a depth less than 60 m (Perreau 1978; Mathur and Juyal 2000). *Terebellum* live on muddy bottom in warm tropical water generally under shallow marine conditions in a depth less than 60 m (Perreau 1978; Mathur and Juyal 2000; Srivastava et al. 2008). *Tetrastomella* lives on sandy to rocky bottom in warm tropical water under marine to brackish water condition (Perreau 1978). The genus *Flemingostrea* can tolerate reduce salinity and there are few genera like *Septifer*, *Cubitostrea*, *Diplodonta*, *Cardium*, *Acanthocardia*, *Loxocardium*, *Vepricardium*, *Tellina*, *Venus*, *Meretris* and *Pitar* can thrive in both marine and brackish water conditions (Perreau 1978). Looking into the whole community it appears that it can be taken as near shore to offshore shelf habitat developed on silt and sandy substrate. The water depth during the deposition of this succession was ranging from 5 to 60 m but more near to 30 m as indicated by the presence of *Noetia*. The sea temperature was of warm with about 20–25 °C temperature as majority of the Gastropods are habitat of tropical warm water.

From the ongoing discussion, it infers that the above community along with the nature of sediments suggestive of open sea conditions with well aerated warm water condition with depth not exceeding euneritic.

Zone-III

This zone is encountered with foraminifers and corals with molluscan genera namely *Nucula (Leionucula)*, *Aviculoperna*, *Chlamys*, *Gryphaeostrea*, *Ostrea*, *Cubitostrea*, *Cardium*, *Trachycardium*, *Nemocardium* (*Discors*), *Tellina* (*Peronidia*), *Meretrix*, *Pitar* (*Calpitaria*), *Callista*, *Corbula*, *Bicorbula*, *Varicorbula*, *Rimella* and *Cassis*, *Noetia*, *Lucina*, *Chama*, *Venus*, *Serratocerithium* and *Globularia*. The genus *Aviculoperna* is an epifaunal free swinging animal attached to objects by means of byssus. *Gryphaeostrea* used to live on these sediments of low energy level commonly associated with clay, marl and limestone. These are mostly live on a sea bottom, well adapted especially on unconsolidated sea floor and the genera *Venericardia* and *Gryphaeostrea* are suggestive of depth less than 60 m. The genera *Flemingostrea* and *Ostrea* reflect mangrove condition which means very shallow sea. However, both the genera are of very small size and lack very coarse ornamentation, therefore, suggestive of comparatively deeper water habitat. The genus *Nemocardium* lives on sandy to muddy bottom under brackish water to marine condition. *Discors* is a suspension feeder just at the interface and had no siphon. *Pitar* is a more mobile and shallow borrower which has a siphon likes the genus *Meretrix* and Nummulitic foraminifers are habitat of tropical and subtropical water in shallower part of neritic depth (Pokorny 1963). According to Pautal (1987) larger foraminifera can occur in inner shelf carbonated bays, protected shelf lagoonal bays etc. Solitary corals or corals as a whole are characteristic of warm, pure, clean and marine water in very shallow depth with gently sloping substratum. These thrived well mainly in seas lying between the tropics. The genus *Lucina* is a semi permanent burrower connected to the surface by inhalant and mucus tubes prefers muddy to sandy bottom in shallow marine tropical sea with water depth less than 60 m (Perreau 1978). *Chama* is an epifaunal bivalve (Murray 1985) initially cemented with the pieces of shell debris but in adult stage prefers free-living condition. It has a thick shell with rugged ornamentation and spines, with the help of spines it is attached to substratum firmly and inspite of strong currents it cannot be uprooted. It is habitat of open shelf. The genus *Venus* is a deep burrower living on sandy to muddy bottom under shallow marine (depth <60 m) to brackish water conditions (Perreau 1978). Amongst gastropods *Serratocerithium* invades a very shallow depth of 4–17 m and *Globularia* is a predator which preyed upon shallow burrowing bivalves in habitat of shallow sea. Corals and foraminifers commonly occur in this succession, thereby, suggestive of open sea condition with well-aerated warm water with water depth is very shallow i.e. sub-littoral habitat perhaps about 10–15 m with sandy substrate.

The open sea condition could not continue for a long time and due to presence of some intermittent barrier and surrounding luxuriant vegetation resulting in deposition of grey shale with limestone bed indicative of alternating open and protected sea condition which is not favourable for thriving of biotic community.

A shallow marine environment of deposition for the Upper Disang Formation of Gelmoul quarry is also implied on the basis of microforaminiferal linings and dinoflagellate cysts in association.

Based on morphological analysis of the dicot leaves, the Oligocene sequences of the Kaina area are indicative of the existence of a moist tropical environment at the time of deposition of the leaves (Guleria et al. 2005). Two species of plant fossil i.e. *Prunus palaeoarmeniaca* and *Dilcherocarpon eocenicum* have been recorded from the Laisong formation (same locality) of Kaina Hills and indicates tropical to subtropical climate (Singh et al. 2012).

Conclusions

The present investigation is mainly confined to upper Disang Formation exposed around Changamdabi and Gelmoul areas of Imphal valley, India. Altogether 80 species of bivalves referable to 51 genera and 28 species of gastropods belonging to 26 genera are recovered from the upper Disang Formation. The molluscan fauna led to recognize three biozones within the Uppar Disang Formation. They are Zone III *Callista (Callista) yawensis-Lucina yawensis*, Zone II *Vulsella pakistanica-Corbula (Varicorbula) daltoni* and Zone I *Caestocorbula (parmicorbula) regulbiensis-Flemingostrea pharoanum aviculina* in the descending order of stratigraphy. The Zones I, II and III range from Late Paleocene-Early Eocene (Thanetian to Ypresian) in age, Middle to Late Eocene (Lutetian to Bartonian) in age and Late Eocene (Priabonian) in age respectively. Therefore, the Upper Disang Formation ranges from Late Paleocene to Late Eocene in age. Two ecological events in the Upper Disang Formation are identified i.e. shallowing of the sea-aerated water corresponding to zones I and open sea conditions with well aerated warm water condition corresponding to zones II and III. This is also supported by the presence of microforaminiferal linings and dinoflagellate cysts and trace fossils. In the higher-up the Laisong Formation has been formed under tropical and subtropical climatic conditions.

Acknowledgements One author (YRS) is grateful to UGC (F.No. 41-1024/2012, SR) and Oil India Limited (Contract No. 6111262) for financial assistance. The authors (YRS & BPS) also thank to the DST, New Delhi for financial assistance (Project No. SR/S4/ES-577/2011 dated 09/03/2012). The authors sincerely acknowledge Shri Ksh Atamajit Singh, Mr. Venus Guruaribam and Miss Reshma Naorem (Research Scholars), Department of Earth Sciences, Manipur for their support during field work.

References

Berggren WA, Kent DV, Swisher CC, Aubury MP (1995). A revised Cenozoic geochronology and chronostratigraphy. In: Berggren WA, Kent DV, Aubury MP, Hardenbol J (eds) Geochronology, time-scales and global stratigraphic correlation, vol 54. Society of Economic Paleontologists and Mineralogists, Special Publication, pp 129–212

Berggren WA, Pearson PN (2005) A revised tropical to subtropical Paleogene planktonic foraminiferal zonation. J Foramin Res 35:279–298

Bukry D (1973) Low-latitude biostratigraphy zonation. Init Rep Deep Sea Drill Proj 34:1–34

Bukry D (1975) Coccolith and silicoflagellate stratigraphy, northwestern Pacific Ocean, Deep Sea Project Leg 32. Init Rep Deep Sea Drill Proj 32:677–701

Dockery III DT (1998) Molluscan faunas across the Paleocene/Eocene Series boundary in the North American Gulf Costal plain. In: Aubry MP, Lucas S, Berggren WA (eds) Late Paleocene-Early Eocene climatic and biotic events in the marine and terrestrial records. Columbia University Press, New Yok, pp 296–322

Guleria JS, Singh RKH, Mehrotra RC, Soibam I, Kishor R (2005) Paleogene plant fossils of Manipur and their paleoecological significance. Palaeobotanist 54:61–77

G.S.I. (2011) Geology and mineral resources of Manipur, Mizoram, Nagaland and Tripura, vol 30. Miscellaneous Publication, Geological Society of India, pp 36–67

Hartman JH, Roth B (1998) Late Paleocene and Early Eocene nonmarine molluscan faunal change in the Bighorn basin, Northwestern Wyomning and South-Central Montana. In: Aubry MP, Lucas S, Berggren WA (eds) Late Paleocene-Early Eocene climatic and biotic events in the marine and terrestrial records. Columbia University Press, New Yok, pp 323–379

Heilmann-Clausen C (1988) The Danish sub-basin. Paleogene dinoflagellates. News fahrbuch fur Geol Palaeontol Abh A101:339–343

Kachhara RP, Soibam I, Jamir NM (2000) Upper age limit of the Disang Group in Manipur. In: XVI Indian Colloquium on Micropalaeontology and stratigraphy vol 37, National Institute of Oceanography, Goa. Special volume ONGC Bulletin, pp 215–218

Martini E (1971) Standard Tertiary and Quarternary calcareous nannoplanktonic zonation, p. 739-785. In: Farinacci A (ed) Proceedings II Planktonic Conference. Italy, Rome, p 2

Mathur NS, Juyal KP (2000) Atlas of Early Paleogene invertebrate fossils of the Himalayan foot hill belt. Wadia Institute of Himalayan Geology, Dehradun, 1–166, pls I–XXXIII

Mathur LP, Evans P (1964) Oil in India. In: Proceedings 22nd International Geological Congress, pp 1–35

Michelsen O, Thomsen E, Danielsen M et al. (1998). Cenozoic sequence stratigraphy in the eastern North Sea, vol 60. Society of Economic Paleontologists and Mineralogists Special Publication, Tulsa, OK: SEPM, pp 91–118

Mishra UK (1990) Palaeontological study of the Disang and Barail sediments in parts of Phek district, Nagaland. Rec Geol Surv India 123:167–168

Murray JW (1985) Atlas of invertebrate macrofossils. Longman, the Palaeontolgical Association, United States of America, p 241

Pandey J, Dave A (1998) Stratigraphy of Indian petroliferous basins. XVI ICMS 2:110–111

Pascoe EH (1912) Traverse across from Dimapur to neighbourhood of Saramati peak. Rec Geol Surv India 42:261

Pautal L (1987) Foraminiferal assemblages of some Early Eocene environments (bays) from the Northern Corbires, France. In: Hart MB (ed) Micropalaeontology of Carbonate environments, pp 75–82

Perreau M (1978) Mollusques: environments et palaéoécologie. In: Baillere JB (ed) Extrait de<<l'Information Scientifique≫, 2, Paris, pp 55–89

Pokorny V (1963) Principles of zoological micropalaeontology. Pergamon Press, p 652

Prashad B (1932) Siboga-Expedite. The Lamellibranchia of the Siboga Expedition, Systematic part 2, Pelecypods (exclusive of Pectenidae). Monograph LIIIC of Uitkosmst. op. Zool. Bot. Oceanogr. en Geol. Gebied, pp 1–358

Pross J, Brinkhuis H (2005) Organic-walled dinoflagellate cysts as paleoenvironmental indicators in the Paleogene; a synopsis of concepts. Palaontol Z 79(1):53–59

Ranga Rao A (1983) Geology and hydrocarbon potential of a part of Assam-Arakan basin and its adjacent region. Pet Asia J 4:127–158

Serra-Kiel J, Hottinger L, Caus E, Drobne K, Ferrandez C, Jauhri AK, Less G, Pavlovec R, Pignatti J, Samso JM, Schaub H, Sirel E, Strougo A, Tambareau Y, Tosquella J, Zakrevskaya E (1998) Larger foraminiferal biostratigraphy of the Tethyan Paleocene and Eocene. Bull Soc Geol du France 169:281–299

Sijagumayum U, Singh YR, Kachhara RP (2011) Some molluscan fossils from the Upper Disang Formation of Changamdabi, East Imphal District, Manipur. J Palaeontol Soc India 56:165–169
Sijagurumayum U, Singh YR, Kachhara RP (2014) Eocene molluscan fossils from the Upper Disang Formation of Imphal valley, Manipur, India. J Palaeontol Soc India 59(2):59–68
Singh BP, Singh YR, Andotra DS, Patra A, Srivastava VK, Guruaribam V, Sijagurumayum U, Singh GP (2016) Tectonically driven late Paleocene (57.9–54.7 Ma) transgression and climatically forced latest middle Eocene (41.3–38.0 Ma) regression on the Indian subcontinent. J Asian Earth Sci 115:124–132
Singh MC, Kushwaha RAS, Srivastava G, Mehrotra RC (2012) Plant remains from the Laisong formation of Manipur. J Geol Soc India 79(3):287–294
Singh RK, Rodriguez-Tovar FJ, Soibam I (2008) Trace fossils of the Upper Eocene-Lower Oligocene transition of the Manipur, Indo-Myanmar range (Northeast India). Turk J Earth Sci 17:821–834
Singh YR, Sijagumayum U, Ranjita DRK (2010) Preliminary studies of fossils from the Palaeogene rocks exposed around Changamdabi area, Manipur. Mem Geol Soc India 75:143–148
Singh YR, Jiangu L, Singh BP, Guruaribam V (2013) Microfloraminiferal linings from the upper part of the Upper Disang Formation at Gelmoul quarry, Churachandpur, Imphal valley and their bearing on palaeoenvironment. Curr Sci 105(9):1223–1226
Singh YR Singh BP, Li J (2015) Hydrocarbon Potential of the Paleogene Disang Group, Manipur Region, India-A Palynological Approach. In: Mukherjee S (ed) Petroleum geosciences: Indian contexts. Springer International Publishing, Switzerland. https://doi.org/10.1007/978-3-319-03119-4_8
Singh YR , Sijagurumayum U, Singh BP (2017) Molluscan fauna from the Upper Disang Formation of Manipur, India: their biostratigraphic implications. Jour Palaeont Soc India 62(1):59–76
Sluijs A, Pross J, Brinkhuis H (2005) From greenhouse to icehouse; organic walled dinoflagellate cysts as paleoenvironmental indicators in the Paleogene. Earth-Sci Rev 68:281–315
Soibam I (2000) Structural and tectonic framework of Manipur. Souvenir, X Manipur Science Congress (March 15–17), pp 26–37
Srivastava DK, Singh AP, Tiwari RP, Jauhri AK (2008) Cassiduloids (Echinoidea) from the Siju Formation (late Lutetian-early Bartonian) of the South Garo Hills, Meghalaya, India. Revue de Paleobiologie, Geneve (Switzerland) 27(2):511–523
Stancliffe RPW (1989) Microforaminiferal linins: their classification, biostratigraphy and paleoecology, with special reference to specimens from British Oxfordian sediments. Micropalaeontology 35(4):337–352
Zachos JCM, Pagani L, Sloan E, Thomas, Billups K, (2001) Trends, rhythms and aberrations in global climate 65 Ma to Present. Science 292:686–693